Lecture Notes in Computer Science   11657

More information about this series at http://www.springer.com/series/7407

Victor Malyshkin (Ed.)

Parallel Computing Technologies

15th International Conference, PaCT 2019
Almaty, Kazakhstan, August 19–23, 2019
Proceedings

 Springer

Editor
Victor Malyshkin ⓘ
Institute of Computational Mathematics
and Mathematical Geophysics SB RAS
Novosibirsk State University,
Novosibirsk State Technical University
Novosibirsk, Russia

ISSN 0302-9743 ISSN 1611-3349 (electronic)
Lecture Notes in Computer Science
ISBN 978-3-030-25635-7 ISBN 978-3-030-25636-4 (eBook)
https://doi.org/10.1007/978-3-030-25636-4

LNCS Sublibrary: SL1 – Theoretical Computer Science and General Issues

This Springer imprint is published by the registered company Springer Nature Switzerland AG
The registered company address is: Gewerbestrasse 11, 6330 Cham, Switzerland

Preface

The 15th International Conference on Parallel Computing Technologies (PaCT 2019) was a four-day event held in Almaty, Kazakhstan. It was organized by the Institute of Computational Mathematics and Mathematical Geophysics of the Russian Academy of Sciences (Novosibirsk) in cooperation with Novosibirsk State University, Novosibirsk State Technical University, al-Farabi Kazakh National University (Almaty, Kazakhstan), and the University of International Business (Almaty).

Previous conferences of the PaCT series were held in various cities of Russia every odd year beginning with PaCT 1991 that took place in Novosibirsk (Akademgorodok). Since 1995, all the PaCT proceedings have been published by Springer in the LNCS series.

The aim of the PaCT 2019 conference was to provide a forum for an exchange of views among the international community of researchers in the field of development of parallel computing technologies. The PaCT 2019 Program Committee selected papers that contributed new knowledge in methods and tools for parallel solution of topical large-scale problems. The papers selected for PaCT 2019:

- Present and study tools for parallel program development such as languages, performance analyzers, automated performance tuners
- Examine and optimize the processes related to management of jobs, data and computing resources at high-performance computing centers
- Look into ways to enhance productivity of those who use high-performance computing resources to solve problems in their application domains
- Propose new models and algorithms in numerical analysis and data processing specifically targeted at parallel computing architectures
- Theoretically study practically relevant properties of distributed systems

Authors from 17 countries submitted 72 papers. The submitted papers were subjected to a single-blind reviewing process. The average number of reviews per submitted paper was 2.6. The Program Committee selected 24 full papers and ten short papers for presentation at the PaCT 2019.

Many thanks to our sponsors: the Ministry of Science and Higher Education of the Russian Federation, Russian Academy of Sciences, JSC Kazakhtelecom, Microsoft, and RSC Technologies.

August 2019   Victor Malyshkin

Organization

The PaCT 2019 was organized by the Institute of Computational Mathematics and Mathematical Geophysics, Siberian Branch of Russian Academy of Sciences (Novosibirsk, Russia) in cooperation with Novosibirsk State University, Novosibirsk State Technical University, al-Farabi Kazakh National University (Almaty, Kazakhstan), and the University of International Business (Almaty, Kazakhstan).

Organizing Committee

Conference Co-chairs

V. E. Malyshkin	ICMMG SB RAS, NSU, NSTU, Novosibirsk, Russia
G. M. Mutanov	KazNU named after al Farabi, Almaty, Kazakhstan
D. Zh. Akhmed-Zaki	UIB, KazNU named after al Farabi, Almaty, Kazakhstan

Conference Secretary

M. A. Gorodnichev	ICMMG SB RAS, NSU, NSTU, Russia

Organizing Committee

S. M. Achasova	ICMMG SB RAS, Russia
S. B. Arykov	ICMMG SB RAS, NSTU, Russia
M. A. Gorodnichev	ICMMG SB RAS, NSU, NSTU, Russia
T. S. Imankulov	KazNU named after al Farabi, Kazakhstan
M. N. Kalimoldayev	Institute of Information and Computational Technologies, Almaty, Kazakhstan
S. E. Kireev	ICMMG SB RAS, NSU, Russia
A. E. Kireeva	ICMMG SB RAS, Russia
A. B. Kydyrbekuly	KazNU named after al Farabi, Kazakhstan
D. V. Lebedev	UIB, KazNU named after al Farabi, Kazakhstan
A. M. Mahmetova	UIB, Kazakhstan
M. E. Mansurova	KazNU named after al Farabi, Kazakhstan
V. P. Markova	ICMMG SB RAS, NSU, NSTU, Russia
Yu. G. Medvedev	ICMMG SB RAS, Russia
V. A. Perepelkin	ICMMG SB RAS, NSU, Russia
T. S. Ramazanov	KazNU named after al Farabi, Kazakhstan
G. A. Schukin	ICMMG SB RAS, NSTU, Russia
V. S. Timofeev	NSTU, Russia
U. A. Tukeyev	KazNU named after al Farabi, Kazakhstan
D. B. Zhakebayev	KazNU named after al Farabi, Kazakhstan

Program Committee

Victor Malyshkin (Co-chair)	Novosibirsk State University, Novosibirsk State Technical University, Russia
Darkhan Akhmed-Zaki (Co-chair)	University of International Business, al-Farabi Kazakh National University, Kazakhstan
Sergey Abramov	Russian Academy of Sciences, Russia
Farhad Arbab	Leiden University, The Netherlands
Jan Baetens	Ghent University, Belgium
Stefania Bandini	University of Milano-Bicocca, Italy
Thomas Casavant	University of Iowa, USA
Pierpaolo Degano	University of Pisa, Italy
Dominique Désérable	National Institute for Applied Sciences, Rennes, France
Victor Gergel	Lobachevsky State University of Nizhni Novgorod, Russia
Bernard Goossens	University of Perpignan, France
Sergei Gorlatch	University of Münster, Germany
Yuri G. Karpov	St.Petersburg State Polytechnic University, Russia
Alexey Lastovetsky	University College Dublin, Ireland
Jie Li	University of Tsukuba, Japan
Thomas Ludwig	University of Hamburg, and German Climate Computing Center, Germany
Giancarlo Mauri	University of Milano-Bicocca, Italy
Igor Menshov	Russian Academy of Sciences, Russia
Nikolay Mirenkov	University of Aizu, Japan
Marcin Paprzycki	Polish Academy of Sciences, Poland
Dana Petcu	West University of Timisoara, Romania
Viktor Prasanna	University of Southern California, USA
Michel Raynal	Research Institute in Computer Science and Random Systems, Rennes, France
Bernard Roux	National Center for Scientific Research, France
Uwe Schwiegelshohn	Technical University of Dortmund, Germany
Waleed W. Smari	Ball Aerospace & Technologies Corp., Ohio, USA
Victor Toporkov	National Research University Moscow Power Engineering Institute, Russia
Carsten Trinitis	University of Bedfordshire, UK and Technical University of Munich, Germany
Roman Wyrzykowski	Czestochowa University of Technology, Poland

Additional Reviewers

Svetlana Achasova
Christian Beecks
Florian Fey
Maxim Gorodnichev
Sergey Kireev
Mikhail Marchenko
Yuri Medvedev
Vladislav Perepelkin

Anastasia Perepelkina
Ari Rasch
Georgy Schukin
Richard Schulze
Aleksey Snytnikov
Oleg Sukhoroslov
Juri Tomak

Sponsoring Institutions

Ministry of Education and Science of the Russian Federation
Russian Academy of Sciences
JSC Kazakhtelecom
Microsoft
RSC Technologies

Contents

Methods and Tools for Parallel Solution of Large-Scale Problems

Data Processing

Cellular Automata

Distributed Algorithms

Programming Languages and Execution Environments

Automated Construction of High Performance Distributed Programs in LuNA System

Darkhan Akhmed-Zaki[1], Danil Lebedev[1] , Victor Malyshkin[2,3,4], and Vladislav Perepelkin[2,3(✉)]

[1] Al-Farabi Kazakh National University, Almaty, Kazakhstan
[2] Institute of Computational Mathematics
and Mathematical Geophysics SB RAS, Novosibirsk, Russia
perepelkin@ssd.sscc.ru
[3] Novosibirsk State University, Novosibirsk, Russia
[4] Novosibirsk State Technical University, Novosibirsk, Russia

Abstract. The paper concerns the problem of efficient distributed execution of fragmented programs in LuNA system, which is a automated parallel programs construction system. In LuNA an application algorithm is represented with a high-level programming language, which makes the representation portable, but also causes the complex problem of automatic construction of an efficient distributed program, which implements the algorithm on given hardware and data. The concept of adding supplementary information (recommendations) is employed to direct the process of program construction based on user knowledge. With this approach the user does not have to program complex distributed logic, while the system makes advantage of the user knowledge to optimize program and its execution. Implementation of this concept within LuNA system is concerned. In particular, a conventional compiler is employed to optimize the generated code. Some performance tests are conducted to compare efficiency of the approach with both previous LuNA release and reference hand-coded MPI implementation performance.

Keywords: Automated parallel programs construction ·
Fragmented programming technology · LuNA system

1 Introduction

Considerable constant growth of supercomputers' capabilities during last decades is accompanied with the increase of complexity of high performance computing hardware usage. This, in turn, makes implementation of large-scale numerical models harder for users of supercomputers. Efficient utilization of modern supercomputers' resources requires an application to be scalable and tunable to hardware configuration. In some cases dynamic load balancing, co-processors support (GPU, FPGA, etc.), fault tolerance and other properties are required. Implementation of such properties is not easy and requires specific knowledge and skills, different from what implementation of the "numerical" part of the program requires.

© Springer Nature Switzerland AG 2019
V. Malyshkin (Ed.): PaCT 2019, LNCS 11657, pp. 3–9, 2019.
https://doi.org/10.1007/978-3-030-25636-4_1

Especially this problem affects users, who develop new numerical models and algorithms, and therefore they are unable to use ANSYS Fluent [1], NAMD [2] and other highly-efficient software tools, optimized by skillful programmers. Strict performance and memory constraints also make unusable most non-programmer friendly mathematical software, such as MathWorks MATLAB [3], GNU OCTAVE [4] or Wolfram Mathematica [5]. The only option to reduce complexity of efficient parallel program development is to employ programming systems [6–11], which automate many low-level error-prone routine jobs and provide higher level means, suitable for various particular cases.

In Charm++ [6] computations are represented as a set of distributed communicating objects called chares. The run-time environment is capable of serializing and redistributing chares, scheduling their execution and performing other execution management tasks in order to optimize program execution. The user is allowed to tune some settings of the execution, including choice of dynamic load balancer. Charm++ achieves high efficiency while freeing the user from a number of complex tasks of parallel programming. In PaRSEC [7] the application domain is limited to a dense linear algebra algorithms class (and some other similar algorithms). In particular, iterations with dynamic conditions are not supported. This and other constraints are used to make particular systems algorithms and heuristics effective, which, in turn, allows to achieve high performance within the application domain. Legion [8] system follows a powerful approach to separately define computations and their execution management as orthogonal parts of the application. With this approach the user is responsible for programming resources distribution, computations scheduling and other execution management tasks, but the means Legion provides allow doing it without the risk of bringing errors into code. LuNA system [9] follows the similar approach, but instead of obliging the user to do the management the system allows automated construction of the management code. Many other systems exist and evolve to study various computational models, system algorithms and heuristics and develop better facilities of parallel programs construction automation [10, 11].

It can be stated, that big effort is put into development of such systems, although much more work has to be done in order to widen their application domains and improve quality of the automation performed.

This paper discusses the approach employed by LuNA system to achieve satisfactory performance of constructed parallel programs. LuNA is a system of automated construction of parallel programs, which implement large-scale numerical models for supercomputers. The system is being developed in the Institute of Computational Mathematics and Mathematical Geophysics, SB RAS.

The next sections present the fragmented programming technology approach upon which LuNA system is based, the implementation of the approach in LuNA system and some performance tests. The conclusion and future works section ends the paper.

2 The Fragmented Programming Technology Approach

In the fragmented programming technology an application algorithm is represented in a hardware-independent form called fragmented algorithm. Fragmented algorithm (FA) is basically a declarative specification, that defines two potentially infinite sets—a set of computational fragments (CF) and a set of data fragments (DF), where each CF is a side-effect free sequential subroutine invocation and each DF being an immutable piece of data. For each CF two finite subsets of DFs are defined to be input and output DFs correspondingly. The CF's subroutine computes values of output DFs provided values of input DFs are available in local memory. FA as an enumeration representation employs a number of operators, which describe DFs and CFs. The representation is based on the definition of computational model [12], i.e. FA is a particular form of computational model, in which exactly one algorithm is deductible.

Fragmented program (FP) is an FA with supplementary information called recommendations. While FA defines computations functionally (i.e. how DFs are computed from other DFs), recommendations affect non-functional properties of the computations, such as computational time, memory usage, network traffic, etc. For example, a recommendation may force two DFs to share the same memory buffer within different time spans in order to reduce memory usage, or a recommendation may define data allocation strategy for a distributed array of DFs, etc.

FA and recommendations are orthogonal in sense that recommendation do not affect the values computed, but only affect how FA entities (DFs and CFs) are mapped into limited resources of a multicomputer in time. Different recommendations cause execution of the same FA to be optimized for different hardware configuration and/or optimization criteria (memory, time, network, etc.). There are two different kinds of recommendations. The first one is informational recommendation, which formulates properties of FA, which are hard to obtain automatically, for example estimated computational complexity of different CFs or the structure of DFs. The second kind of recommendations is prescriptive recommendation, which directs the execution in some way, for example mapping of DFs to computing nodes or order of CFs execution. Neither kind of recommendations is mandatory and even if recommendations are supplied, they can be partially or completely ignored by the system.

Such an orthogonality is common for various programming systems [6–9], since it is the basis, which allows a system to control execution. Let's illustrate some differences in systems' approaches on the example of objects (fragments, jobs, etc.) distribution. In some systems, such as Charm++, the system distributes the objects using system algorithms. In other systems, such as PaRSEC, the user specifies the distribution without programming it, and the system implements it. In systems, such as Legion the user needs to program the distribution using system API.

In LuNA a hybrid approach is employed. If no recommendations are supplied, the system will decide on distribution using system algorithms. If informational recommendations are supplied, a (probably) better distribution will be constructed based on this additional knowledge. If prescriptive recommendations are given, then they will be followed by the system. The prescriptive recommendations are least portable, they are useful until the system is able to automatically construct satisfactory distribution. After that the prescriptive recommendations should be ignored. Informational recommendations are useful in a longer term. They describe significant properties of FA, which are hard to obtain automatically and are used to construct better distribution by knowing the particular case and thus using better particular distribution construction algorithms and heuristics. Once system algorithms of static and dynamic analysis become more powerful, informational recommendations become superfluous. At that point pure FA is sufficient to construct an efficient parallel program.

According to this approach FA is made free of all non-functional decisions, which include multiple assignment (data mutability), order of computations (except informational dependencies), resources distribution, garbage collection and so on. In Charm++, for instance, multiple assignment present, which is currently employed to optimize performance, but later it will become an obstacle for existing Charm++ programs. Recommendations currently play critical role in achieving high performance, because current knowledge in parallel programming automation is not enough to efficiently execute such high performance representations as FA automatically. Recommendations cover the lack of such knowledge and allow to achieve satisfactory performance of FA execution.

3   LuNA System

FP is described in two languages—LuNA and C++. LuNA is used to specify DFs and CFs, as well as recommendations, while C++ is used to define sequential subroutines, which are used to implement CFs in run time. C++ is a powerful conventional language, supported by well-developed compilers and other tools, thus making single jobs —CFs—highly efficient, leaving the system solely with problems of distributed execution.

Older LuNA releases employed the semi-interpretation approach, where FP is interpreted in run time by LuNA run-time system. With this approach the run-time system interprets FP, constructs internal objects, which correspond to CFs and DFs, distributes them to computing nodes, transfers input DFs to CFs and executes CFs once all input DFs are available locally, etc. Current LuNA release employs conceptually the same, but practically more efficient approach. With this approach each CF is considered as a lightweight process, controlled by a program and being executed in a passive run-time environment, accessible via API. Program for each CF is generated automatically by LuNA compiler and usually comprises the following main steps:

- Migrate to another node (if needed), where CF will be executed,
- Request input DFs and wait for them to arrive,
- Perform execution on input DFs with production of output DFs,
- Spawn and initialize new CFs,
- Perform finalization actions.

Finalization actions may include deletion of DFs, storing computed DFs to current or remote computing nodes and so on. Certain steps may vary depending on CF type (single CF execution, subroutine invocation, for- or while- loop, if-then-else operator, etc.), allowed in LuNA language. (Here and below CF's program denotes the program, generated for the CF by LuNA compiler, which should be differentiated from C++ sequential subroutines, which are provided by user as a part of FP.) CF's program also depends on compiler algorithms, recommendations, hardware configuration, etc. Generally, all static decisions on how FP should perform are formulated as CFs' programs. Note, that CFs' programs are not rigid. For instance, the migration step is statically generated, but exact node and route to it may be computed dynamically. Generally, all dynamic decisions are left to run time.

Since CF's programs are generated in C++, they are also optimized by conventional C++ compiler, which takes care of many minor, but important optimizations, such as static expressions evaluation, dead code elimination, call stack optimizations and all other optimizations conventional compilers are good at.

While delegating serial code optimization (sequential CF's implementations and CF's generated programs) to a well-developed C++ compiler, LuNA compiler and run-time system focus on the distributed part of the execution. Based on recommendations, decisions on CFs and DFs distribution to computing nodes, order of CFs execution, garbage collection and others are made statically (in LuNA compiler) and/or dynamically (in run-time system). Consideration of these algorithms is out of scope of the paper and can be found in other publications on LuNA system.

4   Performance Evaluation

To investigate performance of generated programs in comparison with the previous approach a number of tests was conducted. As an application a model 3D heat equation solution in unit cube is considered. This application was studied in our previous paper [13], where more details on the application can be found. The application data consists of a 3D mesh, decomposed in three dimensions into subdomains. The computations are performed iteratively, where each step is solved with pipelined Thomas algorithm [14].

The testing was conducted on MVS-10P supercomputer of the Joint Supercomputer Centre of Russian Academy of Sciences [15]. It comprises $2 \times$ Xeon E5-2690 CPU-based computing nodes with 64 GB RAM each. The following parameters, representative for such applications, were chosen. Mesh size: from 100^3 to 1000^3 with step 100 (in every dimension), number of cores: from 2^3 (8) to 6^3 (216) with step 1 (in each dimension).

The results are shown in Fig. 1. Here LI (LuNA-Interpreter) denotes the previous LuNA release, where run-time interpretation approach is employed, while LC (LuNA-Compiled) denotes the current approach, where CFs' programs are generated. MPI denotes the reference implementation, hand-coded using Message Passing Interface.

From Fig. 1 it can be seen, that current LuNA release produces a much more efficient implementation, than the previous release, although reference MPI implementation outperforms them both. It also can be seen, that the most advantage LC over LI can be observed for smaller fragments sizes, which is expected, since serial code optimization mainly reduces overhead, which is proportional to number of fragments (and not their sizes, for example). The reference MPI implementation is about 10 times

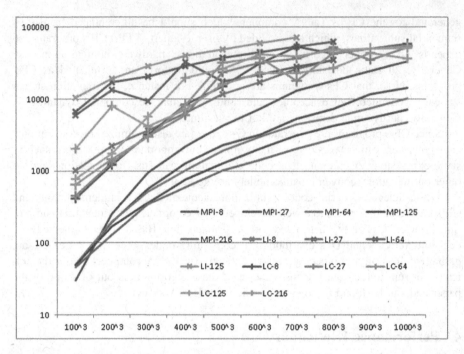

Fig. 1. Program execution time (in seconds). MPI- LI- and LC- are MPI-based, LuNA-Interpreter and LuNA-Compiled implementations correspondingly. The number denotes the number of cores. The X axis is the mesh size.

faster, which means, that more optimizations are required. In particular, network overhead, imposed by run-time system communications, has to be reduced. However, such a slowdown may be tolerable, because, firstly, development of FP required less skill and effort from the user, and, secondly, with system optimization existing FPs become more efficient as a consequence without any need to change.

5   Conclusion

An approach to achieve efficient execution of parallel programs, defined in a high level language, is considered, as well as its implementation in LuNA system for automated parallel programs construction. Performance tests were conducted to compare current LuNA performance with the previous release and reference hand-coded implementation of the same test. In the future both software optimization and development of intelligent system algorithms are required to achieve better performance.

References

1. ANSYS Fluent Web Page. https://www.ansys.com/products/fluids/ansys-fluent. Accessed 01 Apr 2019
2. Phillips, J., et al.: Scalable molecular dynamics with NAMD. J. Comput. Chem. **26**, 1781–1802 (2005)
3. MathWorks MATLAB official web-site. https://www.mathworks.com/products/matlab.html. Accessed 01 Apr 2019
4. GNU Octave Web Site. https://www.gnu.org/software/octave/. Accessed 01 Apr 2019
5. WOLFRAM MATHEMATICA Web Site. http://www.wolfram.com/mathematica/. Accessed 01 Apr 2019
6. Robson, M., Buch, R., Kale, L.: Runtime coordinated heterogeneous tasks in Charm++. In: Proceedings of the Second International Workshop on Extreme Scale Programming Models and Middleware (2016)
7. Wu, W., Bouteiller, A., Bosilca, G., Faverge, M., Dongarra, J.: Hierarchical DAG scheduling for hybrid distributed systems. In: 29th IEEE International Parallel and Distributed Processing Symposium (2014)
8. Bauer, M., Treichler, S., Slaughter, E., Aiken, A.: Legion: expressing locality and independence with logical regions. In: The International Conference on Supercomputing (SC 2012) (2012)
9. Malyshkin, V.E., Perepelkin, V.A.: LuNA fragmented programming system, main functions and peculiarities of run-time subsystem. In: Malyshkin, V. (ed.) PaCT 2011. LNCS, vol. 6873, pp. 53–61. Springer, Heidelberg (2011). https://doi.org/10.1007/978-3-642-23178-0_5
10. Sterling, T., Anderson, M., Brodowicz, M.: A survey: runtime software systems for high performance computing. Supercomput. Front. Innovations Int. J. **4**(1), 48–68 (2017). https://doi.org/10.14529/jsfi170103
11. Thoman, P., Dichev, K., Heller, T., et al.: A taxonomy of task-based parallel programming technologies for high-performance computing. J. Supercomputing **74**(4), 1422–1434 (2018). https://doi.org/10.1007/s11227-018-2238-4
12. Valkovsky, V., Malyshkin, V.: Synthesis of Parallel Programs and Systems on the Basis of Computational Models. Nauka, Novosibirak (1988)
13. Akhmed-Zaki, D., Lebedev, D., Perepelkin, V.: J. Supercomput. (2018). https://doi.org/10.1007/s11227-018-2710-1
14. Sapronov, I., Bykov, A.: Parallel pipelined algorithm. Atom 2009, no. 44, pp. 24–25 (2009). (in Russian)
15. Joint Supercomputing Centre of Russian Academy of Sciences Official Site. http://www.jscc.ru/. Accessed 01 Apr 2019

LuNA-ICLU Compiler for Automated Generation of Iterative Fragmented Programs

Nikolay Belyaev[1,2] and Sergey Kireev[1,2(✉)] ⓘ

[1] ICMMG SB RAS, Novosibirsk, Russia
kireev@ssd.sscc.ru
[2] Novosibirsk State University, Novosibirsk, Russia

Abstract. The work focuses on the application of Fragmented Programming approach to automated generation of a parallel programs for solving applied numerical problems. A new parallel programming system LuNA-ICLU applying this approach was introduced. The LuNA-ICLU compiler translates a fragmented program of a particular type written in the LuNA language to an MPI program with dynamic load balancing support. The application algorithm representation and the system algorithms used in the LuNA-ICLU system are described. Performance comparison results show a speedup compared to the previous implementation of the LuNA programming system.

Keywords: Fragmented programming technology · LuNA system ·
Parallel program generation · Dynamic load balancing

1 Introduction

The problem of efficient parallel implementation of numerical algorithms on supercomputers remains relevant since the advent of supercomputers. Previously, low-level programming of processes or threads with different memory models was mainly used [1]. In recent decades, the growing diversity and complexity of computing architectures and the need to raise the level of programming have made automation of solving system parallel programming problems increasingly important. A number of parallel programming systems was developed in order to simplify the development of parallel programs. An overview of modern parallel programming systems for supercomputers may be found in [2,3]. The following features may characterize them.

– Separation of the application algorithm description from its implementation. A special algorithm representation is usually developed to describe the application algorithm [4–11]. The representation is supported by an API based on an existing language [4–6] (or its extension [7]) or a DSL [8–11]. Efficient

Supported by the budget project of the ICMMG SB RAS No. 0315-2019-0007.

V. Malyshkin (Ed.): PaCT 2019, LNCS 11657, pp. 10–17, 2019.
https://doi.org/10.1007/978-3-030-25636-4_2

execution of the algorithm presented in this way is provided by special system software, a compiler and/or a distributed runtime system.

- Fragmented representation of an algorithm. The complexity of the automatic decomposition of the application algorithm in general case still makes it necessary to perform the decomposition manually. Thus, the algorithm must be represented in a fragmented form [4–11].

A common representation of an algorithm for many parallel programming systems is a set of tasks (fragments of computations) linked by data and control dependencies, forming a graph. The system software provides parallel execution of tasks, while satisfying the dependencies. The task graph can be defined statically [9,11], or be formed dynamically during the execution of the program [5,6,10]. The static representation of the task graph has the advantage that the entire structure of the graph is known before execution, which allows wider scoped compile-time optimizations. Examples of systems with static task graph representation are PaRSEC (DAGuE) [9,10] and LuNA [11]. Compared to PaRSEC, the LuNA language can represent a wider class of algorithms.

LuNA system is an implementation of Fragmented Programming technology being developed at the ICMMG SB RAS in Novosibirsk, Russia. Program in LuNA language (fragmented program) defines a potentially infinite data flow graph, built of single-assignment variables called data fragments (DFs) and single-execution operations called fragments of computation (CFs). Each DF contains one or a portion of application variables. CFs compute some DFs from others. There are two types of CFs in LuNA language: atomic and structured. Atomic CFs are implemented by C/C++ subroutines, while structured CFs are bipartite graphs of CFs and DFs. The LuNA language supports the following structured CFs: conditional CFs ("if" operator), indexed sets of CFs ("for" and "while" operators), and subprograms ("sub" operator). CFs' or DFs' names may contain an arbitrary number of indices, that allow them to be interpreted as arrays.

The current implementation of the LuNA runtime system is a distributed interpreter of LuNA programs. In the process of execution it gradually unfolds a compact notation of a potentially infinite task graph, performing dynamical management of a distributed set of DFs and CFs. However, the use of universal control algorithms in the implementation has led to the fact that the LuNA runtime system has a considerable overhead, which leads to a poor performance on real-world applications [12,13].

The paper presents another approach to the implementation of the LuNA system based on the static translation of a LuNA program into an MPI program. In this approach, the set of supported algorithms was narrowed to a class of iterative algorithms over rectangular n-D arrays, where n is the dimension of the array. The LuNA language was extended by additional high-level constructs in order to ease the program analysis. The implementation of this approach is a new LuNA-ICLU compiler. It provides construction of an MPI program with dynamic load balancing support. Using the example of the particle-in-cell method implementation, it is shown that the performance achieved by the

LuNA-ICLU is better than that of LuNA system and is comparable to the performance of a manually written MPI program.

2   LuNA-ICLU System

To overcome the problems affecting performance of the LuNA system, the LuNA-ICLU system is developed. As described above, performance problems of LuNA system are basically caused by using universal system algorithms of fragmented program execution. The idea of the LuNA-ICLU system is to apply system algorithms that are able to generate automatically a static MPI program from strongly defined class of fragmented programs. So, the applied program developer does not have to solve the system parallel programming problems such as developing of dynamic load balancing algorithms.

To generate a static MPI program from a given fragmented program it is necessary to analyze information dependencies between CFs described in the input fragmented program. Expressions of the LuNA language use CFs and DFs, including the indexed ones, which are parts of fragmented arrays. Index expressions can be complex and difficult to analyze. To overcome this problem a limited class of input fragmented programs is defined. In addition, the LuNA language was extended by certain high-level statements, which are described below.

In the current implementation of the LuNA-ICLU compiler the class of supported algorithms is the following. The fragmented program can contain 1D or 2D fragmented data arrays (arrays of DFs) and iteration processes described via "while" operator. DF values on current iteration are computed from a set of DF values from one or more previous iterations. The dimensions of DF arrays are strictly separated into temporal, over which iterations go, and spatial. Within iteration each element of DF array may be computed by CF from the elements of DF arrays with corresponding spatial dimension indices being the same. For example, DF A[i] can be computed from B[i], but not from B[i+1] or B[i*2]. The sizes of the corresponding spatial dimensions of different arrays must also coincide. The other types of dependencies should be supported in the language and compiler by special operators (see below). Such a class of algorithms is simple enough for compiler to analyze and contains solutions for many applied problems. In this paper a fragmented program for the PIC method solver is described. In future the class of supported input programs can be extended by implementing certain analyzing and code generating modules for compiler.

3   LuNA Language Extension

In order to overcome the problems of the fragmented program static analysis, the LuNA language has been extended by new syntactic constructions.

– The "DFArray" statement defines an array of DFs (its structure and sizes) that should be distributed among the nodes of multicomputer.

– Among the dimensions of the DF arrays, "spatial" and "temporal" dimensions
  are clearly distinguished. A "spatial" dimension is denoted by the symbols
  "[" and "]" and defines a set of DFs that correspond to the same iteration of
  the iterative process. A "temporal" dimension is denoted by the symbols "("
  and ")" and defines different iterations of the iterative process.
– Data dependencies between DF array elements on different iterations of
  "while" loop are specified explicitly in a loop header using expressions such
  as: `<A(i-1), A(i) --> A(i+1)>`.
– The "borders_exchange" and "reduce" operators define frequently met templates of structured CFs over arrays of DFs in order to simplify the process
  of information dependencies analysis and to apply a special optimized implementation in a target program.
– The "dynamic" statement marks a set of CFs in the iteration body that may
  cause a load disbalance.

4   System Algorithms in LuNA-ICLU System

4.1   Control-Building Algorithm

Since the idea of the LuNA-ICLU system is to generate a static MPI-C++
program from a fragmented program written in LuNA-ICLU language, there is
a necessity to design an algorithm that take a fragmented program as input and
convert it to a fragmented program with defined control, i.e. it should define a
partial order relation on a set of CFs.

In this paper, the bulk synchronous parallel (BSP) model for the target MPI
program was considered. Thus, a sequence of CF calls interleaved with communication stages should be built for each MPI process. CFs with spatially distributed
indices are distributed among MPI processes according to a distribution function
(see below), while the calls to the other CFs are duplicated in each MPI process.
The control-building algorithm follows the requirement that each CF must have
all its input DF values computed and stored in the memory of the corresponding
MPI process before it can be executed. The communication stages of the target
MPI program comprise operations such as DF boundaries exchange, reductions,
load balancing, etc.

4.2   Arrays Distribution Algorithm

To generate an MPI program from the fragmented program it is required to
generate a distribution of DFs by MPI processes. In the current implementation
only DFs that are elements of DF arrays are distributed. All other DFs are
duplicated in all MPI processes. Indexed CFs are distributed in accordance with
indexed DFs they produce.

In the target MPI program the distribution is defined by a mapping function that maps spatial coordinates of array elements to MPI processes. Compiler should generate this function and emit it to the target MPI program. The

requirement to the distribution generation algorithm is that it should provide the distribution of DFs that is as close as possible to a uniform. A naive algorithm is applied in the LuNA-ICLU compiler. It considers DFs to be of the same weight, so each DF array dimension is divided by a corresponding size of the Cartesian MPI communicator.

4.3   Dynamic Load Balancing Algorithm

A "dynamic" statement is used by LuNA program developer to tell the compiler that a given subset of CFs can cause a load disbalance on multicomputer nodes at runtime. Compiler should generate the call of load balancing algorithm implementation from LuNA-ICLU runtime library or inline the implementation of some dynamic load balancing algorithm to the output program in order to execute such kind of CFs efficiently.

In the LuNA-ICLU system the dynamic load balancing algorithm is implemented in a runtime library and the compiler inserts calls of corresponding implementation to output program. The load balancing algorithm itself meets the following requirements.

- The algorithm must overcome the load disbalance by changing the mapping function (see Sect. 4.2). At load balancing stage, DFs from overloaded multicomputer nodes are transferred to underloaded ones.
- The algorithm should be parameterized. This requirement is caused by a necessity to tune the algorithm for different applied algorithms and supercomputers. Examples of such parameters are unbalance threshold and frequency of load measurement. In the future versions of the system the execution profile analysis is going to be applied in order to tune the parameters automatically.

In the current implementation a dynamic diffusion load balancing algorithm is applied. In the description below we consider two DFs as neighbors if both DFs are the components of the same DF array and one of their corresponding indices differs by one. We also consider two processes as neighbors if these processes store neighboring DFs. Each process of the target MPI program stores a set of DFs' values that are available locally and a list of each DF's neighbors. The algorithm itself is the following:

1. Each process checks if there is a necessity to call the load balancer (the current iteration number of the iteration process is used).
2. Each process exchanges its current load value (which is basically a measured time spent on execution of CFs specified by the "dynamic" block) with all its neighboring processes.
3. Each process is searching for a neighbor with a maximum load difference compared to itself.
4. If the maximum loads difference is greater than the minimum disbalance threshold (which is basically a parameter of the algorithm), then the process calculates the number of DFs to be sent to the found neighboring process and selects certain DFs.

5. Each process exchanges the information about selected DFs and their neighbors with all neighboring processes.
6. Each process exchanges the values of selected DFs with neighboring processes.
7. Each process updates information about stored DFs and their neighbors.

The considered algorithm has several disadvantages. For example, restriction to local communications may cause a load gradient within a load threshold between neighboring processes, but with a large disbalance between distant processes. In addition, the number of neighboring processes may increase to a large value, which will increase the overhead of load balancing. However, as can be seen from the next section, the algorithm can be applied to resolve the load disbalance appeared when executing fragmented programs.

5   Performance Evaluation

To evaluate the performance of the program obtained by the LUNA-ICLU compiler a test problem of gravitating dust cloud simulation is considered [14]. The simulation algorithm is based on the particle-in-cell method [15]. Parameters of the simulation used in all test runs were the following: mesh size $160 \times 160 \times 100$, number of particles 500 000 000, number of time steps 800. Initial particles distribution was a ball with uniform density located in the center of the simulation domain. The domain decomposition in two directions into 16×16 fragments was applied, so that only several fragments in the center contain particles. Since the main computational load is associated with particles, such problem statement leads to a load imbalance.

Three implementations of the algorithm were developed, using MPI, LuNA and LuNA-ICLU. Moreover, two versions of the programs generated by the LuNA-ICLU compiler were compared: with load balancing and without it. The parameters of the load balancer were the following: the balancing module was invoked every fifth time step, the minimum disbalance threshold was set to 10%. All tests were run using 16 nodes of the MVS-10P Tornado cluster (16 cores per cluster node, 256 cores in total) [16]. The hand-coded MPI program and the MPI program generated by LUNA-ICLU compiler were run using one MPI process per core, whereas the LuNA program was run with one process per node and 16 working threads per process.

Figure 1 shows execution times obtained for different parallel implementations of the considered application algorithm. LuNA-ICLU implementation without load balancing outperforms the LuNA implementation by 10%, whereas with load balancing enabled the execution time decrease is 33%. Hand-written and manually optimizes MPI program even without load balancing outperforms all the other implementations, presumably due to more efficient memory management.

Figure 2 shows the dynamics of time spent by all cores at each time step on useful calculations compared to the time spent on communication operations, including waiting, when running LuNA-ICLU implementations. Without the load balancing enabled, calculations took up only 20% of the total time, whereas load balancing increased this fraction to 45% (60% in the steady state at the end of the simulation).

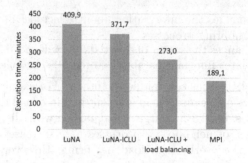

Fig. 1. Execution time for different parallel implementations

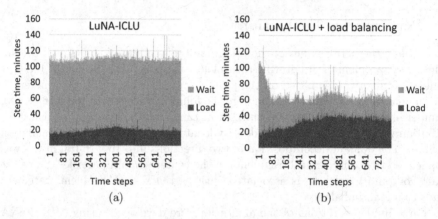

Fig. 2. Dynamics of time spent by all cores at each time step on calculations (Load) and communication operations, including waiting (Wait): LuNA-ICLU implementation without load balancing (a), LuNA-ICLU implementation with load balancing (b)

6   Conclusion

The paper takes a step towards improving the performance of fragmented programs. The problems of the previously developed LuNA system were considered and the prototype of LuNA-ICLU compiler was presented. The results of the performance evaluation are given. It was demonstrated that the performance of LuNA-ICLU system obtained on a PIC method implementation is better than that of the LuNA system and close to the performance of the manually written MPI program. The dynamic load balancing algorithm in the automatically generated MPI program provides a speedup of 1.3 times on the considered problem. The developed fragmented program compiler can be used to automatically generate efficient parallel programs from fragmented programs. In the future, compiler modules can be improved, giving the compiler the ability to support a more complex class of fragmented programs and generate more efficient MPI programs.

References

1. Kessler, C., Keller, J.: Models for parallel computing: review and perspectives. PARS Mitt. **24**, 13–29 (2007)
2. Sterling, T., Anderson, M., Brodowicz, M.: A survey: runtime software systems for high performance computing. Supercomput. Front. Innovations: Int. J. **4**(1), 48–68 (2017). https://doi.org/10.14529/jsfi170103
3. Thoman, P., Dichev, K., Heller, T., et al.: A taxonomy of task-based parallel programming technologies for high-performance computing. J. Supercomput. **74**(4), 1422–1434 (2018). https://doi.org/10.1007/s11227-018-2238-4
4. Legion Programming System. http://legion.stanford.edu. Accessed 23 May 2019
5. HPX - High Performance ParalleX. http://stellar-group.org/libraries/hpx. Accessed 23 May 2019
6. Mattson, T.G., et al.: The open community runtime: a runtime system for extreme scale computing. In: 2016 IEEE High Performance Extreme Computing Conference (HPEC), pp. 1–7 (2016). https://doi.org/10.1109/HPEC.2016.7761580
7. Charm++. http://charm.cs.illinois.edu/research/charm. Accessed 23 May 2019
8. Regent: a Language for Implicit Dataflow Parallelism. http://regent-lang.org. Accessed 23 May 2019
9. Bosilca, G., Bouteiller, A., Danalis, A., Herault, T., Lemarinier, P., Dongarra, J.: DAGuE: a generic distributed DAG engine for high performance computing. In: 2011 IEEE International Symposium on Parallel and Distributed Processing Workshops and Ph.d Forum, Shanghai, pp. 1151–1158 (2011). https://doi.org/10.1109/IPDPS.2011.281
10. PaRSEC - Parallel Runtime Scheduling and Execution Controller. http://icl.utk.edu/parsec. Accessed 23 May 2019
11. Malyshkin, V.E., Perepelkin, V.A.: LuNA fragmented programming system, main functions and peculiarities of run-time subsystem. In: Malyshkin, V. (ed.) PaCT 2011. LNCS, vol. 6873, pp. 53–61. Springer, Heidelberg (2011). https://doi.org/10.1007/978-3-642-23178-0_5
12. Akhmed-Zaki, D., Lebedev, D., Perepelkin, V.: Implementation of a three dimensional three-phase fluid flow ("Oil-Water-Gas") numerical model in LuNA fragmented programming system. J. Supercomput. **73**(2), 624–630 (2017). https://doi.org/10.1007/s11227-016-1780-1
13. Alias, N., Kireev, S.: Fragmentation of IADE method using LuNA system. In: Malyshkin, V. (ed.) PaCT 2017. LNCS, vol. 10421, pp. 85–93. Springer, Cham (2017). https://doi.org/10.1007/978-3-319-62932-2_7
14. Kireev, S.: A parallel 3D code for simulation of self-gravitating gas-dust systems. In: Malyshkin, V. (ed.) PaCT 2009. LNCS, vol. 5698, pp. 406–413. Springer, Heidelberg (2009). https://doi.org/10.1007/978-3-642-03275-2_40
15. Hockney, R.W., Eastwood, J.W.: Computer Simulation Using Particles. IOP Publishing, Bristol (1988)
16. MVS-10P cluster, JSCC RAS. http://www.jscc.ru. Accessed 23 May 2019

Objects of Alternative Set Theory in Set@l Programming Language

Ilya I. Levin[1], Alexey I. Dordopulo[2(✉)], Ivan V. Pisarenko[2],
and Andrey K. Melnikov[3]

[1] Southern Federal University, Academy for Engineering and Technology,
Institute of Computer Technologies and Information Security, Taganrog, Russia
iilevin@sfedu.ru
[2] Supercomputers and Neurocomputers Research Center, Taganrog, Russia
{dordopulo, pisarenko}@superevm.ru
[3] "InformInvestGroup" CJSC, Moscow, Russia
ak@iigroup.ru

Abstract. Software porting between high-performance computer systems with
different architectures requires a major code revision due to the architectural
limitation of available programming languages. To solve the problem, we have
proposed an architecture-independent Set@l programming language based on the
principles of set-theoretic codeview and aspect-oriented programming. In Set@l,
a program consists of a source code, which describes an information graph of a
computational problem, and aspects, which adapt an algorithm to the architecture
and configuration of a computer system. If an algorithm remains unchanged
during its architectural adaptation, calculations and their parallelizing are
described within the Cantor-Bolzano set theory. In the case of algorithm modification, some collections are indefinite, and we can not treat them as traditional
sets with sharply defined elements. To describe indefinite objects, Set@l applies
the alternative set theory developed by P. Vopenka. If collection has indefinite
type and structure at some level of abstraction, it belongs to a "class" type. In
contrast to a class, the indefiniteness of a semiset is an essential and inalienable
attribute. The application of classes, sets and semisets allows to describe various
methods of the algorithm implementation and parallelizing as an entire Set@l
program. In this paper the Jacobi algorithm for the solution of linear equation
systems is considered as an example of the utilization of classes and semisets.

Keywords: Architecture-independent programming ·
Set@l programming language · Alternative set theory ·
Aspect-oriented paradigm

1 Introduction

Nowadays the porting of parallel applications between high-performance computer
systems with different architectures implies the development of a new code due to the
architectural limitations of available programming languages and lack of efficient
methods and tools for architecture-independent description of computational algorithms. Existing approaches to architecture-independent parallel programming have

© Springer Nature Switzerland AG 2019
V. Malyshkin (Ed.): PaCT 2019, LNCS 11657, pp. 18–31, 2019.
https://doi.org/10.1007/978-3-030-25636-4_3

some significant shortcomings: they are based on the specialized translation algorithms (e.g. the Pifagor language of functional programming [1]) or on the fixed parallelization model (e.g. the OpenCL (Open Computing Language) standard [2]). To solve the problem we have proposed an advanced Set@l language of architecture-independent programming [3, 4]. It develops the basic principles of the high-level COLAMO (Common Oriented Language for Architecture of Multi Objects) programming language and set-theory-based SETL (SET Language) programming language which have the following disadvantages. COLAMO [5–7] is oriented only to the structural and procedural organization of computing. Traditional set-theoretical programming languages such as SETL, SETL2 and SETLX [8–10] lack the flexibility of the object- and aspect-oriented approaches, do not use indefinite collections and are not aimed at the description of parallel calculations.

In contrast to the aforementioned programming languages, Set@l classifies collections by parallelism, definiteness and other criteria, including every user's attributes. Furthermore, the Set@l language is based on the paradigm of aspect-oriented programming [11–14]. According to the paradigm, a typical program consists of a source code, which represents an algorithm in the architecture-independent form, and aspects, which specify the features of its implementation on computer system with certain architecture and configuration.

The source code in the Set@l language describes the information graph of a computational problem in terms of sets and relations between them. Architectural independence of the source code is determined by the indefiniteness of collections' types and of their partitions into subsets. The system of aspects specifies the variations of collections' decomposition, completes and redefines their attributes, and adapts an algorithm to the architecture and configuration of a computer system.

The parallelism of elements is one of the essential classification criteria for collections in the Set@l programming language. The language allows to combine collections with various partitions and types of parallelism in order to describe different methods of the algorithm parallelizing.

If aspects do not modify an algorithm during its architectural adaptation, the solution of a computational problem can be described within the Cantor-Bolzano set theory [15] (see example in paper [3]). However, the functionality of aspects is not limited to the parallelization of algorithms. In some cases, it is reasonable to modify an algorithm according to the architectural features of the computer system used for calculations. Then some collections are indefinite and are not sets; so, it is impossible to describe them using the concepts of the Cantor-Bolzano set theory. Another example of the blurred sets phenomena was demonstrated in the paper [3], where we introduced a special type **imp** denoting the indefiniteness of collections by parallelism.

The architecture-independent Set@l programming language describes various implementations of an algorithm in a unified aspect-oriented program. For this purpose we have introduced the classification of collections by the definiteness of their elements [3]. In the Set@l language indefinite collections are described by the special mathematical objects (classes and semisets). The concepts of a class and semiset were proposed by Vopenka within the alternative set theory [16–19]. In this paper we consider the application features of the alternative set theory objects in the Set@l language. To illustrate the aforementioned features, we have chosen the algorithm for the solution of the system of linear algebraic equations (SLAE) by the Jacobi iterative method [20]. As we will discuss in the forthcoming, it is reasonable to modify the algorithm for computer systems with the reconfigurable architecture [21].

2 Approaches to Implementation of Jacobi Algorithm and Their Set-Theoretical Description

There are two basic approaches to the computer-aided solution of a SLAE by the Jacobi iterative method. The first one is shown in Fig. 1a and assumes the verification of a termination condition during each iteration of computing. Figure 1b demonstrates the second approach, which implies one verification after several computational iterations.

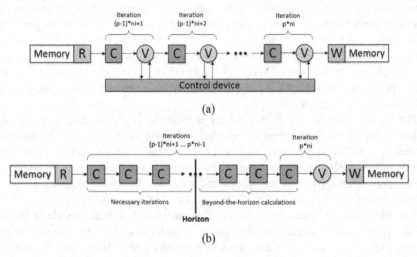

(a)

(b)

Fig. 1. Approaches to the implementation of the Jacobi algorithm for the SLAE solution on reconfigurable computer system: with verification during each iteration (a) and with one verification after several computational iterations (b)

In the case shown in Fig. 1a, each iteration of the Jacobi algorithm for the SLAE solution contains the following operations:

- the calculation of the column-vector of unknown variables (block C);
- the verification of the termination condition (block V) given by $err(k) \leq \delta$, where err is the residual; k is the number of iteration; δ is the fixed value of tolerance.

If the condition is true, the control device transfers data via the untapped blocks C and V and saves the result into a specially allocated area of distributed memory. This variant of implementation completely corresponds to the mathematical description of the Jacobi algorithm. In practice, the considered approach is efficient, but not for all computational architectures. Each verification block V performs the resource-intensive and time-consuming operation of $err(k)$ calculation. The hardware resource required for the implementation of V block is comparable with the C one, and time costs are equivalent too.

For the increase in hardware usage efficiency and reduction of time costs, it is reasonable to modify the Jacobi algorithm for the SLAE solution in case of the reconfigurable architecture. This modification assumes single verification of the termination condition in a cadr (see Fig. 1b). In the case being discussed, a cadr is a set of hardwarily implemented operations. These operations are united into an indivisible computing structure, which performs the functional transformation of input data flows into output ones [22]. If the condition is fulfilled before the operation of verification (in iterations with numbers from $(p-1) \cdot ni + 1$ to $p \cdot ni - 1$), further iterations will not worsen the calculation results. At the same time, the algorithm's modification provides the reduction of time costs: hardware resource freed from V blocks can be used for the placement of additional C blocks. The quantity of C blocks in the cadr is defined by ni parameter. It is worth noting that the both considered variants of the calculations' organization are suitable for multiprocessor computer systems as well as for reconfigurable ones.

The following paragraphs provide some important details on the set-theoretical description of two implementation approaches for the Jacobi algorithm, which are given in Fig. 1.

In the case of verification during each iteration (Fig. 1a), it is possible to detect the number of the last iteration I_m explicitly. In spite of the fact that I_m is unknown in advance, the set K of the algorithm iterations can be sharply defined by the termination condition as follows:

$$K = \mathbf{set}(k \mid k \in \mathbb{N} \,\&\, (k = 1 \text{ or } err(k-1) > \delta)), \qquad (1)$$

where $\mathbb{N}$ is the collection of natural numbers; **set** attribute classifies a collection as a set. In any case, 1-st and I_m-th (when $I_m \neq 1$) iterations belong to K, because the operation of verification always follows the corresponding recalculation procedure. To describe the information graph F of the Jacobi algorithm, we can assign the attributes of calculation (C) and verification (V) to each element of K set:

$$F = \mathbf{set}\left(k^{[C,V]} \mid k \in K\right), \qquad (2)$$

where $[\![\;]\!]$ braces denotes the indefiniteness of collection's type by parallelism. Taking into account statements (1), (2) and decomposition into blocks for the parallelizing by iterations, it is possible to represent K and F as follows:

$$K = [\![[\![1 \ldots ni]\!], [\![ni+1 \ldots 2 \cdot ni]\!], \ldots, [\![(T-1) \cdot ni + 1 \ldots I_m]\!]]\!]; \qquad (3)$$

$$F = \left[\!\!\left[\left[\!\!\left[1^{[C,V]} \ldots ni^{[C,V]} \right]\!\!\right], \left[\!\!\left[(ni+1)^{[C,V]} \ldots (2 \cdot ni)^{[C,V]} \right]\!\!\right], \ldots, \left[\!\!\left[((T-1) \cdot ni + 1)^{[C,V]} \ldots I_m^{[C,V]} \right]\!\!\right] \right]\!\!\right], \qquad (4)$$

where T is the number of the iteration block, in which condition $err \leq \delta$ is fulfilled; ni is the number of iterations in the completed block. According to the arguments given above, the approach to the implementation of the Jacobi algorithm for the SLAE solution with the verification of the termination condition during each iteration is described within the classical Cantor-Bolzano set theory.

The implementation approach shown in Fig. 1b attracts the highest interest, because its set-theoretical description requires the application of indefinite collections. Within the Cantor-Bolzano set theory, only sets with clearly defined elements are considered. Using this theory, we can describe only one special case K^* corresponding to the fulfillment of the termination condition at the iteration with $T \cdot ni$ number:

$$K^* = [\![[\![1 \ldots ni]\!], [\![ni + 1 \ldots 2 \cdot ni]\!], \ldots, [\![(T-1) \cdot ni + 1 \ldots T \cdot ni]\!]]\!], \qquad (5)$$

where T is the number of the last iteration block, and it is defined by stopping criterion $err(T \cdot ni) \leq \delta$. Otherwise, sets K and K^* describe different mathematical objects: collection K^* contains not only necessary but also excessive iterations (see Fig. 1b), and it is impossible to specify the exact location of fulfillment point for termination condition $err \leq \delta$. We can precisely detect that the condition is true at iteration $T \cdot ni$ and is false at iteration $(T-1) \cdot ni$. In contrast to the last element I_m of set K, the last element of K^* indicates the boundary of the subset, to which the iteration of fulfillment belongs. To keep the semantics of K definition and to provide the unified notation of objects in all modules of aspect-oriented programs, we need new set-theoretical methods and description aids, which go beyond the Cantor-Bolzano theory.

An alternative set theory proposed by Czech mathematician P. Vopenka is a field of mathematics dealing with indefinite collections and their set-theoretical description.

According to the Vopenka's theory, *a set* is a sharply defined and definite collection of certain objects. It is characterized by identity and represented as an independent and entire object [16]. For a set we always know exactly if one or another object belongs to it. In fact, plenty of naturally organized collections are not sets, because their elements are not clearly defined. The alternative set theory analyses the phenomenon of indefinite collections with the help of special mathematical objects (classes and semisets).

A class is specified in much the same way as a set, but it does not require the sharp definition of the corresponding collection of objects [16]. However, for each element of a class the belonging concept is considered in its traditional meaning, i.e. it is impossible that the chosen object both belongs and does not belong to a certain class. Since the class is indefinite it is not always feasible to detect precisely if the object belongs to a collection or not. Analogously to a subset, the alternative set theory introduces the concept of a *subclass*, which indicates the relation of inclusion between multiple blurred collections. If a class is a subclass of some clearly defined set, then it is *a semiset* [16]. The concept of a semiset is used for the description of cases, when an indefinite collection is a subset of a definite set.

In the papers of P. Vopenka, the category of *an indefiniteness* (fundamental for the notions of a class and a semiset) is closely connected with the horizon concept. A *horizon* is a special virtual object; if we move closer to it, the indefiniteness of our view appears. In contrast to clearly definite boundaries, the horizon does not have some fixed position and can move on its own. Objects near the horizon are always indefinite; approaching to the horizon increases the indefiniteness of their representation. Indefiniteness of a semiset and a class means that one or several horizons exist and limit our view of the observed objects.

When we implement the Jacobi algorithm for the SLAE solution with one verification in the cadr (see Fig. 1b), the collection of iterations K is a semiset, i.e. a class that has the subset relation $K \subseteq K^*$ with set K^*. The iteration at which the termination condition $err \leq \delta$ becomes true represents a horizon. Due to the implementation features of the algorithm, it is impossible to point out the horizon position precisely (see Fig. 1b). Depending on various factors (e.g., initial approximation or matrix properties), the horizon can move and form different variants of collection K. In the case of the condition fulfillment at iteration $T \cdot ni$, the horizon transforms to a sharp boundary, and semiset K becomes a definite set of operations, which coincides with set K^*. In general, the set difference of K^* and K corresponds to the semiset of special iterations. During these iterations, the condition $err \leq \delta$ is true, but calculations are not terminated because it is impossible to check the condition. The aforementioned semiset describes *beyond-the-horizon calculations*, which are not necessary from the mathematical point of view, but do not lead to the degradation of results. These calculations occur due to the features of the considered approach to the Jacobi algorithm implementation.

In order to describe semiset K as a mathematical object, we have to declare superset K^* and assign the corresponding subset relation between collections:

$$K_{sub}^*(k) = \textbf{set}(k_1 \ldots k_2 \mid k_1 = (k-1) \cdot ni + 1 \, \& \, k_2 = k \cdot ni); \qquad (6)$$

$$K^* = \textbf{set}\big(K_{sub}^*(k) \mid k \in \mathbb{N} \, \& \, (k = 1 \, \textbf{or} \, err((k-1) \cdot ni) > \delta)\big); \qquad (7)$$

$$\textbf{sm}(K) \subseteq \textbf{set}(K^*), \qquad (8)$$

where $K_{sub}^*(k)$ is the k-th subset of K^* corresponding to the k-th block of iterations; $\textbf{sm}$ attribute classifies the collection type as "semiset". The graph F of the Jacobi method implementation with one verification after ni computational iterations is described by the following formula of the relation calculus:

$$F = \textbf{sm}(P(k) \mid k \in K \, \& \, (\textbf{mod}(k, ni) = 0 \rightarrow P = [\![C, V]\!]) \, \& \, \ldots \\ \ldots \, \& \, (\textbf{mod}(k, ni)! = 0 \rightarrow P = [\![C]\!]), \qquad (9)$$

where $\textbf{mod}$ is the reminder of a division. It is worth noting that F contains the elements of collection K and is a semiset too. Each element of F is supplied with the attribute of calculation (C) or verification (V). Taking into account the partition into iteration blocks, semisets K and F have the following form:

$$K = [\![[1 \ldots ni]\!], [\![ni + 1 \ldots 2 \cdot ni]\!], \ldots, [\![(T-1) \cdot ni + 1 \ldots \textbf{?}\textbf{?}; \qquad (10)$$

$$F = [\![[1^{[C]} \ldots ni^{[C,V]}]\!], [\![(ni+1)^{[C]} \ldots (2 \cdot ni)^{[C,V]}]\!], \ldots, [\![((T-1) \cdot ni + 1)^{[C]} \ldots \textbf{?}\textbf{?}, \qquad (11)$$

where $?$ designates the indefiniteness of collection caused by the presence of a horizon, which does not allow to identify the number of the last required iteration exactly. According to expressions (10) and (11), T-th subclasses of collections K and F are semisets, and their union with sharply defined subsets of complete iteration blocks are semisets too.

As we have discussed in paper [3], every collection in the architecture-independent Set@l programming language have one of the following type attributes: a set (**set**), a semiset (**sm**) or a class (**cls**). These attributes describe the definiteness of collections' elements. "Set" and "semiset" types correspond to definite and indefinite collections, respectively, and "class" type identifies collections with the type that can not be specified explicitly. The application of classes provides the unification of objects' names in all units of an aspect-oriented program in Set@l. For example, in the Jacobi algorithm for the SLAE solution, iteration collection K and graph F can be sets (formulas (1)–(4)) as well as semisets (expressions (6)–(11)). It depends on the approach to the algorithm's implementation, which is described in the aspect of processing method. When we develop the source code of the program, the implementation details are still unknown. Therefore, we can not define the type of collection K unambiguously, and it is marked as a class.

To specify a collection as a class in the Set@l programming language, one has to assign **cls** attribute to this collection and give possible variants of its typing in any module of a program:

```
cls(<name of a class>);
typing(<name of a class>): <type 1> or <type 2> or … ;
```

A class is the most general and universal type of collections in Set@l. If it is necessary to introduce a collection in some module of a program, but it is impossible to explicitly define the type and structure of this collection on the current level of abstraction, then the collection can be declared as a class and can be used in the program analogously to a classical set. Owing to the extension of collection's definition in one of the aspects, the collection type and decomposition are concretized during the program's translation. In fact, the indefiniteness of collections by parallelism (denoted by **imp** attribute [3] in the Set@l language) can be described with classes.

3   Description of Jacobi Algorithm in Set@l Programming Language

Typing of collections by various criteria is one of the special features of the architecture-independent Set@l programming language. In addition to the typing of collections by parallelism and indefiniteness [3], programs can include every user's attributes declared by `attribute` keyword as follows:

```
attribute <name of attribute >(<collection or element>):
  <description of attribute>;
end(<name of attribute>);
```

An attribute is assigned to a collection or an element. It defines the methods of their processing by translator or specifies the relations between various objects of a program. Like any other objects in the Set@l language, attributes can form collections with appropriate types of parallelism and definiteness.

For the algorithm of the SLAE solution by the Jacobi iterative method, it is reasonable to declare the following attributes assigned to the elements of iterations' collection K:

- the attribute of calculation C connecting iteration number $k \in K$ with the set-theoretical description of the information graph for the following basic computational operation of the Jacobi method:

$$x_i^{(k+1)} = \frac{1}{a_{i,i}} \cdot \left(b_i - \sum_{1 \leq j \leq n, j \neq i} a_{i,j} \cdot x_j^{(k)} \right); \quad i = 1, 2, \ldots, n, \qquad (12)$$

where i, j are the indexes of row and column numbers; n is the size of SLAE; a is the matrix of the SLAE coefficients; b is the right-hand side vector; $x^{(k)}$ is the vector of unknown variables at k-th iteration;

- the attribute of verification V which sets up the correspondence between the iteration number and collection describing the information graph of $err(k)$ calculation, further checking of condition $err(k) \leq \delta$ and termination of computing in case of the condition fulfillment. The residual is calculated as follows:

$$err(k) = \max\left(\left| x_i^{(k+1)} - x_i^{(k)} \right| \right), \quad i = 1, 2, \ldots, n. \qquad (13)$$

In fact, attributes C and V describe the relation between iteration collection K and set-theoretical representation F of the information graph for the Jacobi algorithm decomposed into subsets by iterations.

Figure 2 shows the code fragment in the Set@1 language that describes the attributes of calculation C and verification V. A comment starts with a double slash symbol "//", and Set@1 treats all the information after it on a line as a comment. The attributes are assigned to the number of iteration k, where k is the element of iterations' collection K (see lines (1) and (7) in Fig. 2). The initial approximation is saved in a special set x_init. Before the first iteration of the Jacobi algorithm, x_init is assigned to the first subset of set x (line (2)). Symbol "*" means the entire selection of elements in a set or subset. To form the conditional statement in line (2) we us the logical operation of implication denoted by "->". The universal quantifier `forall` (lines 3–5) loops over the collection of row numbers I. Collection J contains the numbers of columns in the matrix of the SLAE coefficients. The summation of elements of J (line (4)) is described in terms of the relation calculus as follows:

```
sum(<term of sum>|<predicate for summation index>).
```

In line (8), the maximum of the specific set is searched; this set is formed by the differences between the corresponding elements of unknown vectors at $(k + 1)$-th and k-th iterations. Line (9) describes the verification of the termination condition. If it is true, the last approximation of unknown vector is saved in set x_res.

```
(1)      attribute C(element(k) in K):          // calculation attribute;
(2)          k=1 -> x(k,*)=x_init(*);
(3)          (forall i in I):
(4)              x(k+1,i)=( b(i)-sum(a(i,j)*x(k,j)|j in J and j!=i) )/a(i,i);
(5)          end(forall);
(6)      end(C);
(7)      attribute V(element(k) in K):          // verification attribute;
(8)          err(k)=max(abs(x(k+1,i)-x(k,i))|i in I);
(9)          err(k)<=delta -> x_res(*)=x(k+1,*);
(10)     end(V);
```

Fig. 2. Declaration of the attributes of calculation (*C*) and verification (*V*) in the Set@1 programming language

The fragment of the source code of the Jacobi algorithm for the SLAE solution in Set@1 is given in Fig. 3. In addition to this fragment, the source code contains the description of *C* and *V* attributes (see Fig. 2).

```
(1)      set(I);              // rows;
(2)      set(J);              // columns;
(3)      cls(K);              // iterations of matrix processing;
(4)      cls(F);              // information graph of the Jacobi algorithm;
(5)      graph(F):
(6)          x_res=F(x_init,a,b);        // relation described by class F;
(7)      end(F);
(8)      CI=sub(prod(C,K));   // the subclass of calculation iterations;
(9)      VI=sub(prod(V,K));   // the subclass of verification iterations;
(10)     F=union(CI,VI);      // F is a union of subclasses CI and VI;
```

Fig. 3. The fragment of the source code for the Jacobi algorithm in the Set@1 programming language

Lines (1)–(4) in Fig. 3 declare the collections of rows (*I*) and columns (*J*) of matrix, collection of iterations of the matrix processing (*K*) and collection *F* which represents the information graph of the Jacobi algorithm in a set-theoretical manner. Collections *I* and *J* of rows and columns are sets, because their elements are sharply defined by the size of a coefficient matrix. According to the chosen approach to implementation, collections *K* and *F* are sets or semisets. Therefore, in the source code we declare them as classes. In lines (5)–(7), information graph *F* is described as a relation between the triplet of sets (*x_init*, *a*, *b*) and set *x_res*, where *x_res* is the specially allocated set for the saving of calculation results. Lines (8)–(10) declare the generalized algorithm *F* as a union (union) of two subclasses (sub) *CI* and *VI*. These subclasses include the numbers of the calculation and verification iterations with appropriate attributes *C* and *V* (Fig. 2). In lines (8) and (9), the Cartesian product of collections is denoted by the keyword prod.

The code fragment of the aspect of processing method for the Jacobi algorithm is shown in Fig. 4. This aspect contains two main sections: the decomposition of sets (set construction, line (1)–(5)) and description of approaches to the algorithm's implementation (implementation method, lines (6)–(19)). The approach to implementation is declared by imp_method variable (line (7)). It can take two values: 'oneC_oneV' (the algorithm is implemented with the verification of the termination condition during each iteration according to Fig. 1a) and 'manyC_oneV' (the algorithm is implemented with one verification after several computational iterations (see Fig. 1b)). With regard to the chosen implementation method, class of iterations K can be a set or semiset (line (8)); corresponding branches of the program are described by the case statement (lines (9) and (14)).

```
(1)   set construction::
(2)       I=imp(BL(i)|BL(i)={set,imp}((i-1)*s+1 … i*s) and i in set(1..N));
(3)       J=imp(BC(j)|BC(j)={set,imp}((j-1)*c+1 … j*c) and j in set(1..M));
(4)       K*={set,imp}(IT(k)|IT(k)={set,imp}((k-1)*ni+1 … k*ni) and (k=1 or
                      err((k-1)*ni)>delta));
(5)   end(set construction);

(6)   implementation method::
(7)       imp_method=<variable describing the approach to implementation>;
(8)       typing(K)='set' or 'sm';
(9)       case(imp_method='oneC_oneV'):      // verification during each iteration;
(10)         type(K)='set';
(11)         K=(k|k in K* and (k=1 or err(k-1)>delta));
(12)         F=(conc(C,V)(k)|k in K);
(13)      end (case);
(14)      case (imp_method='manyC_oneV'): // one V block after ni C blocks;
(15)         type(K)='sm';
(16)         K=sub(K*);
(17)         F=(P(k)|k in K and (mod(k,ni)=0 -> P=conc(C,V)) and (mod(k,ni)!=0 ->
                   P=C));
(18)      end(case);
(19) end(implementation method);
```

Fig. 4. The aspect of processing method for the Jacobi algorithm in Set@1

Lines (2) and (3) of the set construction section (Fig. 4) declare the sets of matrix rows (I) and columns (J). These sets are used for the description of calculation (C) and verification (V) attributes (see Fig. 2). Row set I is decomposed into N subsets $BL(i)$, and each subset includes s rows. Column set J is decomposed into M subsets $BC(j)$, and each subset includes c rows. As a results, I and J are given as follows:

$$I = \left[\left[\underbrace{[1...s]}_{BL(1)}, \underbrace{[s+1...2\cdot s]}_{BL(2)}, ..., \underbrace{[(N-1)\cdot s+1...N\cdot s]}_{BL(N)}\right]\right]; \qquad (14)$$

$$J = \left[\left[\underbrace{[\![1\ldots c]\!]}_{BC(1)}, \underbrace{[\![c+1\ldots 2\cdot c]\!]}_{BC(2)}, \ldots, \underbrace{[\![(M-1)\cdot c+1\ldots M\cdot c]\!]}_{BC(M)}\right]\right]. \tag{15}$$

Different variants of parallelization by rows, columns and cells are specified by parameters s, c, N and M in the architectural aspect of the program [3]. The definiteness of a collection and parallelism of its elements are independent typing criteria. Therefore, the corresponding attributes (e.g., set and imp) form parallel-independent set enclosed into curly braces ({set,imp}) in the program code.

In line (4) of the set construction section (see Fig. 4), set K^* is declared. Its decomposition corresponds to the expression (5) and describes the parallelization by iterations typical for computer systems with the reconfigurable architecture. The subset $IT(k)$ of K^* consists of ni iterations and corresponds to k-th iteration block. The amount of subsets in K^* is calculated with the help of the termination condition. The supplementary set K^* is necessary for the declaration of semiset K, which appears in the case of implementation with one verification after several computational iterations (see expressions (6)–(11) and Fig. 1b). In contrast to the semiset K, set K describes the implementation variant characterized by the verification during each iteration (see expressions (1)–(4) and Fig. 1a). It is worth noting that set K is also a subset of collection K^*.

The approach to the implementation of the Jacobi algorithm with the verification of termination condition during every iteration (oneC_oneV, see Fig. 1a) is described by lines (9)–(13) in Fig. 4. In line (10), the typing of class K is refined: in the case being considered, it is a definite set (set). Line (11) forms the structure of set K using previously declared set K^*. For this purpose, only some iterations are selected from K^*. In these iterations, the termination condition is not fulfilled or is fulfilled for the first time. In line (12), the attributes of calculation C and verification V are assigned to every element of set K. The attributes form a parallel dependent set (conc) due to the sequential execution of corresponding operations. Iteration set K is applied for the declaration of collection F that describes the information graph of the Jacobi algorithm. Therefore, F inherits the typing of K ({set,imp}) and decomposition into blocks by iterations.

Lines (14)–(18) in Fig. 4 describe the implementation of the Jacobi algorithm for SLAE solution with one verification of the termination condition after several computational iterations (manyC_oneV, see Fig. 1b). In this case, class K is a semiset (line (15)) and a subclass of sharply defined set K^* (line (16)). K has the K^*-like structure that assumes parallelizing by iterations. Collection F contains the vertices of the information graph and inherits the types ({sm,imp}) and decomposition of semiset K (line (17)). Variable attribute P in the left side of relational expression (see line (17)) takes the following values:

- conc(C,V) or calculation and further verification of the termination condition;
- C or only calculation of unknown vector.

The value of P depends on the remainder (mod) of the division of iteration index k by ni.

Figure 5 demonstrates the code fragment of the architectural aspect of the Set@l-program for the SLAE solution by the Jacobi iterative method.

```
(1)     case(architecture_type='RCS'):        case(architecture_type='MP'):

                            // Rows(I):
(2)         s=1;                                    s=q1;
(3)         N=n;                                    N=n/s;
(4)         type(I)='pipe';                         type(I)='seq';
(5)         type(sub(I))='pipe';                    type(sub(I))='par';

                            // Columns(J):
(6)         c=1;                                    c=q2;
(7)         M=m;                                    M=m/c;
(8)         type(J)='pipe';                         type(J)='seq';
(9)         type(sub(J))='pipe';                    type(sub(J))='par';

                            // Iterations(K):
(10)        case(imp_method='oneC_oneV'):           case(imp_method='oneC_oneV'):
(11)            ni=floor(R/(Rc+Rv));                    ni=1;
(12)        end(case);                              end(case);

(13)        case(imp_method='manyC_oneV'):          case(imp_method='manyC_oneV'):
(14)            ni=floor((R-Rv)/Rc);                    ni=<number of iterations>;
(15)        end (case);                             end (case);

(16)        type(K)='pipe';                         type(K)='seq';
(17)        type(sub(K))='conc';                    type(sub(K))='seq';

(18)    end(case);                              end(case);
```

Fig. 5. The code fragment of the architectural aspect for the Jacobi algorithm

In the case of the reconfigurable architecture (RCS, left column in Fig. 5), the processing by iterations is used: collection K is decomposed into subclasses, and each subclass contains ni iterations. The following parameters of computer system's configuration are utilized during architectural adaptation:

- the available computational resource (R);
- the hardware costs for one iteration of calculation (Rc) and verification (Rv).

These parameters are declared in the special aspect for the configuration of computer system.

In the case of the multiprocessor architecture (MP, right column in Fig. 5), the processing by cells is used. The size of a cell depends on the amount of processors $q_1 \times q_2$. It is worth to note that the code being considered provides the implementation with one verification after several computational iterations (Fig. 1b) on multiprocessor computer systems. To choose this option, one should set integer value of ni, which exceeds unity, and additional parallelizing by iterations will be performed during the implementation of the algorithm.

4   Conclusions

In contrast to available architecture-specialized aids for the parallel programming of high-performance computer systems, the Set@l language allows to describe a computational algorithm and the methods of its modification and parallelizing as separate modules (a source code and aspects) of a unified architecture-independent program.

The description of indefinite collection according to the alternative set theory of P. Vopenka is one of the distinctive features of the Set@l programming language. In addition to typing by parallelism, Set@l provides the classification of collections by the definiteness of their elements. If the type and partition of a collection are not sharply defined on the current level of abstraction, it is declared as a class and is used in code analogously to traditional sets. The structure and typing of a class can be specified and re-declared in other aspects of a program. In addition to classes, semisets are introduced into Set@l. The indefiniteness is the essential characteristic of a semiset, and it can not be eliminated in aspects of a program. Using classes, sets and semisets, one can describe various implementation methods for algorithms in a unified aspect-oriented Set@l-program. The proposed approach to the programming of high-performance computer systems offers new possibilities in the development of architecture- and resource-independent software.

References

1. Legalov, A.I.: Functional language for creation of architecture-independent parallel programs. Comput. Technol. **10**(1), 71–89 (2005). (in Russian)
2. OpenCL: The open standard for parallel programming of heterogeneous systems. https://www.khronos.org/opencl/
3. Levin, I.I., Dordopulo, A.I., Pisarenko, I.V., Mel'nikov, A.K.: Approach to architecture-independent programming of computer systems in aspect-oriented Set@l language. Izv. SFedU. Eng. Sci. **3**, 46–58 (2018). https://doi.org/10.23683/2311-3103-2018-3-46-58. (in Russian)
4. Levin, I.I., Dordopulo, A.I., Mel'nikov, A.K., Pisarenko, I.V.: Aspect-oriented approach to architecture-independent programming of computer systems. In: Proceedings of the 5th All-Russia Conference on Supercomputer Technologies (SCT-2018), Izdatel'stvo YUFU, Taganrog, vol. 1, pp. 181–183 (2018). (in Russian)
5. Kalyaev, I.A., Levin, I.I., Semernikov, E.A., Shmoilov, V.I.: Reconfigurable Multipipeline Computing Structures. Nova Science Publishers, New York (2012)
6. Guzik, V.F., Kalyaev, I.A., Levin, I.I.: Reconfigurable Computer Systems. Izdatel'stvo YUFU, Taganrog (2016). (in Russian)
7. Dordopulo, A.I., Levin, I.I., Kalyaev, I.A., Gudkov, V.A., Gulenok, A.A.: Programming of hybrid computer systems in the programming language COLAMO. Izv. SFedU. Eng. Sc. **11**, 39–54 (2016). https://doi.org/10.18522/2311-3103-2016-11-39-54. (in Russian)
8. Stroetmann, K., Herrmann, T.: SetlX – A Tutorial. Research Gate Website. https://www.researchgate.net/publication/236174821_SetlX_-_A_Tutorial. Accessed 21 Jan 2019
9. Cantone, D., Omodeo, E., Policriti, A.: Set Theory for Computing: From Decision Procedures to Declarative Programming with Sets. Springer-Verlag, New York (2001). https://doi.org/10.1007/978-1-4757-3452-2

10. Dewar, R.: SETL and the evolution of programming. In: Davis, M., Schonberg, E. (eds.) From Linear Operators to Computational Biology, pp. 39–46. Springer, London (2013). https://doi.org/10.1007/978-1-4471-4282-9_4

11. Dessi, M.: Spring 2.5 Aspect-Oriented Programming. Packt Publishing Ltd., Birmingham (2009)

12. Kurdi, H.A.: Review on aspect oriented programming. Int. J. Adv. Comput. Sci. Appl. **4**(9), 22–27 (2013)

13. Podorozhkin, D.Yu., Kogaj, A.R., Safonov, V.O.: Application of aspect-oriented programming methods for development of software systems. J. Comput. Sci. Telecommun. Control Syst. **126**(3), 166–171 (2011)

14. Rebelo, H., Leavens, G.T.: Aspect-oriented programming reloaded. In: Proceedings of the 21st Brazilian Symposium on Programming Languages, SBLP 2017 (2017). Art. no. 10. https://doi.org/10.1145/3125374

15. Hausdorff, F.: Set Theory. AMS Chelsea Publishing, Providence (2005)

16. Vopenka, P.: Alternative Set Theory: A New Look At Infinity. Izdatel'stvo Instituta matematiki, Novosibirsk (2004). (in Russian)

17. Holmes, M.R., Forster, T., Libert, T.: Alternative set theories. In: Gabbay, D.M., Kanamori, A., Woods, J. (eds.) Handbook of the History of Logic: Sets and Extensions in the Twentieth Century, vol. 6, pp. 559–632. Elsevier (2012)

18. Gabrusenko, K.A.: Philosophical foundations of Georg Cantor and Petr Vopenka set theories. Tomsk State Univ. J. **339**, 32–25 (2010). (in Russian)

19. Vopenka, P.: The philosophical foundations of alternative set theory. Int. J. Gen. Syst. **20**(1), 115–126 (1991)

20. Bahvalov, N.S., ZHidkov, N.P., Kobel'kov, G.M.: Numerical methods. BINOM. Laboratoriya znanij, Moscow (2017). (in Russian)

21. Ebrahimi, A., Zandsalimy, M.: Evaluation of FPGA hardware as a new approach for accelerating the numerical solution of CFD problems. IEEE Access **5**, 9717–9727 (2017). https://doi.org/10.1109/ACCESS.2017.2705434

22. Kalyaev, I.A, Levin, I.I., Semernikov, E.A., SHmojlov, V.I.: Reconfigurable Multipipeline Computing Structures, 2nd edn. Izdatel'stvo YUNC RAN, Rostov-on-Don (2009). (in Russian)

Mathematical Abstraction in a Simple Programming Tool for Parallel Embedded Systems

Fritz Mayer-Lindenberg[(✉)]

Technical University of Hamburg, Hamburg, Germany
mayer-lindenberg@tuhh.de

Abstract. We explain the application of a mathematical abstraction to arrive at a simple tool for a variety of parallel embedded systems. The intended target systems are networks of processors used in numeric applications such as digital signal processing and robotics. The processors can include mixes of simple processors configured on an FPGA (field programmable gate array) operating on various number codes. To cope with such hardware and to be able to implement numeric computations with some ease, a new language, π-Nets, was needed and supported by a compiler. Compilation builds on a netlist identifying the processors available for the particular application. It also integrates facilities to simulate entire many-threaded applications to analyze for the precision and the specified timing. The main focus of the paper will be on the language design, however, that firmly builds on mathematical considerations. The abstraction chosen to deal with the various codes is to program on the basis of real numbers, and to do so in terms of predefined operations on tuples. A separate step is then needed to execute on some processor. To deal with errors, the number set is enlarged to also contain 'invalid' data. Further simplification is through the generous overloading of scalar operations to tuples e.g. used as complex signal vectors. Operating on the reals also fits to high-precision embedded computing or performing computations on one or several PCs. To these features, π-Nets adds simple, non standard structures to handle parallelism and real time control. Finally, there is a simple way to specify the target networks with enough detail to allow for compilation and even modeling configurable, FPGA based components in an original way. The paper concludes by a short presentation of an advanced target and by a funny example program.

Keywords: Real computing · Tuple data · Substitutions · Processor networks · Compilation

1 Introduction

The following exposition explains the design of a programming language and the related selection of methods and paradigms. Some are standard, a few appear to be novel, e.g. the transition from the abstract computation of the reals to the operations of the processors on number codes, the simple type system with its capabilities to change the base ring for vector and polynomial operations, and the compiled multi-threading based on a

V. Malyshkin (Ed.): PaCT 2019, LNCS 11657, pp. 32–50, 2019.
https://doi.org/10.1007/978-3-030-25636-4_4

course grained overlay of the data flow. There has been a sustained effort to keep things as simple as possible, using a simple syntax and avoiding unnecessary choices. It would have been still simpler to compile for a single processor of a fixed type but it is not the intended targets mentioned in the abstract that shall be restricted. For subsequent exposition there is the difficulty that our language and the compiler, albeit simple, try to be comprehensive such that a full description would have to touch many different aspects. So we will need to focus on the most essential, non-standard ingredients, the unified syntax for programs and data, but be short on details of the compilation and the thread management, and the run time environment including details on the supported FPGA processors and the hardware infrastructure they attach to.

There is the event of the SoC chips holding several standard processor cores and a fairly complex FPGA part with enough resources to configure dozens of specialized additional processors on it, and fast interfaces to expand them off chip. To fully access their resources one has to work through a technical documentation of many thousands of pages. For such a single chip it is mandatory to address heterogeneous networks, and a head start through a simple tool is most desirable. The present work can be seen as a practical investigation whether such head start is possible, with the by-product of an entry level environment to parallel and distributed programming to try out the concepts. Our paradigm to deal with FPGA programming for the intended applications is to build on a family of separately defined FPGA processors [2] attached to predefined hardware infrastructure, reducing it to defining or specifying a coarse grained network of processors further programmed through software. It is not intended to use FPGA overlays of similar processors as discussed in [22] but to use FPGA networks with application-specific mixes of processors for different number types.

The language π-Nets has evolved since some years and only now attained kind of a stable state. An early version has been described in [21]; a mix of simplifications (e.g., renouncing on an extra machine oriented data type) and functional extensions has been applied since. π-Nets has a predecessor language 'Fifth' used more than 30 years ago by engineers programming for small networks of the microprocessors of that time including the Transputer [5]. It was hardware oriented and based on implementing fairly efficient stack processors for every type of processor to be supported, thereby abstracting from hardware details. Programs could be run and tested interactively. Even today, one will hardly find tools for programming heterogeneous networks of processors at the system level (as opposed to one-by-one) not bound to a particular manufacturer – and small and simple. Therefore, the effort to design and implement a concise language for the intended targets and applications on contemporary processors and FPGA was accepted, taking up ideas from Fifth. The SoC with its external interfaces can actually be considered as a modern counterpart of the Transputer.

Programming languages, so π-Nets are classified based on of the underlying programming paradigms. In our understanding the main task of a programming language is to formally describe finite computations composed of many elementary operations (e.g., '+' and '*'), such that programs can be processed to automatically control a digital computer to actually perform the computation. In an abstract description the number of operations need not be explicit or data independent in which case the programmer has to care for it to be acceptable. In view of large applications, programs should be well-structured and understandable for the human computer user and hide purely technical detail. The existing imperative, functional and object oriented languages are of this kind [9, 10]. In our case, the mathematical orientation suggests a functional view. The only data to be handled in a program will be tuples of real numbers, i.e. tables of functions on finite index sets. Then there are 'program functions' (algorithms composing real operations) that operate on the tuple entries. While several elements of functional languages are used (e.g. the single assignment of data names), function tables and program functions are well distinguished. Program functions are statically defined only and not a type of data to be operated on. This distinction is crucial for being able to compile. Operations on functions and evaluations of constant tuples defined through algorithms are allowed at compile time. At the run time of a compiled program, fixed functions transform variable tuple input to tuple output. Algorithms on tuples don't need memory. For an embedded system performing computing processes in real time, data must be transported through time, however. To handle memory beyond distinguishing data names from memory locations holding them, a notion of automaton is added used borrowing some structural concepts of object oriented languages.

For embedded systems, the most common language is the imperative 'C' as functional and object oriented languages cause overheads for the processing of dynamic data structures. If the target contains several processors, they are usually programmed separately. The object-oriented C++, could be (and has been) used to define an abstract data type of real numbers, yet on top of types representing the numbers in the different codes what we want to avoid. Software tools tend to become complex when an operating system is required for the target, and libraries for time control and communications. The Occam language is an example for a language integrating synchronous communications. Several existing languages including C (e.g. within Arduino), Occam and the more recent Python have been claimed by their inventors and users to be simple languages. Our demands go further, and we need to support other targets.

Besides its basic structure and syntax, also the lexicographic features determine the appeal of a language, e.g. the way how to integrate comments, the keywords, and the allowed characters. For the handling of comments, π-Nets take up Knuth's idea of 'literate programming', interleaving the program with a textual software description [15]. π-Nets programs break up into series of definitions each beginning by one of four keywords only and having a well-defined end. All text lines not beginning by one of the keywords skipped as comments. Only after detecting a starting keyword the compiler switches on its textual analysis up to the end of the definition. With this feature, a program text can look like a page in a textbook with some formula lines in between (the program lines). Literate Programming has been promoted as and is believed to be a way to arrive at more thoughtful programs with fewer errors. Another lexicographic feature is to avoid of pseudo-natural key words such 'begin', 'end', 'if', and 'else'. Finally, π-Nets support the use of some Greek characters, subscript and superscript characters, and some more, depending on a program editor capable of inputting and displaying UTF-8 codes. More text processing features like page formatting and printing, formula editing and more choices for the characters would be helpful. The π-Nets compiler defines ASCII equivalents to the special characters and starts by replacing them by these equivalents before further analysis. To adapt to an editor using more textual features, the expansion into ASCII equivalents would have to be changed.

After these remarks we start by a discussion of the data handled by π-Nets, and the operations selected for them. The next chapter discusses a technique of substitutions of operations that is key to the execution of program functions on physical processors using encoded data, and to extend the mathematical capabilities of the language. We then turn to the functions, automata and processes, to parallelism, communications and timing control, and to the mapping of processes to processors, and conclude on how to implement π-Nets on the PC and other processors, and possible extensions.

2   Real Tuples with Invalid Entries

In this section we define the basic data and operations provided by the language, and a small calculus on how to construct tuples and to address their entries with a slightly non-standard notation. The choice of the reals as the basic data set to build on is motivated as an abstraction of the multitude of number encodings for computers and at the same time to clearly distinguish between a number and its various codes. Most algorithms used on computers were developed in a mathematical context composing real operations and functions and are not bound to particular number codes. We postpone the question of whether and how to 'implement' real operations to support the operation of the compiler and how to transition to the intended target processors operating on fixed length codes. Real numbers are limits of sequences of rational ones, and a computer can at best deliver rational approximations to certain reals within given error bounds. They also provide the abstraction of being able to arbitrarily increase the precision of finite computations. The complexity of real computation depending on the allowed errors is a topic of current research [12, 13]. Mathematical software systems supporting non-integer data typically provide a type of floating-point numbers, implicit and non-parametric in the worst case, and not intended for embedded targets.

After focusing on the reals, we need two extensions to arrive at the π-Nets data sets. The first is related to the processing of errors due to calling functions outside their domains. It consists in adding a set of non-numbers to form the data set $R = IR \cup \{nn's\}$. For both numbers and non-numbers literals are provided. For numbers, they are the usual decimal fixed and floating point literals like 14, 14.0, or $1.4 \cdot 10^{\wedge}1$ for the real number 14. The formats of the alternative real literals don't hint at particular ways to encode the numbers; they do encode precision. Without the decimal point, the number is supposed to be exact, otherwise its last digit is supposed to be rounded. Exact rationals are input as quotients of integers, if needed. Non-numbers are input as strings. The second extension is to the family of tuple sets $R_n = \{(x_0, \ldots, x_{n-1}) | x_i \in R\}$, identifying R and R_1. Tuples may thus have invalid entries. Tuple literals derive from the number literals by listing component literals within a pair of brackets. We note that tuples don't store numbers like a variable does, but are the data that could be stored in a linear array for reals. The R_n are the only data sets used in π-Nets; there are no pointers, no Boolean values, and indexes are just special reals. Tuples always have specific sizes to support static memory allocation and overloading by size.

Operations and functions map one or more input tuples of specific sizes to single output tuples but need not be defined for all combinations of input values. If called outside their domains, they are extended to deliver tuples of non-numbers. They also extend to invalid input by then delivering invalid output as well. Their original domain is the set of input tuples yielding a valid result. As a consequence, all functions defined in a π-Nets program are defined for every input and must terminate after a finite number of steps. Invalid data need not stop an application; applications to signal processing often tolerate some invalid output. The invalid data in R can further be used to transport information about the failure to deliver valid ones. When the numbers are eventually encoded for a machine, the non-numbers have to be mapped to at least one 'invalid' code; otherwise the indication of a result to be invalid would be lost.

An n-tuple $(x_i)_i \in R_n$ is defined to be table of a function $x: I_n \rightarrow R$, with $I_n = \{0, ..., n-1\}$. The natural order of I_n is used to list the table entries without their indexes. As a first operation of tuples, the pair of brackets (..) is used to construct a tuple from a list of numbers or other tuples by concatenating them, as already in the case of the tuple literals. The tuples are 'flat' and do not include the information on how they were concatenated. $((1, 2), (3, 4))$ and $(1, 2, 3, 4)$ are literals for the same tuple. The inverse to concatenation, i.e. breaking a long tuple into parts is implicit only and occurs, when a tuple x is named by a list of names for the sub tuples, 'x $\leftarrow$ a, b, c'. π-Nets functions are called with their argument tuple in the form 'f(x, y, z,...)', i.e. concatenating the arguments into a single tuple. In this case, the concatenation is formal only as the function starts by de-concatenating its argument into the previous parts again, allowing to check the number and sizes of the components. The function would also accept a single, large input tuple defined otherwise. For a single input tuple the call reads 'f(x)' or, alternatively, 'f x', which is common in mathematics and as easy to understand.

There are a few more operations only concerned with accessing the components of their tuple arguments:

x.i	or x_i	(apply to index argument, instead of the common x[i]),
x.y		(compose x,y as functions, y must be index valued),
x:m.i		(apply n*m tuple x as a function $I_n \rightarrow R_m$),
x:m.(i,j)	or x_{ij}	(double indexing for an n*m tuple x),
x:s:m.(i,j)		(double indexing of an m valued n*s tuple, etc.).

The tuple x remains the same, 1D or 2D addressing just being different access operations. The '.' default is 1D indexing or using ':1.'. It can be changed to ':m.' or 's:m.' etc. when x is named. x:s:(i, j) would then become 'x.(i, j)', and x a function $I_r \times I_s \rightarrow R_m$. Tuple sizes are currently limited to 2^{16}. Larger tuples need to reside in memory and can only be accessed by reading sub tuples from there.

The index sets for tuples are small compared to the value set R. Indexes are integers typically computed from integer indexes but not from variable input data except for interpolating from a table. The index ranges are defined at compile time, and index computations are compositions with constants. In contrast, functions composed of arithmetic operations and other predefined ones computing new tuple values for their result typically have infinite domains and are applied to variable data, too. Tuple operations often arise by performing the same scalar operation on every entry. The tuple operation then eliminates a conventional loop and contributes to have shorter, more abstract programs. π-Nets have the built-in feature to automatically extend every function f this way to larger tuples. LISP e.g. uses an explicit operator for special cases of this. If the arguments off are tuples of sizes r, s, t,…, then some or all can be replaced by m-tuples of r, s, t…-tuples for the same m (i.e. using the same index set) to produce an m-tuple of results, e.g.

$$f(x, y{:}s, z{:}t) = (f(x,y_0,z_0),\ldots,f(x,y_{m-1},z_{m-1})).$$

The extension actually applies to the concatenation operator (..), and to the indexing by a number to yield the indexing by a tuple of indexes. A function on m-tuples also applies to m scalars and vice versa. Indexing is different for both versions, however, and needs an extra selection if it is invisible.

The choice of the provided scalar operations is quite conventional, apart from dealing with ideal real operations and their immediate extensions to vector operations:

$+, -, *, /, \%, //$ (int.divide), $<, <=, <>, =, =>, >$,nni(int.test)   sqrt, ld, $2^\wedge$, sin, cos, atan

i32, x16, x35, v144, f32, f64, g45 (rounding operations).

The inclusion of the roundings is a unique, important feature of π-Nets. The concept was introduced in [1]. The roundings correspond to the supported number encodings. They are defined as the compositions of their coding and decoding maps and share their domains. 'x16', 'i32', 'f32', 'f64' are standard formats while 'x35', 'v144', 'g45' are non-standard for FPGA-based processors. 'v144' is a floating point vector variant of the 'x35' fixed point coding that encodes tuples differently from the usual tuples of scalar codes, using a common exponent for a vector of mantissas. As a rounding it operates differently than just by components [2]. The encodings thus become individual operations instead of extra data types. More operations and thereby encodings can be added without otherwise affecting the language. For the evaluation of constants at compile time the compiler employs a virtual machine (VM) 'implementing' its real operations in such a way that all needed roundings can be obtained from rounding the VM reals. As long as no codes of arbitrary precision are to be supported on the processors the VM can still use fixed word size codes for the individual operations. Computing constants, however, also applies to tuple operations and evaluating functions (arbitrary finite compositions of operations), and rounding errors will accumulate. Therefore the VM uses an encoding that is significantly more precise than required just for rounding individual operations, namely a double-double quad precision type [17, 18]. The methods in [20] can be applied to determine whether a composite operation is faithfully rounded. Double floating point codes can be attractive and supported on FPGA processors, too [21]. π-Nets could make use of an arbitrary precision floating point library like [19] to select the real operations of the VM to be of any desired precision to further extend its applications to high performance high precision ones.

The choice of predefined tuple operations includes extensions of the scalar ones and some more. Their selection determines the expressiveness of programs but must also take care to cover the most common cases only to keep things simple:

+, −, *	(std. vector sum and difference, product of functions),
sum, min, max, ‖.‖	(having single tuple arguments),
x y	(dot product of n-tuples x,y, matrix*vector for x size n*m),
x:m y, x/\y	(matrix*matrix, tensor product, vector/Clifford product),
x § y, x §* y	(apply as polynomial function, 1-16 variables, multiply),
x ipl y	(interpolate from table of samples, 1-16 variables),
x find y , order x	(selected set operations).

The notation 'x y' is similar to applying a function writing 'f x'. For scalars it becomes the standard product, now written without the '*' character. A few more operations to solve linear and polynomial equations are under consideration. Sampling and interpolating functions are complementary similarly to the coding and decoding of numbers. The selection has been made after evaluating a number of benchmark programs.

Applications usually add extra, composite operations. The next section explains, how and how data types like complex numbers can be defined and used.

3 Functions, Substitutions, Encoded Execution

In this section, 'function' means 'algorithm based on the predefined operations' and does not refer to the tuples as function tables. The first level of composing operations is the expression. Expressions use infix notation for the arithmetic operations and prefix function calls, and are right associative on the same priority level. The (..) brackets override the infix priorities, thus reappear in another role. Expressions evaluate into single result tuples and don't give access to sub expressions. For such, and for adding program control (branches), a structure of nested blocks of expressions is provides by further extending the options of the (..) brackets. Not having to use different brackets for program control and defining data tuples simplifies the language. The early LISP also worked with a single pair of brackets. Brackets holding programs are only distinguished from data by holding constant expressions only evaluated at compile time whereas variable expressions are evaluated only later to data, and on demand. The options are

- define computed constants
  (3+1/7)
- listing several expression to concatenate their results
  (h+1, h−1)
- listing named sub expressions for referencing, not necessarily output
  ($a^2 − 4$ ←h h+1, h−1)
- using formal parameters to be able to compute with varying data
  (←a $a^2 − 4$ ←h h+1, h−1)
- naming a top-level bracket to be able to call it as a named function
  fct abc (←a $a^2 − 4$ ←h h+1, h−1)

– permit nested open branches to one of several closing brackets
(... expr0 ... x < y ? ... expr1 .) ... expr2)
– indexing, end recursion, and recursive named functions (example in ch.6).

In all cases (except for the naming of a global constant or function), the (..) structure can figure as a data operand within an expression. Data names including formal parameters are valid within the (..) only, disallowing references to tuple values therein. The use of formal parameters makes sense for unnamed sub blocks as the automatic extension to tuples applies. The branching syntax is similar to the 'x?y:z' of 'C'. If the condition ('x<y' in the example) holds, expr1 is computed and output at the exit '.)', otherwise expr2. The usual keywords for branches are thus replaced by '?' and '.)'. There can be no mutual references to values of expressions in the different branches. '?' can also be used test for the result of an expression to be valid. Several conditions can be listed and then stand in conjunction. Without a second branch the block returns the tuple with all entries invalid if the '?' condition fails, or becomes a condition itself. Indexed blocks produce a long, concatenated result tuple. This last option can also be used to define constant tuples, writing e.g. (12: 0) for an all zero 12-tuple. A few more options exist for the process control blocks discussed in the next section.

π-Nets provide an additional structure to the set of functions defined in a program that is primarily intended as a simple substitute for data type definitions, and to reducing the number of symbols through further overloading. A full data type definition would define some particular data set and operations on it. In our case, no data sets can be added, such that only a set of operations remains. The simplified structure is called and declared as a 'function type' with an optional default size for the tuples the member functions apply to. All members of a given function type 'abc' get double names composed of the type name and a selector, e.g. 'abc fgh'. This double naming can be used just to have more expressive names. The overloading comes in when an expression or a block is prefixed with the type name. Then type members are no more selected by their full names but by their selectors only which then overload possible previous meaning of the selectors. A function type 'cpx' of complex operations can e.g. define a member function 'cpx *' (complex multiplication):

$$cpx * (\leftarrow a, b, c, d \quad a\,c - b\,d, \quad a\,d - b\,c).$$

Then all 4-input multiplies in a prefixed block 'cpx (...)' then become complex multiplications.

A unique feature to the knowledge of the author is that through overloading the basic arithmetics to defining some particular ring, this change automatically extends to a change of base ring for the *predefined* linear and polynomial operations as well. 'A v' becomes the multiplication of a complex matrix (a $2n^2$-tuple) to a complex vector (a 2n-tuple), and polynomials have complex coefficients and evaluate a complex argument (a 2-tuple) to a complex result. Multiplying the real '1' (a 1-tuple) to a real or complex number remains unchanged. As a result, reals need not be 'converted' to complex ones by appending a '0', and even complex polynomials evaluate correctly on real input. Many interesting real algebras can be implemented similarly as function types. Ambiguities are resolved by simple rules or restrictions. In the embedded block

$$\ldots \text{expr} \leftarrow 8 \text{ h}\ldots\text{cpx}(\ldots \leftarrow x\ldots h\S x \leftarrow y\ldots)\ldots$$

the argument x of the polynomial 'h§' is complex but the polynomial is one with real coefficients as the tuple h was defined outside the block. h§x correctly evaluates to the complex number y. Also, there is the option to overload '+' and '*' with the integer modulo operations to deal with finite fields. We note that the overloading of the arithmetic operations for other data is very common in mathematics, and not at all a source of misunderstandings. Even if the listing of expressions within the unified brackets may appear oversimplified, this is compensated by the ease to switch to different data types, to use expressive double names, and e.g. to program on the level of complex tuple operations. Not to construct additional data sets reflects that the tuples sets and the supported mapping of tuple indexes are rich enough for most numeric applications.

Another mathematical feature has been selected for π-Nets to support applications to robotics, namely the variant of automatic differentiation to derive an algorithm to compute partial derivatives of a function from an algorithm just for the function. By integrating this into the compiler, there is nearly no complication for the programmer. The function f: $R_n \rightarrow R_m$ is simply given the composite name T'f in its definition along with the block defining f. It can then be called both as 'f' and as 'T'f' which is then a function $R_n \times R_n \rightarrow R_m \times R_n$ of two tuple arguments, mapping $(u, v) \rightarrow (f(u), Df(u)$ v) where Df denotes the full differential of f. Similarly, a tuple of n-tuples named as T'U is a tuple of 2n-tuples that can also be referenced by U as a tuple of n-tuples. Automatic differentiation is a formal operation not actually determining Df as a linear approximation to f. If T'U is obtained by sampling a differentiable curve c then T'f maps T'U correctly to the samples of f°c. In other words, T'f correctly composes with sampled tuple functions T'U. The idea of automatic differentiation is not new [8], but the place to integrate it may be. Another useful variant of automatic differentiation is to derive the Hamiltonian vector field of a function and compute solution curves [7].

There is a common basis for all of the indicated automatic features, the extension of functions to tuples, the switch from real to complex operations, and replacing function and operations be their tangential maps. It is to exchange the operations in an algorithm with other operation yet maintaining the composition scheme to arrive at an algorithm for different data, or defining algorithms through a composition scheme and a compatible assignment of operations [3]. For the extension to tuples the change is from the original operations in the program to their individual extensions. The replacement of real by complex operations or the like already occurs during the textual analysis and the construction of the intermediate code by the compiler. A substitution also occurs for the eventual execution of a program on some processor using some particular number encoding. The processor performs 'encoded' versions of the real operation on the data codes to produce result codes. These don't decode to the true result of the real operation but to the rounded one, the rounding being the composition of first coding then decoding. The switch is here from real to encoded operations or, for the processing on the reals, the change of replacing every operation by the one obtained by performing it and rounding its result. In this case, the composition of real operations no longer corresponds to the composition of the encoded operations in general, and the machine delivers worse approximations to the true results than a unique final rounding. Using this substitution the VM can simulate the encoded operations.

4 Parallelism, Communications and Timing

The pure functions discussed above are complex SW building blocks. Whether they are really executed depends on the processes defined for the application. There may be many processes; all start in parallel when the application is started and perform encoded operations on processors of the target. Processes use additional building blocks, in particular variables and automata with a state memory. Variables are pre-defined automata offering read and write operations of tuples of real numbers, and serve to define automata with hidden state memories. They are organized by automata types defining variables and associated functions. The access functions are, apart from per-forming read and write operations, similar to the pure functions before. Each automaton of the same type has its own instances for the variables. We skip the details.

An application process is defined by first listing its variables and sub automata and then providing a control block starting by '#' to indicate that a new thread is described:

$$\text{apc pnm, x, 4 y}\quad(\# \ldots \ldots)$$

would define a process 'pnm' with a scalar variable x and a variable y holding 4-tuples. Optionally, initial values can be specified for them. The control block is similar to the algorithmic block used for the pure functions, can contain branches, sub blocks etc. It differs by having no arguments and result, and allowing a few extra, non functional, sequential operations (read, write, send, receive) and structures (#, $$). This completes what is needed and provided to abstractly describe parallel real-time systems. The extras are confined to the top level processes which also invoke pure functions and access functions on their sub automata.

The write operation to a variable x is reserved to the process for which it is declared and reads

$$\text{expr} \gg x$$

where 'expr' is a tuple expression for the data to be written. The chosen syntax for this is thus quite different from the naming of an intermediate tuple item, 'expr←nme'. The read operation is simply 'x' like calling a function, this time actually calling to the automaton 'x'. A process 'p' can also read variables of another process 'q' writing 'q x' which is called 'state sampling'. State sampling involves no synchronization but is a first way to let the processes communicate. Handshaking and synchronous communi-cation can be derived from this. The send operation from p to q sends a list of tuples,

$$r, s, t, \ldots \gg q.$$

The tuples are received by q one by one simply writing 'p'. This is stream com-munications and involves synchronization. Stream communications are supposed to be buffered such that only the receiver has to wait for data. The similarity of the syntax to accessing a variable is intentional. Variables and sub automata can be viewed as primitive processes communicated with. The syntax also fully abstracts from how communication is implemented on the PC or an embedded target.

Sampling and stream communications also serve for general input and output (IO), name by sending, receiving, and sampling with predefined or external processes. Screen output is through sending tuples to the predefined host terminal process 'HTC'. File IO is handled similarly. When an embedded system is programmed for, the processes will eventually be executed by the target processors using dummy processes to describe IO. Alternatively, the environment can be modeled by giving appropriate process definitions for it, and executing them in a comprehensive simulation.

In order to be able to easily define many processes for a highly parallel target, or to group processes with a related timing or at least with related control, processes can break up ('be dissected') into several threads. The dissection is by means of the '#' control already used at the start. It can be used almost anywhere (not within expressions) to coarsely cut the control block into sections which are labeled to indicate that sections belong to the same thread. Writing to a specific variable and sending to a specific process is always bound to a specific thread. The pattern

$$(\#S \ \dots \ \text{expr} \leftarrow y \ \dots \dots \ \#T \ \ \dots \dots \ f(y) \leftarrow z \ \dots \dots \ \#S \ \dots \dots \ z >> u \ \dots \ )$$
$$\quad\quad\quad \text{(send y)} \quad\quad\quad\quad\quad\quad \text{(rec y) (send z)} \quad\quad\quad \text{(rec z)}$$

is for a process subdivided into two threads S,T. The threads are processes of their own and might eventually run on different processors. The shown communications are then needed but fully abstracted from by simply referencing data computed in another thread (as T references y in the call of f). In the example the result of f is sent back to S which is similar performing a remote function call from S to T. The dissection does not affect the data flow of the full process but simply defines a distribution of the workload. This also holds for the control flow. A branch carried out in one thread also carries over to another if it has sections in one or both branches. In

$$(\#S \dots \text{cond?} \dots \#T \dots \ .) \dots \ \#T \ \dots)$$

thread S computes a branch condition and branches accordingly. T takes over in the first branch (after the '?'). S also executes the first half of the second branch (after '.)') until T takes over again. The branch condition has not been computed by T but T must branch, too. This implies that the information whether the branch condition holds or not will be communicated to T. As a result communications between the threads can be compiled as all follow the same control path such that one can define unique sequences of send and receive operations between any two threads.

Dissection is also made compatible with the structure of sub blocks for the case that the block also contains sections of another than the calling thread. Flow control within the block is constrained to it, i.e. a branching within the block does not affect outside threads which then have independent control flows for a while. Also, there is no communication between insides the block and outsides but communications with other threads may be through explicit send and receive operations. Threads of the block which already started before synchronize at their entry and each time the block is repeated. The application processes could all be packed as sub processes inside a large container process yet with no particular benefit. Blocks with threads can be indexed and deliver a tuple result to the calling thread. The following creates 100 threads computing the components of a vector in parallel:

$$\ldots (100 : \ \# \ldots \ \ldots r, s, t) \leftarrow 100 : 3 \ v \ldots$$

The dissection into threads is particularly simple to define and to change as it only concerns the labeling of the '#' dissectors. The finest granularity would be to have just one expression in every section, each being a thread of its own except those performing stream communications and writing variables. Single operations, even tuple operations or function calls, cannot be further dissected. As the dissections respect the control flow, they can be viewed as a coarse-grained overlay to the data flow with the purpose to group the expressions to be evaluated into threads.

The last, still missing element is timing control. A single command is in use for this. It causes the calling thread to wait for a given time after the previous wait command or the start of the actual process or sub process before performing the next output to a variable or process. It is denoted '$$d' for waiting for d seconds. The cyclic process (cyclic due to the '<' at the end)

$$(\# \ \ 0 \gg x \ \ \$\$1 \ \ 1 \gg x \ \ \$\$1 \ \ <)$$

repetitively outputs to a variable 'x' that can be sampled by an external process as a square wave (period 2 s). The parameter of '$$' need not be constant. The control depends on a predefined process outputting the real time from the start of an application in seconds to a variable RT. The command can also be used to define timeouts for one or more alternative receive operations.

'$$d' has another important use. In a sequence

$$' \$\$c \ \ f(x) \leftarrow h \ \$\$d \ h \gg y'$$

waiting at '$$d' only occurs when the execution time to compute 'f(x)' on a processor is less than d. This is a real time condition which can be checked by simulation.

5   Simulation and Execution on the PC, and Code Generation

The PC is not required to be able to execute the real-number algorithms compiled on it. For the task of compiling for simple processors under real time restrictions it is enough to compute at a level of precision from which the computations of the processors using codes of finite word sizes can be obtained through rounding. The PC disposes of the virtual machine already needed by the compiler to compute constants. It is used to interpret the real operations in the intermediate code output by the compiler, too, and can do this fairly efficiently as it performs precompiled tuple operations using fairly precise yet still finite word size codes. The availability of the VM thus plays an important triple role. It also allows the PC to figure within the target network and take over some processing (and, typically, the user interfacing). If the PC runs as part of an application it can interactively sample variables on other processors which is useful for control and for debugging. And it can be used to simulate entire applications including the processes in its environment, their timing and testing for real time conditions. The VM executes the many application threads with context switches each time a thread has to wait as prescribed by π-Nets. The π-Nets approach also works for networks of PCs connected by an LAN, each running a VM of its own.

Native code generation is being worked on for an ARM based micro controller and will extend to the ARM processors in SoC chips with fairly complex FPGA parts. On an FPGA, entire processors can be configured performing scalar and non-scalar operations using standard or non-standard number codes of unusual lengths. Processors can but generally need not run an operating system (OS). The compiler then attaches a small run time kernel instead. FPGA OS have started to emerge, too [11]. A family of FPGA based processors attached to a dedicated hardware infrastructure for external communications, memory interfacing and providing sequential controllers has been designed in related projects, combining a controller with a number of arithmetic units of different complexities to choose from [2]. They execute from small memory blocks within the FPGA in the range of a few 10 kB only and rely on DMA supported software caching for more. Each runs up to four threads, providing separate registers for them and performing context switches in zero time. The design tries to maximize ALU efficiency through parallel memory and control operations. An FPGA can be configured to hold several to many of them depending on the complexities of the arithmetic circuits, each bringing in some 10^8 operations per second. External memory connects to the FPGA and provides the storage for large programs and data.

Ongoing work is to port the processors with some needed changes to a more recent FPGA platform. This is particularly motivated not only by hoping to attract applications, but also by the design of an experimental processor system that links 50 SoC nodes into a powerful parallel computer. It can make use of all π-Nets facilities and is the flexible hardware counterpart to what may prove a flexible tool for its use [24]. This system contains 100 ARM processors, up to a thousand FPGA processors for running thousands of threads, 100 GB of DRAM and 800 GB of nonvolatile storage. Memory is strictly distributed to the nodes. Every node connects to 7 neighboring nodes though high-speed interfaces managed by the infrastructure. The system is wired up for the famous Hoffman-Singleton graph [23] with a diameter of two only.

To define the target network for an application with just enough detail to permit compilation, π-Nets provides the syntax to enter a netlist for it through a series of simple commands, building on a hierarchy of modules containing processors of some types, networking and memory nodes, and sub systems built from them. The command to define an additional target component has the general form

node (type name) component name ← components linked to it

and defines for every component the previous ones it is linked to. Indexed sets of components of the same type can be defined using an extra parameter, and parameters like memory sizes be specified. Optionally, code distribution can be specified through one or several boot trees. There are predefined nodes, the 'HOST' computer of the PC type connected to the 'LAN' node. 'LAN' is a non-computational node of the B type ('bus'). Another non-computational type is the 'M' node just providing memory. An M node can be connected to several processor nodes, and variables of the attached processors can be allocated on it. M nodes are used to model the external memories attached to an FPGA and to which the simple FPGA processors are linked. If the definition of the target by defining node types and nodes can be considered as a sort of programming, then it is a purely structural one. The targets definition can also be understood as a target specification, and the definition of a network of FPGA processors could be further processed to actually derive the FPGA configuration data.

The component hierarchy includes some special support for dealing with FPGA based processors. The FPGA is presented as a virtual component type with an internal network, memory nodes and fixed function automata (i.e. a hardware infrastructure), including hardwired processors if there are any. Various configurations adding sub networks of processors configured on the FPGA are then defined as types inheriting from the virtual type. The link to the encodings supported by the rounding operations of the language is that the processors implementing the encoded operations are selected by the name of the encoding; this works on the basis of the compatible processors only differing by their arithmetic unit circuits. A node type definition inheriting from a type SoC with memory nodes M0 and M1 and attaching 8 FPGA processors of the type 'f45' and 4 of the type 'v144' reads

$$\text{node type (SoC)} \quad \text{(f45)} \; 8R \leftarrow M0 \quad \text{(v144)} \; 4\,S \leftarrow M1$$

The target definition for the 50-node computer mentioned above reads

$$\text{node (SoC.x)} \; 50 \; P \leftarrow \; LAN, hs \; P.$$

It declares a set of 50 SoC nodes of sub types x derived from the virtual type by defining various combinations of processors on them. 'x' stands for a tuple/table to select from a set of configurations for every P node. Thus the single tuple constant x selects one of the large set of possible heterogeneous configurations of the entire system. The computer is linked to the 'LAN' and thereby to the host processor for code downloads etc., and the P nodes are connected to each other according to the interconnection list 'hs', another tuple constant holding the table for the Hoffman-Singleton graph. Depending on the contents of 'x', and also due to the presence of ARM processors in every SoC node and the PC, the system is highly heterogeneous. The configuration can be dynamically changed on individual nodes.

A modern PC with several processor cores and a graphics subsystem with a separate memory could be described similarly to the SoC chip as another parallel target. We skip further details on this and content ourselves to state that even with the extras to specify the target, the simplicity of the language is not compromised as most details are resolved by the compiler and the runtime environment, including code distribution and FPGA configuration from an attached memory. It remains to be said how the hardware is actually employed for executing an application program. The required assignments are manual by annotating the program text. This and some other choices described before would profit from being automated, deriving e.g. a target network specification from the network of application processes [4]. Editing the thread and processor assignments is easy, however, simulation helps to further optimize it, and gives control to the user. The thread and processor assignments actually stand in tables within the intermediate code and could be optimized automatically without recompiling the intermediate code; only code generation must be performed.

The workload assignment and the transition to the executing processors proceeds on a per-thread basis and involves two selections. This only regards the processes to be performed by the embedded system, not to those just modeling the environment. First an encoding for executing the thread (or the entire process) must be prescribed. Every thread uses a single encoding, but different threads can choose independently. If a thread communicates data to another one using a different encoding, automatic code conversions occur. Just selecting the encodings already permits simulation. The VM interprets the intermediate code and executes by automatically adding the selected rounding to every real operation. This can, of course, lead to new program errors.

The second step is to select executing processors from the processors of the target. The processors need to be available for executing and need to support the selected data encoding. The annotation is confined to the entries into the threads. To select the encoding and the processor one annotates

$$apc \dots (\#S \dots \dots \#T \ f64 \ on \ host \dots \dots).$$

After making the assignments, the program can finally be compiled, this time for every processor and the threads selected for it only. The selection of the processors for the threads also determines the distribution of the variables the write to.

The processors will generally execute the predefined tuple operations from compiled subroutines (as does the VM). For a most precise simulation by the VM, the target processors must do so using the same algorithms as the VM. There may be choices between more than one algorithm performing similarly using the VM's highest precision but differently for different target codes. Context switches between threads running on the same processor can be managed by an operating system. On a simple processor, conditional indirect jumps and a single return variable suffice. As function calls are not interrupted, no return stacks have to be switched. For processors executing interrupt routines at different priority levels, the described switching of threads would be implemented for every interrupt level needed. For the FPGA processors, the code generator builds on the hardware infrastructure and 'knows' about its communications structures and protocols. For multi-threaded processes, communications can be compiled up to the point of performing compile-time routing.

6 A Funny Example

This entire chapter including its headline is an example of a π-Nets program. All text lines but the program lines are skipped by the π-Nets compiler as comments. The process 'mus' defined below outputs several sequences of tones via the MIDI interface of the PC as sequences of pitch codes. Here, the sequences are derived from the Fibonacci sequence mod(25) generated by a recursive function named 'fib'. Every number of the sequence is the sum of the two previous ones. The first two stand at the beginning of the block performing the calculation and must be integers in the range 0..24, in order to be used as tones. The recursion always breaks for the applied non-negative integer input. Without the mod operation '%25' the sequence would approach infinity very fast, and the sequence of tones would leave the range of audible frequencies if large numbers were accepted at all. So it rests within a range of two octaves, but would repeat periodically after a while. The definition is

```
fct fib ( 0,1 ← a,b,i   i>0 ?   b,  a+b %25,  i-1 <)   a )
```

The computation by 'fib' uses an end recursion indicated by '<)' instead of '.)'. The π-Nets block offers the feature that the argument 'i' of the function call is extended by constants whereas the recursive recalls must deliver the full argument a,b,i. 'i' is used as the iteration count of the recursion. The branch behind '<)' is the breaking branch and delivers the result. 'a+b %25' computes (a+b)%25, '%' having a low priority.

The process 'mus' controls three automata of the predefined type MDO (midi output channels) using a separate thread for each of them. Each is set to a particular instrument and a desired volume such that the threads generate their separate melodies during 14 measures of 2/4 beats and run and out-put in parallel. They play a monophone melody, an accompaniment by dual tones and a rhythmic pattern on an instrument called 'wood', and finish by a 'drum' sound. The threads with the names of the MDO automata can use the constants defined for that type and output unassigned numbers or results of expressions to them. Constants of the form \16 (1/16 note) set tone durations by forcing a synchronization delay and then end the tones started previously. 'c\4' thus generates a quarter note 'c' on the instruments of the thread, and '\1' w/o previously started tones a pause of a full measure (2 s). The output synchronization replaces explicit wait commands, and enhances the readability of the tone sequence (cf. the thread 'vc3' below). The add operations like 'fis + fib(..)' etc. transpose the fib values to another starting tone (here 'fis').

```
apc mus on host,(mdo) vc1    (pia,77),
                   (mdo) vc2    (band,55),
                   (mdo) vc3    (wood,88)
(
   #vc1(14:←i Fis+fib(3i) B+fib(3i+1)\38 c+fib(3i+2)\8 )
        c, e, g, c' \2
   #vc2 \1 (24:←i fis+fib(2i)\16   fis+fib(2i+1)\316 )
   #vc3 (7: c'\316 c'\316 c'\4 c'\316 c'\316 )
        drum 100    c'\38
)
```

The indexed blocks '(24: ...)' etc. are repeated sequentially in the order of the index values. The index is named 'i' in the first two blocks. The musical output ends, once the last thread has finished (vc2 by its final accord). Note that the program out-put is generated by the process in physical time. The output is invariable, however, could be directed to a file and be read back as a timed tuple constant.

'mus' can be used as a starting point to experiment with other algorithms to generate melodies [16]. The threads vc1-3 could be mapped to different PC computers as spatially separated sources of sound.

7 Summary and Conclusion

Our discussion has focused on π-Nets as a language for numeric computations, emphasizing its simplicity in spite of the needs of the parallel and distributed targets. At its basis, there are the predefined tuple sets and abstract, non scalar operations on them. Tuples are used as coordinates for the numeric treatment of almost all mathematical structures, even coordinate-free structures that are defined through sets of equivalent coordinates and tracking changes of coordinates, things which appear more difficult from a less abstract starting point [14]. Transforms like the FFT or discrete exterior calculus [6] profit from the available tuple compositions. The structures to describe parallelism and real time control have proven to be versatile; they may appear primitive but operate on a high level. The same holds for the mathematical algorithms that call to non-trivial compiler operations.

As to the algorithmic language, our quest for simplicity is due to the belief that a tool should be sharp and effective, but at the same time easy to use in order not to distract time and concentration from the applications to be made. As for every task, for the design a simple programming language there are many choices from the basic paradigms to details of the syntax, and many ways to go. One may e.g. argue about the strange '$x \leftarrow n$' naming which is our way to distinguish between variables, names for referencing, and the equality relation. Indexing by subscripts or the '$x.i$' may be easier to accommodate to. The massive overloading to other tuple sizes and types may be considered dangerous but stays close to what is usual in mathematics. The lack of data set constructions is compensated by the predefined rich types of n-tuples and sub sets of them and the ability to perform tuple compositions. The lack of an extra type of integers is compensated by the fact that the compiler can track index computations with an integer result and support these with precise integer codes. For an efficient execution on FPGA based processors, there is the large selection of number codes. The most important and unique feature of π-Nets may be the use of the reals as basic data while staying grounded up to the point of to be able to compile for very simple processors. Numbers and their codes remain clearly distinguished. Criticism is certainly in place regarding several aspects of the present prototypical implementation.

There are several steps to be taken to improve the quality and the scope of the compiler implementation which at its present state mainly serves to evaluate the language and adjust it if needed, to support a few target systems for purposes of their special support, to demonstrate the ease to compile applications, and to experiment with implementing or supporting additional mathematical methods. The operation of the VM will be extended to higher precisions typically not needed on embedded processors but beneficial for the quality of constant foldings and simulation, and for supporting additional high precision PC based computing applications exploiting current multi-core and GPU hardware. The support for the embedded targets could be enhanced in various ways, too, by automating workload distribution and optimizing for efficiency, and performing software caching and reconfiguration automatically instead of using the existing explicit commands. The target support conceptually offered by the present π-Nets tool already exceeds what most other programming tools offer.

References

1. Mayer-Lindenberg, F.: A management scheme for the basic types in high level languages. In: Vojtáš, P., Bieliková, M., Charron-Bost, B., Sýkora, O. (eds.) SOFSEM 2005. LNCS, vol. 3381, pp. 390–393. Springer, Heidelberg (2005). https://doi.org/10.1007/978-3-540-30577-4_46

2. Mayer-Lindenberg, F.: A modular processor architecture for high-performance computing applications on FPGA. In: Conference on Computer Design, CDES 2012, Las Vegas, USA (2012). https://134.28.202.18/t3resources/ict/dateien/Mitarbeiter/f-mayer-lindenberg/Las_Vegas.pdf

3. Mayer-Lindenberg, F.: Dedicated Digital Processors: Methods in Hardware/Software System Design. Wiley, London (2004)

4. Mayer-Lindenberg, F.: High-level FPGA programming through mapping process networks to FPGA resources. In: 2009 International Conference on Reconfigurable Computing and FPG as ReConFig 2009, Cancun, Quintana Roo, Mexico, pp. 302–307 (2009). https://doi.org/10.1109/reconfig.2009.73

5. Mayer-Lindenberg, F.: Fifth on the transputer. Microprocessing Microprogramming **19**(5), 367–373 (1987). https://doi.org/10.1016/0165-6074(87)90248-1

6. Desbrun, M., Hirani, A.N., Leok, M., Marsden, J.E.: Discrete exterior calculus. https://arxiv.org/abs/math/0508341v2

7. Marsden, J.E., West, M.: Discrete mechanics and variational integrators. Acta Numerica **10**(1), 357–514 (2001). https://doi.org/10.1017/S096249290100006X

8. Bücker, H.M., Corliss, G., Hovland, P., Naumann, U., Norris, B.: Automatic Differentiation: Applications, Theory, and Implementations. Lecture Notes in Computational Science and Engineering. Springer, Heidelberg (2006). https://doi.org/10.1007/3-540-28438-9

9. Sebesta, R.W.: Concepts of Programming Languages. The Benjamin/Cummings Series in Computer Science. Benjamin/Cummings, Redwood City (1989)

10. Krishnamurthy, E.V.: Parallel Processing: Principles and Practice. Addison-Wesley, Sydney (1989)

11. Eckert, M., Meyer, D., Haase, J., Klauer, B.: Operating system concepts for reconfigurable computing: review and survey. Int. J. Reconfigurable Comput. **2016**, 1–11 (2016). https://doi.org/10.1155/2016/2478907. Article No. 2478907

12. Kawamura, A., Ota, H., Rösnick, C., Ziegler, M.: Computational complexity of smooth differential equations. In: Rovan, B., Sassone, V., Widmayer, P. (eds.) MFCS 2012. LNCS, vol. 7464, pp. 578–589. Springer, Heidelberg (2012). https://doi.org/10.1007/978-3-642-32589-2_51

13. Blum, L., Cucker, F., Shub, M., Smale, S.: Complexity and Real Computation. Springer, New York (1998). https://doi.org/10.1007/978-1-4612-0701-6

14. Padula, A.D., Scott, S.D., Symes, W.W.: A software framework for abstract expression of coordinate-free linear algebra and optimization algorithms. ACM Trans. Math. Softw. **36**(2), 1–36 (2009). https://doi.org/10.1145/1499096.1499097. Article No. 8

15. Knuth, D.E.: Literate Programming. Comput. J. **27**(2), 97–111 (1984). https://doi.org/10.1093/comjnl/27.2.97

16. Nierhaus, G.: Algorithmic Composition. Paradigms of Automated Music Generation. Springer, Wien (2009). https://doi.org/10.1007/978-3-211-75540-2

17. Knuth, D.: The Art of Computer Programming, Volume 2. The: Seminumerical Algorithms 4.2.3, 3rd edn. Addison-Wesley, Sydney (1998)

18. Hida, Y., Li, S., Bailey, D.: Library for double-double and quad-double arithmetic (2008). https://www.researchgate.net/publication/228570156_Library_for_Double-Double_and_Quad-Double_Arithmetic

19. Fousse, L., Hanrot, G., Lefèvre, V., Pélissier, P., Zimmermann, P.: MPFR: a multiple-precision binary floating-point library with correct rounding. ACM Trans. Math. Softw. (TOMS) **33**(2), 1–14 (2007). https://doi.org/10.1145/1236463.1236468. Article No. 13

20. Lange, M., Rump, S.: Faithfully rounded FP computations. preprint, vol. **1**, no. 1 (2017). https://urldefense.proofpoint.com/v2/url?u=http-3A__www.ti3.tuhh.de&d=DwIBaQ&c=vh6FgFnduejNhPPD0fl_yRaSfZy8CWbWnIf4XJhSqx8&r=phx_h-t0CpJpXIoE7Nt7XzoVuWOl1rYzfCfuZFItYqZo5lxViGBLk_fC3J092Uza&m=as-6aujO3oSMlwe2QObpbwos2EV580rwz45Btb2diZk&s=x4bV2sQwvF2OiCfTJI5P0v04JsZK2ZzuyxOXxY7-8JY&e

21. Mayer-Lindenberg, F., Beller, V.: An FPGA-based floating-point processor array supporting a high-precision dot product. In: 2006 IEEE International Conference on Field Programmable Technology, Bangkok, Thailand, pp. 317–320. IEEE (2006). https://doi.org/10.1109/fpt.2006.270337

22. Li, X., Phung, C.F., Maskell, D.L.: FPGA overlays: hardware-based computing for the masses. In: Proceedings of the Eighth International Conference on Advances in Computing, Electronics and Electrical Technology - CEET 2018, pp. 25–31. SEEK Digital Library (2018). https://doi.org/10.15224/978-1-63248-144-3-12

23. Hafner, P.R.: On the graphs of Hoffman-Singleton and Higman-Sims. Electron. J. Comb. **11**(1), 1–33 (2004). Article No. R77

24. Parallelrechner ER-4, Technische Universität Hamburg. www.tuhh.de/ict/forschung/parallelrechner-er-4.html

Improving the Accuracy of Energy Predictive Models for Multicore CPUs Using *Additivity* of Performance Monitoring Counters

Arsalan Shahid(✉)⬛, Muhammad Fahad⬛, Ravi Reddy Manumachu⬛,
and Alexey Lastovetsky⬛

School of Computer Science, University College Dublin, Belfield, Dublin 4, Ireland
{arsalan.shahid,muhammad.fahad}@ucdconnect.ie,
{ravi.manumachu,alexey.lastovetsky}@ucd.ie

Abstract. Energy predictive modelling using performance monitoring counters (PMCs) has emerged as the leading mainstream approach for modelling the energy consumption of an application. Modern computing platforms such as multicore CPUs provide a large set of PMCs. The programmers, however, can obtain only a small number of PMCs (typically 3–4) during an application run due to the limited number of hardware registers dedicated to storing them. Therefore, selection of a reliable subset of PMCs as predictor variables is crucial to the prediction accuracy of online energy models. State-of-the-art methods for selecting the PMCs are largely based on their correlation with energy consumption.

Recently, *Additivity* is introduced as a property of PMCs that appears to have significant impact on the accuracy of energy predictive models. It is based on an experimental observation that energy consumption of serial execution of two applications is equal to the sum of the energy consumption of those applications when they are run separately. In this work, we demonstrate how the accuracy of energy predictive models based on three popular techniques (Linear regression, Random forests, and Neural networks) can be improved by selecting PMCs based on a property of *additivity*.

Keywords: Performance monitoring counters · Energy consumption · Energy modelling · Multicore CPU · Energy predictive models

1 Introduction

Energy is now a first-class design constraint along with performance in all computing settings. It is a critical limitation for battery-operated mobile systems. Energy-proportional designs [1] in servers are crucial to the operational efficiency of data centres. According to a 2010 DOE Office of Science report [3], it is the leading concern for High Performance Computing (HPC) system designs.

© Springer Nature Switzerland AG 2019
V. Malyshkin (Ed.): PaCT 2019, LNCS 11657, pp. 51–66, 2019.
https://doi.org/10.1007/978-3-030-25636-4_5

Energy consumption in computing contributes nearly 3% to the overall carbon footprint and is now a serious environmental concern [24].

Energy efficiency in computing is driven by innovations in hardware represented by the micro-architectural and chip-design advancements, and software that can be grouped into two categories: (a). System-level energy optimization, and (b). Application-level energy optimization. System-level optimization methods aim to maximize energy efficiency of the environment where the applications are executed using techniques such as DVFS (dynamic voltage and frequency scaling), Dynamic Power Management (DPM), and energy-aware scheduling. Application-level optimization methods use application-level parameters and models to maximize the energy efficiency of the applications.

Accurate measurement of energy consumption during an application execution is key to energy minimization techniques at software level. There are three popular approaches to providing it: (a). System-level physical measurements using external power meters, (b). Measurements using on-chip power sensors, and (c). Energy predictive models.

While the first approach is known to be accurate, it can only provide the measurement at a computer level and therefore lacks the ability to provide fine-grained component-level decomposition of the energy consumption of an application. This is a serious drawback. Consider, for example, a computer consisting of a multicore CPU and an accelerator (GPU or Xeon Phi), which is representative of nodes in modern supercomputers. While it is easy to determine the total energy consumption of a hybrid application run that utilizes both the processing elements (CPU and accelerator) using the first approach, it is difficult to determine their individual contributions. This decomposition is critical to energy models, which are key inputs to data partitioning algorithms that are critical building blocks for optimization of the application for energy. Without the ability to determine accurate decomposition of the total energy consumption, one has to employ an exhaustive approach (involving huge computational complexity) to determine the optimal data partitioning that optimizes the application for energy.

The second approach has no definitive research works proving its accuracy.

The third approach of energy predictive modelling emerged as the pre-eminent alternative. The existing models predominantly use performance monitoring counters as predictor variables for modelling energy consumption. Performance monitoring counters are special-purpose registers provided in modern microprocessors to store the counts of software and hardware activities. We will use the acronym PMCs to refer to software events, which are pure kernel-level counters such as *page-faults*, *context-switches*, etc. as well as micro-architectural events originating from the processor and its performance monitoring unit called the hardware events such as *cache-misses*, *branch-instructions*, etc. They have been developed primarily to aid low-level performance analysis and tuning. While remarkably PMCs have not been used for performance modelling, they have been speedily adopted for energy predictive modelling and have come to dominate its landscape over the years. The energy predictive models are, however, trained and validated using system-level physical measurements of energy consumptions of the training and test applications. The most common approach proposing an energy predictive model is to determine the energy consumption of a hardware component based on linear regression of the performance events occurring in the hardware component during an application run. The total energy consumption is then calculated as the sum of these individual energy consumptions. Therefore, this approach constructs component-level models of energy consumption and composes them using summation to predict the energy consumption during an application run.

We focus in this work on energy predictive modelling using PMCs. Modern computing platforms such as multicore CPUs provide a large set of PMCs. The most popular tools that can be used to gather the values of the PMCs for a platform include Likwid [25], PAPI [18], Intel PCM [11], and Linux *perf* [19]. The programmers, however, can obtain only a small number of PMCs (typically 3–4) during an application run due to the limited number of hardware registers dedicated to storing them. Consider, for example, the Intel Haswell server whose specification is shown in Table 1. *Likwid* tool provides 167 PMCs for this platform. To obtain the values of the PMCs for an application, the application must be executed about 53 times since only a limited number of PMCs can be obtained in a single application run.

Table 1. Specification of the Intel Haswell and Intel Skylake multicore CPUs

Technical Specifications	Intel Haswell Server	Intel Skylake Server
Processor	Intel E5-2670 v3 @2.30 GHz	Intel Xeon Gold 6152
OS	CentOS 7	Ubuntu 16.04 LTS
Micro-architecture	Haswell	Skylake
Thread(s) per core	2	2
Cores per socket	12	22
Socket(s)	2	1
NUMA node(s)	2	1
L1d cache/L1I cache	32 KB/32 KB	32 KB/32 KB
L2 cache	256 KB	1024 KB
L3 cache	30720 KB	30976 KB
Main memory	64 GB DDR4	96 GB DDR4
TDP	240 W	140 W
Idle Power	58 W	32 W

Since only 3–4 PMCs can be collected in a single application run, selecting such a reliable subset as predictor variables is crucial to the prediction accuracy of online energy models.

We classify techniques for selecting the PMCs into following four categories:

- Techniques that consider all the PMCs offered by a tool for a platform with the goal to capture all possible contributors to energy consumption. To the best of our knowledge, we found no research works that adopt this approach.
- Techniques that are based on a statistical methodology such as correlation, principal component analysis (PCA) etc. [15, 28].
- Techniques that use expert advice or intuition to pick a subset (that may not necessarily be determined in one application run) and that, in experts' opinion, is a dominant contributor to energy consumption [8].
- Techniques that select parameters with physical significance based on fundamental laws such as energy conservation of computing [21].

Shahid et al. [21] introduced a new property of PMCs that appear to have significant impact on the accuracy of energy predictive models. It is based on an experimental observation that dynamic energy consumption of serial execution of two applications is equal to the sum of the dynamic energy consumption of those applications when they are run separately. The property, therefore, is based on a simple and intuitive rule that if the parameter is intended for a linear predictive model, the value of a PMC for a serial execution of two applications should be equal to the sum of its values obtained for the individual execution of each application. The PMC is branded as *non-additive* on a platform if there exists an application for which the calculated value differs significantly from the value observed for the application execution on the platform. The use of *non-additive* PMCs in a model impairs its prediction accuracy. The authors show by employing a detailed statistical experimental methodology on a modern Intel Haswell multicore server CPU that while many PMCs are potentially *additive*, a considerable number of PMCs are not. Some of the *non-additive* PMCs are widely used in energy predictive models as key predictor variables.

In this work, we study how the criterion of *additivity* can be used to select PMCs to improve the accuracy of the following types of models: Linear regression (*LR*), Random forests (*RF*), and Neural networks (*NN*). We observe that a large number of energy predictive models in the literature (Sect. 3) is based on these three methods. In a linear regression, we solve a linear model by estimating the regression coefficients. The *RF* is a decision tree based non-linear model build by constructing many linear boundaries. A linear transfer function is used to train our *NN*. *Additivity* property has been envisioned to be useful for selection of PMCs to use as predictor variables in linear energy predictive models. In this paper, we first validate it using detailed experimental evaluation on two modern multicore platforms: (1). Intel Haswell and (2). Intel Skylake. We further investigate its applicability on non-linear modelling techniques such as *RF* and *NN*. We analyze these techniques in terms of the PMCs employed in them and make sure that they appear as *additive* linear parameters. We demonstrate that *additivity* is highly applicable to non-linear methods that employ linear functions for composition of models.

We perform three classes of experiments: Class A, Class B, and Class C. For Class A, we use a dual-socket Intel Haswell multicore server (Table 1).

We select six PMCs which are common in the state-of-the-art models [4,8,14,27] and which are highly correlated with dynamic energy consumption. We build three sets of models. The first set, ({$LR1, LR2, ..., LR6$}, contains linear regression models (*LRS*). The second set, {$RF1, RF2, ..., RF6$}, contains random forest models (*RFS*). The third set, {$NN1, NN2, ..., NN6$}, contains neural network models (NNS). In each set, the models contain decreasing number of *non-additive* PMCs. Consider, for example, the first set. Model LR1 employs all the selected PMCs as predictor variables. Model LR2 is based on five most *additive* PMCs. Model LR3 uses four most *additive* PMCs and so on until Model LR6 containing the highest *additive* PMC.

The predictions of the models are compared with system-level physical measurements using power meters ([9]), which we consider to be the ground truth. Our results show that the removal of *non-additive* PMCs improves the average prediction accuracy of *LR* from 31.2% to 18.01%. Similarly, the average prediction accuracy for *RF* is improved from 38% to 24%, and for *NN* from 30% to 24%.

We find no PMC to be *additive* for all categories of applications within a tolerance of 5%. For Class B and Class C experiments, we use a single-socket Intel Skylake server (Table 1) to study the application specific energy predictive models. We choose two highly optimized scientific kernels offered by Intel math kernel library (MKL): (a). Fast Fourier transform (FFT) and (b). Dense matrix-matrix multiplication application (DGEMM). We identify a set of nine most *additive* PMCs (*PA*) common for both the applications and a set of nine PMCs that are *non-additive* (*PNA*) but which are used in state-of-the-art energy predictive models. For Class B, we build three models, {LR-A,RF-A,NN-A}, based on *PA* and three models, {LR-NA,RF-NA,NN-NA}, based on *PNA*. We show that the models based on *PA* demonstrate notably better prediction accuracy.

For Class C, since only four PMCs can be collected in a single application run, we compose two sets of PMCs, *PA4* and *PNA4*. *PA4* contains four highly energy correlated PMCs selected from *PA*, and *PNA4* contains four most correlated PMCs selected from *PNA*. Models that use *PA4* demonstrate noteworthy improvement in average prediction accuracy in comparison with models composed using *PNA4*. We also observed that higher correlation with energy when applied to *non-additive* PMCs does not improve their prediction accuracy. The models based on *PNA4* perform even worse than those based on *PNA*.

We conclude, therefore, that correlation with dynamic energy consumption alone is not sufficient to provide good prediction accuracy but should be combined with methods such as *additivity* that take into account the physical significance of the parameters originating from fundamental laws such as energy conservation of computing.

To summarize, the main contribution of this work is a study of the impact of *additivity* on the accuracy of mainstream PMCs-based energy predictive modelling techniques.

The rest of this paper is organized as follows. Section 2 present the terminology related to power and energy followed by related work in Sect. 3. Section 4 explains the *additivity* criterion of PMCs and its implications for energy predictive models. In Sect. 5, we present our experimental methodology including setup and design of the three classes of experiments. Section 5 presents the experimental results. Finally, Sect. 6 concludes the paper.

2   Terminologies

There are two types of power consumptions in a component: dynamic power and static power. Dynamic power consumption is caused by the switching activity in the component's circuits. Static power or idle power is the power consumed when the component is not active or doing work. From an application point of view, we define dynamic and static power consumption as the power consumption of the whole system with and without the given application execution. From the component point of view, we define dynamic and static power consumption of the component as the power consumption of the component with and without the given application utilizing the component during its execution.

There are two types of energy consumptions, static energy and dynamic energy. We define the static energy consumption as the energy consumption of the platform without the given application execution. Dynamic energy consumption is calculated by subtracting this static energy consumption from the total energy consumption of the platform during the given application execution. If P_S is the static power consumption of the platform, E_T is the total energy consumption of the platform during the execution of an application, which takes T_E seconds, then the dynamic energy E_D can be calculated as, $E_D = E_T - (P_S \times T_E)$.

In this work, we consider only the dynamic energy consumption. We describe the rationale behind using dynamic energy consumption in the section 1 of supplemental [22].

3   Related Work

This section presents a brief literature survey of some important tools widely used to obtain PMCs, notable research on energy predictive models, and research works that provide a critical review of PMCs.

Tools to obtain PMCs. Perf [19] can be used to gather the PMCs for CPUs in Linux. PAPI [18] and Likwid [25] allow obtaining PMCs for Intel and AMD microprocessors. *Intel PCM* [11] gives PMCs of core and uncore components of an Intel processor.

Notable Energy Predictive Models for CPUs. Initial Models correlating PMCs to energy values include [6,10,12,13]. Events such as integer operations, floating-point operations, memory requests due to cache misses, component access rates, instructions per cycle (IPC), CPU/disk and network utilization, etc. were believed to be strongly correlated with energy consumption. Simple linear models have been developed using PMCs and correlated features to predict energy consumption of platforms. Rivoire et al. [20] study and compare five full-system real-time power models using a variety of machines and benchmarks. They report that PMC-based model is the best overall in terms of accuracy since it accounted for majority of the contributors to system's dynamic power. Other notable PMC-based linear models are [2,8,23,26]. Manila [15] construct a densely populated multi-dimensional space of PMCs and predict the energy consumption of platform using a nearest neighborhood search algorithm. Zhuo et al. [28] present a PMC-based energy consumption models for task characteristics in cloud data center using regression algorithms.

Critiques of PMCs for Energy Predictive Modelling. Some attempts where poor prediction accuracy of PMCs for energy predictive modeling has been critically examined include [5,7,16,17]. Researchers highlight the fundamental limitation to obtain all the PMCs simultaneously or in one application run and show that linear regression models give prediction errors as high as 150%.

4 *Additivity* of PMCs

The property of *additivity* is based on a simple and intuitive rule that if a PMC is intended as a parameter in a linear term of the energy predictive model then its value for a compound application should be equal to the sum of its values for the executions of the base applications constituting the compound application. It is based on the experimental observation that the dynamic energy consumption of a serial execution of two applications is the sum of dynamic energy consumptions observed for the individual execution of each application.

We now present a test to determine if a PMC is *non-additive* or potentially *additive*. It comprises of two stages. A PMC must pass both stages to be pronounced *additive* for a given compound application on a given platform.

In the first stage, we determine if the PMC is deterministic and reproducible.

In the second stage, we examine how the PMC of the compound application relates to its values for the base applications. At first, we collect the values of the PMC for the base applications by executing them separately. Then, we execute the *compound* application and obtain its value of the PMC. Typically, the core computations for the compound application consist of the core computations of the base applications programmatically placed one after the other.

If the PMC of the *compound* application is equal to the sum of the PMCs of the base applications (with a tolerance of 5.0%), we classify the PMC as potentially *additive*. Otherwise, it is *non-additive*.

For each PMC, we determine the maximum percentage error. For a *compound* application, the percentage error (averaged over several runs) is calculated as follows:

$$Error(\%) = (|\frac{(\overline{e_{b1}} + \overline{e_{b2}}) - \overline{e_c}}{\overline{e_{b1}} + \overline{e_{b2}}}|) \times 100 \tag{1}$$

where $\overline{e_c}, \overline{e_{b1}}, \overline{e_{b2}}$ are the sample means of predictor variables for the compound application and the constituent base applications respectively. The maximum percentage error is then calculated as the maximum of the errors for all the *compound* applications in the experimental testsuite.

We automated the determination of a PMC's *additivity* using a tool called *AdditivityChecker* (see section 3 of the supplemental [22]).

5   Experimental Results

The experiments are carried out on two modern multicore platforms: (1). an Intel Haswell based dual-socket server and (2). an Intel Skylake based single-socket server. The specifications for both are given in Table 1. We choose a diverse set of benchmarks in our test suite (section 4 of supplemental [22]) with highly memory bound and compute bound scientific computing applications such as DGEMM and FFT from Intel math kernel library (MKL), scientific applications from NAS Parallel benchmarking suite, Intel HPCG, *stress*, non-optimized and non-scientific applications. Apart from reducing bias, one other reason to compose a diverse test suite is to have a range of PMCs for different executions of applications on the platform.

For an application execution, we measure the following: (1). the dynamic energy consumption, (2). the execution time and (3). PMCs. The dynamic energy consumption of the platform is provided by WattsUp pro power meter and the readings are obtained programatically using a detailed statistical methodology employing HCLWattsUp API [9]. The power meters are periodically calibrated using an ANSI C12.20 revenue-grade power meter, Yokogawa WT210. To ensure the reliability of our results, we follow a statistical methodology where a sample mean for a response variable is obtained from several experimental runs. We follow a strict statistical methodology to ensure the reliability of our experiments (see section 3 of supplemental [22]).

We use *Likwid* package [25] to obtain the PMCs. It offers 164 PMCs and 385 PMCs on Intel Haswell and Intel Skylake platform, respectively. We eliminate PMCs with counts less than or equal to 10. The eliminated PMCs have no significance on modeling the dynamic energy consumption of our platform since they are non-reproducible over several runs of the same application on our platform.

The reduced set contains 151 PMCs for Intel Haswell and 323 for Intel Skylake. The collection of all of them takes a huge amount of time since only four PMCs can be obtained in a single application run. This is because of a limited number of hardware registers dedicated for storing them. We also notice that some PMCs can only be collected individually or in sets of two or three for single execution of an application. Therefore, we observe that each application must be executed about 53 and 99 times on Intel Haswell and Intel Skylake platform, respectively, to collect all the PMCs.

We select three predictive models for our experiments: (1). Linear Regression Model (*LR*), (2). Random Forest (*RF*), and (3). Neural Networks (*NN*). We explain them in detail in section 1 of supplemental [22]. In all these models, PMCs appear as parameters in linear terms, and therefore must be *additive*.

We now divide our experiments into three classes, class A, class B and class C, as follows:

1. Class A: we show the improvements in the average prediction accuracy of the three modeling techniques by the *additivity* of PMCs. A diverse set of applications (see section 4 of supplemental [22]) on a dual socket Intel Haswell multicore server is used in these experiments.
2. Class B: we study the impact of the *additivity* of PMCs on prediction accuracy of application-specific energy predictive models. Two highly memory bound and compute bound scientific computing applications such as DGEMM and FFT from Intel MKL, are used in these experiments.
3. Class C: we compare the accuracy of two four parameter models. Both models employ subsets of parameters from the original selected set. The only difference is that one subset include higher energy correlated parameters, and the other contains the most *additive* parameters.

5.1 *Class A*: Improving Prediction Accuracy of Energy Predictive Models Using *Additivity*

We conduct the Class A experiments on the dual-socket Intel Haswell multicore server (see Table 1). We choose six PMCs (X_1 to X_6 in Table 2), which are widely used in energy predictive models. We build a dataset of 277 points as *base* applications by executing the applications from our test suite with different problem sizes. This dataset is used to train the models. We build a test dataset containing points for 50 *compound* applications which are composed up of serial executions of *base* applications. Each point contains the dynamic energy consumption and PMCs for the execution of an application. We apply *additivity* test with allowed error percentage of 5% and found no PMC to be *additive*. We list the PMCs and their *additivity* error percentages in Table 2.

Table 2. List of selected PMCs for modelling with their *additivity* test errors (%).

Selected PMCs	Additivity test error (%)
X_1: IDQ_MITE_UOPS	13
X_2: IDQ_MS_UOPS	37
X_3: ICACHE_64B_IFTAG_MISS	36
X_4: ARITH_DIVIDER_COUNT	80
X_5: L2_RQSTS_MISS	14
X_6: UOPS_EXECUTED_PORT_PORT_6	10

We build three sets of models, LRS = {LR1, LR2, LR3, LR4, LR5, LR6}, RFS = {RF1, RF2, RF3, RF4, RF5, RF6}, and NNS = {NN1, NN2, NN3, NN4, NN5, NN6}. In each set, the models contain decreasing number of *non-additive* PMCs. Consider, for example, the first set. Model LR1 employs all the selected PMCs as predictor variables. Model LR2 is based on five most *additive* PMCs. PMC X_4 is removed because it has the highest *non-additivity*. Model LR3 uses

Table 3. Linear predictive models (LR1-LR6) using zero intercepts and positive coefficients with their minimum, average, and maximum prediction errors.

Model	PMCs	Coefficients	Percentage prediction errors (min, avg, max)
LR1	$X_1, X_2, X_3, X_4, X_5, X_6$	3.83E−09, 3.67E−10, 5.30E−07, 0, 5.56E−08, 0	(6.6, 31.2, 61.9)
LR2	X_1, X_2, X_3, X_5, X_6	3.83E−09, 3.67E−10, 5.30E−07, 0, 5.56E−08	(6.6, 31.2, 61.9)
LR3	X_1, X_3, X_5, X_6	3.75E−09, 5.34E−07, 5.58E−08, 0	(2.5, 25.3, 62.1)
LR4	X_1, X_5, X_6	4.00E−09, 5.59E−08, 0	(2.5, 23.86, 100.3)
LR5	X_1, X_6	4.60E−09, 1.46E−09	(2.5, 18.01, 89.45)
LR6	X_6	1.60E−09	(2.5, 68.5, 90.5)

Table 4. Random forest (RF) regression based energy predictive models (RF1-RF6) with their minimum, average, and maximum prediction errors.

Model	PMCs	Percentage prediction errors (min, avg, max)
RF1	$X_1, X_2, X_3, X_4, X_5, X_6$	(2.78, 37.8, 185.4)
RF2	X_1, X_2, X_3, X_5, X_6	(2.5, 30.4, 199.6)
RF3	X_1, X_3, X_5, X_6	(2.5, 30.02, 104)
RF4	X_1, X_5, X_6	(2.5, 23.68, 59.3)
RF5	X_1, X_6	(2.5, 43.4, 174.4)
RF6	X_6	(2.5, 57.7, 172.1)

four most *additive* PMCs and so on until Model LR6 containing the highest *additive* PMC, which is X_6.

We compare the predictions of the models with system-level physical measurements using HCLWattsUp, which we consider to be the ground truth. The minimum, average, and maximum percentage prediction errors for the models in the sets LRS, RFS, and NNS are given in Tables 3, 4 and 5.

Table 5. Neural Networks based energy predictive models (NN1-NN6) with their minimum, average, and maximum prediction errors.

Model	PMCs	Percentage prediction errors (min, avg, max)
NN1	$X_1, X_2, X_3, X_4, X_5, X_6$	(2.5, 30.31, 192.3)
NN2	X_1, X_2, X_3, X_5, X_6	(2.5, 26.32, 201.2)
NN3	X_1, X_3, X_5, X_6	(2.5, 24.14, 160.1)
NN4	X_1, X_5, X_6	(2.5, 24.06, 180.3)
NN5	X_1, X_6	(2.5, 40.21, 202.45)
NN6	X_6	(2.5, 45.05, 180.5)

Since we are modelling dynamic energy consumption, the linear models in Table 3 are built using penalized linear regression using *R programming* interface that forces the coefficients to be non-negative. All the models also have zero intercept. One can see that the accuracy of the models improves as we remove the highest *non-additive* PMCs one by one until Model $LR5$, which exhibits the least average prediction error of 18.01%. We observe that $LR6$ has the worst average prediction error of 68.5% due to poor linear fit.

Table 4 shows the same trend for random forest models in RFS until Model $RF4$, which has the least average prediction error of 23.68%. Table 5 also shows the same trend for neural network models in NNS until Model $NN4$ with the least average prediction error of 24.06%.

It can be seen that improvements in average prediction accuracy due to *additivity* are less for RF and NN models compared to linear models where we are certain that *additivity* is crucial. The maximum prediction error percentages for RF and NN models are particularly bad. We will investigate in our future work how *additivity* can be used to reduce the maximum error percentage for the three types of models. One can see, however, that the average prediction error percentages of the best RF and NN models are close to the average prediction accuracy of the best linear model suggesting that the RF and NN models exhibit a relationship close to linearity.

5.2   *Class B*: Impact of *Additivity* on the Prediction Accuracy of Application-specific Energy Predictive Models

In this section, we study the accuracy of application specific energy predictive models built using LR, RF, and NN techniques. We choose a single-socket Intel Skylake server (Table 1) for the experiments. We found no PMC to be *additive* within tolerance of 5% for the application suite (see section 4 of supplemental [22]). However, we discover that some PMCs are highly *additive* for two highly optimized scientific kernels: Fast Fourier Transform (FFT) and Dense Matrix-Multiplication application (DGEMM), from Intel Math Kernel Library (MKL).

We build a dataset of 50 *base* applications using different problem sizes for DGEMM and FFT and apply the *additivity* test. The range of problem sizes for DGEMM is 6500×6500 to 20000×20000, and for FFT is 22400×22400 to 29000×29000. We select this range because of reasonable execution time (>3 s) of the applications. We also build a dataset of 30 *compound* applications from these *base* applications.

The *Additivity* test based on the two datasets reveals that there are a number of PMCs which are commonly *additive* for both applications. We select nine PMCs that are highly *additive* with *additivity* test errors of less than 1%. We also select nine PMCs which are *non-additive* for both the applications but which have been employed as predictor variables in energy predictive models given in literature (Sect. 3). We check the correlation of all PMCs with dynamic energy consumption. The selected PMCs with their correlations are given in Table 6.

We denote the set of *additive* PMCs by PA and *non-additive* PMCs by PNA. We build a dataset containing 801 points representing DGEMM and FFT for a range of problem sizes from 6400×6400 to 38400×38400 and 22400×22400 to 41536×41536, respectively, with a constant step sizes of 64. We record the dynamic energy consumption and the selected PMCs (Table 6) for each application. We split the dataset into training and test datasets. Training dataset contains 651 points used to train the three energy predictive models. Test dataset contains 150 points.

We build two linear models, {LR-A,LR-NA}, two random forest models, {RF-A,RF-NA}, and two neural network models, {NN-A,NN-NA}. The models {LR-A,RF-A,NN-A} are trained using PMCs belonging to *PA* and the models {LR-NA,RF-NA,NN-NA} are trained using PMCs belonging to *PNA*. Table 7a show the prediction error percentages of the models. One can see that the models based on *PA* have better average prediction accuracy than the models based on PNA.

Table 6. *Additive* and *non-additive* PMCs highly correlated with dynamic energy consumption. 0 to 1 represents positive correlation of 0% to 100%.

	Additive PMCs	Correlation
$X1$	UOPS_RETIRED_CYCLES_GE_4_UOPS_EXEC	0.992
$X2$	FP_ARITH_INST_RETIRED_DOUBLE	0.993
$X3$	MEM_INST_RETIRED_ALL_STORES	0.870
$X4$	UOPS_EXECUTED_CORE	0.993
$X5$	UOPS_DISPATCHED_PORT_PORT_4	0.870
$X6$	IDQ_DSB_CYCLES_6_UOPS	0.981
$X7$	IDQ_ALL_DSB_CYCLES_5_UOPS	0.972
$X8$	IDQ_ALL_CYCLES_6_UOPS	0.993
$X9$	MEM_LOAD_RETIRED_L3_MISS	−0.112
	Non-additive PMCs	
$Y1$	ICACHE_64B_IFTAG_MISS	0.960
$Y2$	CPU_CLOCK_THREAD_UNHALTED	0.600
$Y3$	BR_MISP_RETIRED_ALL_BRANCHES	0.992
$Y4$	MEM_LOAD_L3_HIT_RETIRED_XSNP_MISS	−0.020
$Y5$	FRONTEND_RETIRED_L2_MISS	0.806
$Y6$	ITLB_MISSES_STLB_HIT	0.111
$Y7$	L2_TRANS_CODE_RD	0.860
$Y8$	IDQ_MS_UOPS	0.99
$Y9$	ARITH_DIVIDER_COUNT	0.986

5.3 *Class C*: Comparison of the Impact of Energy Correlation and *Additivity* of PMCs on the Accuracy of Energy Predictive Models

Since only four PMCs can be collected in a single application run, selection of such a reliable subset is crucial to the prediction accuracy of online energy models. The Intel Skylake server (Table 1) is used for the experiments. We use

Table 7. Prediction accuracies of *LR*, *RF*, and *NN* models. **(a)** Class B experiments using nine PMCs. **(b)** Class C experiments using four PMCs.

Model	PMCs	Prediction Errors (%) [Min, Avg, Max]	Model	PMCs	Prediction Errors (%) [Min, Avg, Max]
LR-A	*PA*	(0.005, 35.32, 225.5)	LR-A4	*PA4*	(0.024, 25.12, 87.25)
LR-NA	*PNA*	(0.449, 85.61, 4039)	LR-NA4	*PNA4*	(0.449, 85.61, 4039)
RF-A	*PA*	(.0001, 29.39, 157.4)	RF-A4	*PA4*	(0.005, 22.73, 207.7)
RF-NA	*PNA*	(0.004, 36.90, 1682)	RF-NA4	*PNA4*	(0.035, 38.06, 1628)
NN-A	*PA*	(0.001, 15.43, 104.2)	NN-A4	*PA4*	(0.003, 11.46, 152.2)
NN-NA	*PNA*	(0.003, 21.04, 170.3)	NN-NA4	*PNA4*	(0.016, 21.32, 227.5)

        (a)   (b)

PA and *PNA* from Class B experiments to build two sets of four most energy correlated PMCs. The first set PA4, $\{X1, X2, X4, X8\}$, is constructed using *PA* and the second set *PNA4*, $\{Y1, Y3, Y8, Y9\}$, using *PNA*.

We build two linear models, {LR-A4,LR-NA4}, two random forest models, {RF-A4,RF-NA4}, and two neural network models, {NN-A4,NN-NA4}. The models {LR-A4,RF-A4,NN-A4} are trained using PMCs belonging to *PA4* and the models {LR-NA4,RF-NA4,NN-NA4} are trained using PMCs belonging to PNA4. The training and test datasets are the same as those for Class B experiments.

Table 7b shows the prediction error percentages of the models. Model NN-A4 has the least average prediction error of 11.46%. We can see that models {LR-NA4,RF-NA4,NN-NA4} built using highly correlated but *non-additive* PMCs do not demonstrate any improvement in average prediction accuracy compared to models {LR-NA,RF-NA,NN-NA} based on nine *non-additive* PMCs.

The models based on *PA4* containing four most *additive* and highly correlated PMCs have better average prediction accuracy than the models based on the set of *non-additive* PMCs, PNA4.

We conclude, therefore, that correlation with dynamic energy consumption alone is not sufficient to provide good average prediction accuracy but should be combined with methods such as *additivity* that take into account the physical significance of the parameters originating from fundamental laws such as energy conservation of computing.

6   Conclusion

The ability of PMC-based predictive models to provide fine-grained decomposition of energy consumption during the execution of an application makes them ideal fundamental building blocks for several application-level energy optimization techniques. Modern computing platforms such as multicore CPUs provide a large set of PMCs. However, only a limited number of PMCs (typically 3–4) can be obtained during an application run. Therefore, selection of a reliable subset of 3–4 PMCs is crucial to the prediction accuracy of online energy predictive models. The existing techniques select the PMCs based on their correlation with total energy consumption and construct models employing data analytical approaches such as linear regression, random forests, and neural networks. They do not consider the physical significance of a PMC parameter arising from fundamental laws such as energy conservation of computing.

In this work, we demonstrated how the accuracy of energy predictive models based on three popular techniques (Linear regression, Random forests, and Neural networks) can be improved by selecting PMCs based on a criterion of *Additivity*, which is derived from the application of energy conservation law for computing.

We showed that the removal of *non-additive* PMCs from the list of predictor variables in energy predictive models improved their accuracy. We illustrated that using highly *additive* PMCs resulted in notable improvements in the average prediction accuracy of application-specific models compared to application-specific models employing *non-additive* PMCs. Finally, we studied how a reliable subset of 3–4 PMCs can be constructed for employment in *online* energy predictive models. We showed that using correlation based PMC selection methods to *non-additive* PMCs do not improve the average prediction accuracy of energy models. We demonstrated that using highly correlated PMCs but which are also highly *additive* significantly improves the average prediction accuracy of the models.

In our future work, we will focus on theoretic framework explaining why additivity, which is based on a fundamental physical law of energy conservation, improves the prediction accuracy for the three types of models.

Acknowledgement. This publication has emanated from research conducted with the financial support of Science Foundation Ireland (SFI) under Grant Number 14/IA/2474.

References

1. Barroso, L.A., Hölzle, U.: The case for energy-proportional computing. Computer **12**, 33–37 (2007)
2. Basmadjian, R., Ali, N., Niedermeier, F., de Meer, H., Giuliani, G.: A methodology to predict the power consumption of servers in data centres. In: 2nd International Conference on Energy-Efficient Computing and Networking. ACM (2011)
3. DOE: The opportunities and challenges of exascale computing (2010). http://science.energy.gov/~/media/ascr//pdf/reports/Exascale_subcommittee_report.pdf
4. Dolz, M.F., Kunkel, J., Chasapis, K., Catalán, S.: An analytical methodology to derive power models based on hardware and software metrics. Comput. Sci.-Res. Dev. **31**(4), 165–174 (2016)
5. Economou, D., Rivoire, S., Kozyrakis, C., Ranganathan, P.: Full-system power analysis and modeling for server environments. In: In Proceedings of Workshop on Modeling, Benchmarking, and Simulation, pp. 70–77 (2006)
6. Fan, X., Weber, W.D., Barroso, L.A.: Power provisioning for a warehouse-sized computer. In: 34th Annual International Symposium on Computer architecture, pp. 13–23. ACM (2007)

7. Hackenberg, D., Ilsche, T., Schöne, R., Molka, D., Schmidt, M., Nagel, W.E.: Power measurement techniques on standard compute nodes: a quantitative comparison. In: 2013 IEEE International Symposium on Performance Analysis of Systems and Software (ISPASS), pp. 194–204. IEEE (2013)

8. Haj-Yihia, J., Yasin, A., Asher, Y.B., Mendelson, A.: Fine-grain power breakdown of modern out-of-order cores and its implications on skylake-based systems. ACM Trans. Archit. Code Optim. (TACO) **13**(4), 56 (2016)

9. HCL: HCLWattsUp: API for power and energy measurements using WattsUp Pro Meter (2016). http://git.ucd.ie/hcl/hclwattsup

10. Heath, T., Diniz, B., Horizonte, B., Carrera, E.V., Bianchini, R.: Energy conservation in heterogeneous server clusters. In: 10th ACM SIGPLAN Symposium on Principles and Practice of Parallel Programming (PPoPP), pp. 186–195. ACM (2005)

11. IntelPCM: Intel® performance counter monitor - a better way to measure cpu utilization (2012). https://software.intel.com/en-us/articles/intel-performance-counter-monitor

12. Isci, C., Martonosi, M.: Runtime power monitoring in high-end processors: methodology and empirical data. In: 36th Annual IEEE/ACM International Symposium on Microarchitecture, p. 93. IEEE Computer Society (2003)

13. Kansal, A., Zhao, F.: Fine-grained energy profiling for power-aware application design. ACM SIGMETRICS Perform. Eval. Rev. **36**(2), 26 (2008)

14. Li, T., John, L.K.: Run-time modeling and estimation of operating system power consumption. In: ACM SIGMETRICS Performance Evaluation Review, vol. 31, pp. 160–171. ACM (2003)

15. Mair, J., Huang, Z., Eyers, D.: Manila: using a densely populated pmc-space for power modelling within large-scale systems. Parallel Comput. **82**, 37–56 (2019)

16. McCullough, J.C., Agarwal, Y., Chandrashekar, J., Kuppuswamy, S., Snoeren, A.C., Gupta, R.K.: Evaluating the effectiveness of model-based power characterization. In: Proceedings of the 2011 USENIX Conference on USENIX Annual Technical Conference. USENIXATC 2011. USENIX Association (2011)

17. O'Brien, K., Pietri, I., Reddy, R., Lastovetsky, A., Sakellariou, R.: A survey of power and energy predictive models in HPC systems and applications. ACM Comput. Surv. **50**(3), 37 (2017)

18. PAPI: Performance application programming interface 5.4.1 (2015). http://icl.cs.utk.edu/papi/

19. Perf Wiki: perf: Linux profiling with performance counters (2017). https://perf.wiki.kernel.org/index.php/Main_Page

20. Rivoire, S., Ranganathan, P., Kozyrakis, C.: A comparison of high-level full-system power models. In: Proceedings of the 2008 Conference on Power Aware Computing and Systems, HotPower 2008. USENIX Association (2008)

21. Shahid, A., Fahad, M., Reddy, R., Lastovetsky, A.: Additivity: a selection criterion for performance events for reliable energy predictive modeling. Supercomput. Front. Innovations **4**(4), 50–65 (2017)

22. Shahid, A., Fahad, M., Reddy Manumachu, R., Lastovetsky, A.: Supplemental: Improving the accuracy of energy predictive models for multicore cpus using Additivity of performance monitoring counters (2019). https://github.com/ArsalanShahid116/SLOPE-PMC/blob/master/PaCT-2019-Additivity-supplemental.pdf

23. Singh, K., Bhadauria, M., McKee, S.A.: Real time power estimation and thread scheduling via performance counters. SIGARCH Comput. Archit. News **37**(2), 46–55 (2009)

24. Smarr, L.: Project greenlight: optimizing cyber-infrastructure for a carbon-constrained world. Computer **43**(1), 22–27 (2010)
25. Treibig, J., Hager, G., Wellein, G.: LIKWID: a lightweight performance-oriented tool suite for x86 multicore environments. In: 2010 39th International Conference on Parallel Processing Workshops (ICPPW), pp. 207–216. IEEE (2010)
26. Wang, H., Jing, Q., Chen, R., He, B., Qian, Z., Zhou, L.: Distributed systems meet economics: pricing in the cloud. In: Proceedings of the 2nd USENIX Conference on Hot Topics in Cloud Computing. USENIX Association (2010)
27. Wang, S.: Ph.d thesis: Software power analysis and optimization for power-aware multicore systems (2014)
28. Zhou, Z., Abawajy, J.H., Li, F., Hu, Z., Chowdhury, M.U., Alelaiwi, A., Li, K.: Fine-grained energy consumption model of servers based on task characteristics in cloud data center. IEEE Access **6**, 27080–27090 (2018)

An Experimental Study of Data Transfer Strategies for Execution of Scientific Workflows

Oleg Sukhoroslov[✉]

Institute for Information Transmission Problems
of the Russian Academy of Sciences, Moscow, Russia
sukhoroslov@iitp.ru

Abstract. The paper studies the impact of data transfer strategies on the execution of scientific workflows. Five strategies are described, which define when and in what order data transfers are performed during the workflow execution. The strategies are experimentally evaluated by means of simulation using a realistic network model. It is demonstrated that the execution time of data-intensive workflows significantly depends on the used strategy. In particular, Eager and Lazy strategies, often used in theory and practice of workflow scheduling, demonstrate the poor results in most cases. The alternative strategies provide up to 36% makespan improvement by overlapping communications and computations, prioritizing data transfers and reducing network contention.

Keywords: Scientific workflows · Data-intensive computing · Task scheduling · Data management · Simulation

1 Introduction

Workflows is an important class of loosely coupled parallel applications that consist of multiple tasks with control or data dependencies. Such applications are widely used for automation of complex computational and data processing pipelines in science and technology [16]. The tasks in workflows run independently by exchanging data only through their input and output files.

Workflows are well suited for parallel execution on distributed computing systems such as clusters, grids and clouds. However, the efficiency of workflow execution in a system critically depends on the methods used to schedule the tasks among the system nodes which is an active area of research [21,22]. The typical objective is to minimize the workflow execution time or cost, possibly subject to additional constraints such as a fixed budget or a deadline.

The explosive growth of data observed in many domains has led to the proliferation of workflows that consume and produce large amounts of data that has to be transferred between the tasks during the workflow execution. For example, in the survey of scientific workflows from several domains [9] the size of data files

© Springer Nature Switzerland AG 2019
V. Malyshkin (Ed.): PaCT 2019, LNCS 11657, pp. 67–79, 2019.
https://doi.org/10.1007/978-3-030-25636-4_6

varied from 2 GB to more than 200 TB. In some cases, the CPU time allocated to I/O operations exceeded the time spent by computations.

Data-intensive workflows have specific runtime requirements [10]. Besides the task scheduling, the execution of such workflows requires a careful choice of data management strategies, since data transfers can significantly impact the execution time. A multitude of approaches have been proposed over the last decade that consider the data location and transfers when scheduling workflows in distributed systems. These works make different assumptions on the data sharing and transfer models, however the trade-offs among the different strategies and their impact on the workflow execution are poorly studied. Also, previous work almost completely ignores the network contention caused by concurrent data transfers which can significantly impact the data transfer times.

We argue that the efficient execution of data-intensive workflows requires considering not only the placement of data files (space dimension), but also the scheduling of data transfers (time dimension). As a first step in this direction, we investigate several data transfer strategies in isolation from the task scheduling algorithm. We describe five strategies and evaluate their impact on the execution of scientific workflows by means of simulation using a realistic network model.

We demonstrate that the execution time of data-intensive workflows significantly depends on the used data transfer strategy. In particular, Eager strategy, assumed in many theoretical studies of workflow scheduling, performs the worst in most cases. Similarly, Lazy strategy, used in many practical implementations, never achieves the best results. The alternative strategies provide up to 36% makespan improvement by overlapping communications and computations, prioritizing data transfers and reducing network contention. The relative performance of the studied strategies depends on the workflow properties.

The paper is structured as follows. Section 2 discusses the related work. Section 3 describes the workflow scheduling problem along with the used workflow and system models. Section 4 describes and discusses the studied data transfer strategies. Section 5 presents the results of simulation experiments. Section 6 concludes and discusses the future work.

2 Related Work

A multitude of approaches have been proposed that consider the data location and transfers when scheduling workflows in distributed infrastructures.

Heterogeneous Earliest Finish Time (HEFT) [18] is a well-known list scheduling heuristic that takes into account the workflow graph (data dependencies) and data transfer times. However, HEFT and other similar heuristics do not consider the data placement during the scheduling, i.e the locality of data files is not explicitly explored. More recent works on scheduling workflows in clouds [1,5,13] take into account VM provision, financial costs and deadlines, but also do not explore the data assignment. Works [11,20,23] focus on execution of workflows across multiple data centers and propose data placement strategies to reduce the data transfers between data centers. However, the task scheduling and minimization of execution time are not deeply treated in such approaches, e.g. tasks are simply assigned to the data center which stores the most of input data. In [7] an integrated task and data placement algorithm based on graph partitioning is proposed with the goal of minimizing data transfers. Bryk et al. [4] propose a dynamic scheduling algorithm for workflow ensembles in clouds that minimizes the number of transfers by taking advantage of data caching and file locality. Finally, works [15,17] explicitly treat the data and task assignment problems together and propose scheduling algorithms aiming at minimizing both data transfers and the total execution time of the workflow.

These works make different assumptions on the data sharing and transfer models. For example, some works assume direct data transfers between the execution nodes [17,18], while others rely on a shared storage system (e.g. Amazon S3) for exchanging data between the workflow tasks [4,15]. Similarly, the data staging strategies range from the eager transfers of input data as soon as it is available [18] to the lazy transfers tightly coupled with task execution [4]. At the same time, the trade-offs among the different strategies and their impact on the execution of data-intensive workflows are poorly studied. Bharathi et al. [2] analyzed different data staging strategies based on the degree of interaction between the workflow manager and the data placement service. However, the studied strategies rely on the centralized management of data transfers and explore only simple sequential ordering of data transfers.

Also, previous work almost completely ignores the network contention which can significantly impact the data transfer times [12]. For example, HEFT and other static heuristics use the full network bandwidth for estimating the data transfer times, since modifying such algorithms to take into account the interference of data transfers is very challenging. Indeed, the transfers associated with a just scheduled task can impact the transfers and start times of the previously scheduled tasks. This assumption is also used in dynamic scheduling algorithms, which at least can try to adapt to the introduced inaccuracies. Similarly, when evaluating the proposed algorithms authors often rely on simulators with flaws in their network model [19]. Interestingly, Bryk et al. [4], while using simple transfer time estimates in their scheduling algorithm, actually simulate the bandwidth sharing in the modeled system and serialize data transfers to mitigate the congestion effects. This confirms the importance of employed data transfer strategy.

This paper addresses the aforementioned issues by analyzing the impact of different data transfer strategies on the execution of data-intensive workflows while taking into account the network contention. In contrast to related works, direct data transfers between the execution nodes are assumed without the use of a shared storage or a centralized data transfer manager.

3 Problem Description

In this section, we describe the main assumptions of the used workflow and system models, introduce the workflow scheduling problem and HEFT algorithm.

A *workflow* is modeled as a directed acyclic graph, $W = (T, D)$, where T is the set of t vertices (tasks) and D is the set of d edges (dependencies) between the tasks. Each task t_i has a weight w_i equal to the required amount of computations. Each edge $(i, j) \in D$ represents a precedence constraint, such that task t_i should complete before task t_j starts, and has a weight $d_{i,j}$ equal to the amount of data required to be transmitted between the tasks. A task without any parent is called an *entry* task and a task without any child is called an *exit* task.

The distributed computing system is modeled as a set N of n nodes connected via a network. Each node n_i is characterized by its performance p_i which allows to estimate the task execution times. It is assumed that each node can execute one task at a time and the task execution is nonpreemptive. The network has a star topology where each node is connected to a central backbone via a dedicated link l_i characterized by its bandwidth B_i and latency L_i. The rate of communication between a pair of nodes is determined only by the characteristics of the corresponding links. The link bandwidth is shared between concurrent data transfers using the realistic model [19].

The entry and exit tasks are executed on a dedicated *master* node, which does not participate in the execution of ordinary tasks. This node corresponds to the machine which stores the workflow input data and where the output data should be placed after the workflow execution. The intermediate data produced by the tasks is stored on the nodes that executed the corresponding tasks. The data required for a task execution is transferred directly from the corresponding nodes. It is also assumed that task execution can be overlapped with data transfers.

The workflow scheduling problem is to find the optimal assignment of workflow tasks to system nodes with respect to a given criterion. In this work, we consider minimizing the workflow execution time (makespan) and use the well-known Heterogeneous Earliest Finish Time (HEFT) algorithm [18]. This algorithm takes into account the data dependencies between tasks by employing the following list scheduling heuristics. The tasks are scheduled in descending order of their rank computed as

$$rank(t_i) = \overline{w_i} + \max_{t_j \in children(t_i)} \left(\overline{c_{i,j}} + rank(t_j) \right),$$

where $\overline{w_i}$ is the average execution time of task t_i and $\overline{c_{i,j}}$ is the average communication time between tasks t_i and t_j. Each task is scheduled to a node with a minimum earliest finish time for this task.

HEFT and other static scheduling algorithms require a priori estimates of task execution and data transfer times. In this work, it is assumed that the former are exact, while the latter are obtained using the full network bandwidth as in related works. The network contention caused by concurrent data transfers during the workflow execution can significantly impact the transfer times and invalidate the assumptions made by the algorithm. This can lead to the performance degradation of produced schedules [12]. However, the amount of network contention can depend on a strategy used for scheduling of data transfers.

4   Data Transfer Strategies

Data transfer strategy defines when and in what order data transfers, corresponding to edges in a workflow DAG, are performed during the workflow execution. For each data transfer, the source task is called *producer* and the destination task is called *consumer*. In this study, the following strategies are considered.

Eager: In this strategy, the data transfer starts immediately after the data is ready, i.e. the producer is completed, and the destination node is known, i.e. the consumer is scheduled to some node. In case of static scheduling, the latter information can be made available for all nodes before the workflow execution, so that the data transfers are started as earliest as possible. This strategy is often implicitly assumed in theoretical and simulation studies of workflow scheduling algorithms, because it looks effective and is simple to model analytically. However, in practice, this strategy can cause severe network contention for data-intensive workflows, thereby delaying the execution of upcoming tasks and resulting in significant divergence from the original static schedule.

Lazy: In this strategy, the data transfer is performed when the destination node is ready to execute the consumer task, subject to readiness of the data. Lazy strategy is the opposite to Eager strategy, since it delays the data transfer to the latest time possible, i.e. when the data is actually needed, by tightly coupling the data transfer and task execution. The obvious drawback of this approach is that it does not allow to overlap communications and computations, since the node is idle when it waits for the data transfer to complete. Nevertheless, this approach is often used in real systems along with dynamic scheduling of ready tasks to idle nodes. In this case, the destination node is not known beforehand, and Lazy is the only applicable strategy. For static or forward dynamic scheduling, the node schedule is known completely or for some time ahead, which enables the use of more advanced strategies for overlapping communications and computations.

Eager and Lazy strategies form the two opposite sides of the spectrum of possible data transfer strategies, each with its shortcomings. The following strategies try to address these shortcomings by prefetching task input data and prioritizing data transfers according to the workflow execution schedule.

Prefetch: In this strategy, the data transfer is scheduled when the destination node begins to execute a task immediately preceding the consumer task. This approach is similar to prefetching technique widely used in computer science where the data expected to be needed soon is loaded in advance. In contrast to Lazy strategy, this approach allows to reduce the node idle time by overlapping the data transfer with task execution. However, the idle time can not be completely eliminated if the data transfer takes more time than the execution of preceding task, or if the data transfer is delayed because the data is not ready. This approach requires the information about the current and the next task scheduled on the node, i.e. the use of static or forward dynamic scheduling.

Queue: In this strategy, data transfers on each destination node are scheduled sequentially in the order of planned execution of consumer tasks on this node. In comparison to Prefetch strategy, this approach allows to more flexibly load data for upcoming tasks in advance, i.e. before the execution of preceding task. In contrast to Eager strategy, this approach prioritizes data transfers, so that the tasks soon to be executed receive their data before the tasks far in the schedule. Also, since incoming data transfers on each node are performed sequentially, this strategy can reduce network contention, though outgoing transfers are not limited. This strategy requires a task schedule on each node, i.e. the use of static or forward dynamic scheduling. When only a single next task is known, this strategy is equivalent to Prefetch. A possible drawback of this strategy is that it can delay data transfers, and consequently task execution, by introducing additional dependencies on data transfers of the preceding task. In particular, if some input data for a given task has become ready earlier than the data for the preceding task, the former cannot be downloaded before the latter.

QueueECT: In this strategy, data transfers on each destination node are scheduled sequentially in the order of expected completion time of producer tasks, breaking the ties with the order of planned execution of consumer tasks as in Queue. The intuition behind this strategy is to avoid the mentioned drawback of Queue by prioritizing data transfers for data that is expected to be ready earlier. While this approach allows to better utilize network by avoiding delays of data transfers, it has two drawbacks. First, it can delay execution of a task due to interfering data transfers of succeeding tasks. Second, it requires information about the expected completion time of each task, while other strategies require only a list of scheduled tasks on each node. This also makes this strategy sensitive to inaccuracies in estimates during the workflow execution.

5   Experimental Study

The impact of described data transfer strategies has been studied by means of simulation using *pysimgrid*[1], an open source framework for studying scheduling in distributed computing systems. This framework is implemented on the base of mature SimGrid toolkit [6] which includes a verified network model [19]. *pysimgrid* implements a thin Python wrapper around the native SimGrid C API and provides convenient interfaces for implementation of scheduling algorithms and running simulations. The framework includes implementations of several scheduling algorithms along with tools for generation of synthetic systems and applications, batch execution of experiments and analysis of produced results.

The studied data transfer strategies have been implemented in *pysimgrid* by using the SimDAG library from SimGrid. It maintains an internal DAG representation of the workflow, which explicitly treats data transfers as special tasks, i.e. vertices that are connected with producer and consumer tasks. Data transfer tasks are automatically started by SimDAG when the producer is completed and the consumer is scheduled. This behavior corresponds to Eager strategy. Other strategies have been implemented by adding extra dependencies between the

---

[1] https://github.com/alexmnazarenko/pysimgrid.

data transfer and compute tasks. For example, Lazy strategy is implemented by adding dependencies between data transfer tasks and preceding compute tasks.

To model diverse workflow structures, the following workflows based on real-world scientific applications are used in experiments: CyberShake, Epigenomics, Inspiral, Montage [3], Montage1.5 [8], 1000Genome [14]. The majority of workflows consist of 100 tasks, except Montage1.5 (472 tasks) and 1000Genome (52 tasks). The workflows have been converted from DAX to DOT format used by *pysimgrid*. During the conversion, multiple data transfers between the same pair of tasks were replaced by a single data transfer with total data size, since *pysimgrid* do not support multigraphs. To model different levels of data intensity, multiple workflow instances were produced by scaling all data transfer sizes to meet the specified CCR (communication to computation ratio) values. CCR is reported as the ratio of the sum of data transfer times, disregarding network contention, to the sum of task execution times, using the mean node performance.

The systems used in experiments consist of 5 or 10 worker nodes with performance randomly distributed between 1 and 4 GFLOPS. Each node is connected with others (via central backbone) by a network link with 100 MB/s bandwidth and 100 us latency. 100 random systems are generated for each node count.

HEFT implementation from *pysimgrid* is used for scheduling of workflow tasks. The choice of static algorithm is motivated by the use of task schedules in many strategies. Also, this allows to investigate the influence of data transfer strategies on degradation of static schedule due to network contention. The algorithm implementation is modified to take into account both Eager and Lazy strategies when computing data transfer estimates. Taking into account other strategies inside HEFT is much harder and is left for future work.

The described workflows have been executed in simulated heterogeneous systems using the studied data transfer strategies. Makespan is used as the base performance metric. For each workflow-system pair we perform runs using each of studied data transfer strategies and then normalize their makespans to the makespan of the baseline strategy, Eager. To reduce variance, we compute the mean of normalized makespans across all systems and report these values in the tables. The complete experimental setup is published on GitHub[2].

Table 1 contains the results for execution of workflows with varying CCR on systems with 5 and 10 nodes (the best results for each configuration are marked with *). As expected, for small CCR the effect of data transfers on the makespan is minimal, and all strategies perform similar. However, when CCR is increasing, the results of different strategies increasingly diverge. Prefetch and Eager strategies demonstrate the best and the worst results in the majority of cases (79% and 67%) respectively, except Montage1.5 discussed later. Lazy strategy never achieves the best results and is worst in 21% of cases. Queue strategy performs the best in 44% of cases with results close to Prefetch. QueueECT shows mixed results with good results for Montage and 1000Genome, and poor ones for Epigenomics and Inspiral.

---

[2] https://github.com/osukhoroslov/pysimgrid-experiments/tree/master/pact2019.

Table 1. Normalized makespan for systems with 5 (left) and 10 (right) nodes

CCR, %	Eager	Lazy	Pfetch	Queue	QECT	CCR, %	Eager	Lazy	Pfetch	Queue	QECT
CyberShake						CyberShake					
1	1.000	0.998	0.997*	0.997*	0.997*	1	1.000	0.996	0.995*	0.995*	0.995*
5	1.000	0.991	0.983*	0.984	0.985	5	1.000	0.985	0.978*	0.980	0.981
10	1.000	0.985	0.971*	0.971*	0.974	10	1.000	0.977	0.965*	0.969	0.971
20	1.000	0.975	0.953*	0.955	0.963	20	1.000	0.970	0.949*	0.954	0.965
Epigenomics						Epigenomics					
1	1.000	0.974	0.964*	0.964*	1.000	1	1.000	0.963	0.954*	0.954*	0.999
5	1.000	0.899	0.850*	0.850*	1.000	5	1.000	0.893	0.827*	0.827*	0.998
10	1.000	0.853	0.750*	0.750*	1.000	10	1.000	0.880	0.737*	0.737*	1.000
20	1.000	0.821	0.641*	0.647	1.000	20	1.000	0.898	0.761*	0.761*	1.000
Inspiral						Inspiral					
1	1.000	0.968	0.957*	0.957*	0.996	1	1.000	0.942	0.933*	0.934	0.992
5	1.000	0.878	0.839	0.837*	0.996	5	1.000	0.827	0.802*	0.819	0.990
10	1.000	0.818	0.757	0.749*	0.995	10	1.000	0.790	0.767*	0.797	0.986
20	1.000	0.782	0.722	0.691*	0.995	20	1.000	0.810	0.807*	0.811	0.984
Montage						Montage					
1	1.000	1.006	0.998*	1.001	0.998*	1	1.000	1.006	0.997*	1.002	0.998
5	1.000	1.033	0.990*	1.001	0.990*	5	1.000	1.026	0.976*	1.002	0.986
10	1.000	1.062	0.969*	0.991	0.982	10	1.000	1.050	0.940*	0.990	0.997
20	1.000	1.045	0.874*	0.899	0.984	20	1.000	1.036	0.851*	0.916	1.005
Montage1.5						Montage1.5					
1	1.000*	1.005	1.004	1.003	1.000*	1	1.000*	1.002	1.000*	1.002	1.000*
5	1.000*	1.063	1.079	1.078	1.001	5	1.000*	1.040	1.064	1.061	1.006
10	1.000*	1.133	1.205	1.215	1.005	10	1.000*	1.136	1.237	1.251	1.017
20	1.000*	1.260	1.448	1.479	1.017	20	1.000*	1.231	1.511	1.548	1.079
1000Genome						1000Genome					
1	1.000	0.973	0.966*	0.966*	0.966*	1	1.000	0.966	0.962*	0.962*	0.962*
5	1.000	0.898	0.859*	0.859*	0.859*	5	1.000	0.894	0.864*	0.864*	0.864*
10	1.000	0.853	0.768*	0.768*	0.768*	10	1.000	0.883	0.802*	0.802*	0.803
20	1.000	0.824	0.729*	0.734	0.734	20	1.000	0.905	0.838*	0.838*	0.839

Figure 1 contains the Gantt charts for execution of Epigenomics workflow with CCR = 20% on a 5-node system. In this case, Prefetch strategy reduced the workflow makespan by 36% in comparison to Eager (Queue, excluded from the figure, has similar schedule). This workflow has large input data, which is required by 26 of 100 tasks, and relatively small intermediate and output data. Therefore, Eager strategy quickly saturates the network with input data transfers and significantly delays task execution. Lazy strategy manages to decrease the makespan by 18% by reducing the network contention, but creates noticeable idle gaps between the task executions. Prefetch and Queue further decrease the makespan by 18% by overlapping data transfers with task execution.

The results for Montage1.5 stands out from the rest of experiments, since Eager strategy consistently outperforms other strategies, while Prefetch and Queue perform the worst. Figure 2 contains the workflow structure and Gantt charts for execution of Montage1.5 instance with CCR = 20% on a 5-node system. These results can be explained by the workflow structure and task sizes. The first layer consists of 48 *mProjectPP* tasks (colored yellow), the second

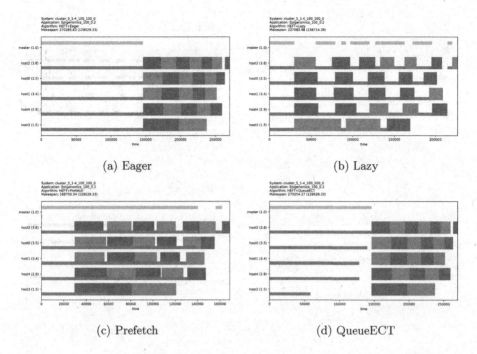

(a) Eager (b) Lazy

(c) Prefetch (d) QueueECT

Fig. 1. Gantt charts for execution of Epigenomics CCR = 20% on a 5-node system (blue - task execution, green - data upload, red - data download) (Color figure online)

layer consists of 320 *mDiffFit* tasks (blue) and the fifth layer consists of 48 *mBackground* tasks (green). The task executions in Gantt charts are colored according to these task groups, and data transfers are colored in black and gray. The execution of *mProjectPP* tasks dominates the run time, while *mDiffFit* and *mBackground* tasks, requiring data produced by *mProjectPP* tasks, take significantly less time. Eager strategy manages to transfer the required data to subsequent tasks the earliest by overlapping transfers with *mProjectPP* tasks without impacting their execution. Prefetch strategy starts to transfer required data late and fails to overlap data transfers with execution of small tasks. Queue strategy starts early, but since it serializes data transfers according to the task execution order, it can severely delay some transfers while waiting for data for preceding tasks if such data is produced in a different order. Indeed, QueueECT strategy, which serializes data transfers according to expected data readiness, performs close to Eager. Interestingly, Lazy strategy performs slightly better than Prefetch and Queue.

In contrast to Montage1.5, the Montage instance with 100 tasks has balanced task sizes across all layers and is executed the fastest with Prefetch strategy. CyberShake workflow demonstrates the lowest speedup, since its makespan is dominated by the large data transfers to two initial tasks, which leaves less room for optimizations. For 1000Genome workflow, the results of Prefetch, Queue and

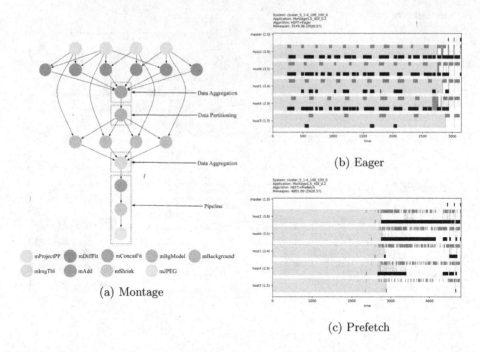

(a) Montage

(b) Eager

(c) Prefetch

Fig. 2. Montage workflow structure (a) and Gantt charts (b-c) for execution of Montage1.5 CCR = 20% on a 5-node system

QueueECT are almost identical, since all significant data transfers are concentrated on the first layer of the workflow.

The ratio of simulated makespan to makespan expected by HEFT algorithm for experiments on 5-node systems is presented in Table 2. As expected, the error caused by inaccurate data transfer time estimates made in the algorithm increases with CCR. However, the resulting error significantly depends on the

Table 2. The ratio of simulated makespan to makespan expected by HEFT for runs on 5-node systems

CCR, %	Eager	Lazy	Pfetch	Queue	QECT
CyberShake					
1	1.023	1.020	1.020	1.020	1.020
20	1.366	1.302	1.301	1.304	1.315
Epigenomics					
1	1.048	1.010	1.011	1.011	1.048
20	1.926	1.319	1.233	1.245	1.926
Inspiral					
1	1.049	1.005	1.004	1.004	1.045
20	2.008	1.320	1.450	1.386	1.997

CCR, %	Eager	Lazy	Pfetch	Queue	QECT
Montage					
1	1.002	1.004	1.001	1.003	1.001
20	1.188	1.143	1.036	1.066	1.169
Montage1.5					
1	1.001	1.005	1.005	1.004	1.001
20	1.056	1.266	1.531	1.563	1.074
1000Genome					
1	1.048	1.011	1.012	1.012	1.012
20	1.918	1.365	1.398	1.407	1.407

used data transfer strategy. For example, the use of Eager strategy for Epigenomics, Inspiral and 1000Genome results in 100% error, i.e. the real makespan is twice the expected, while the use of other strategies allows to significantly reduce this error (up to 23%). Note that the error reported for Lazy strategy is not consistent with results from the Table 1 since the HEFT implementation was modified to take into account this strategy when computing estimates.

6   Conclusion and Future Work

In this paper, several data transfer strategies for execution of scientific workflows have been described and experimentally evaluated by simulating execution of different workflows based on real-world scientific applications. It is demonstrated that the execution time of data-intensive workflows significantly depends on the used data transfer strategy. In particular, the commonly used Eager and Lazy strategies demonstrate the poor results in the most of cases. Prefetch and Queue strategies performed the best by overlapping communications and computations, prioritizing data transfers and reducing network contention, which resulted in up to 36% makespan improvement. Nonetheless, as was also demonstrated, there are cases where Eager can outperform other strategies, so the relative performance of these strategies depends on the workflow properties.

An obvious limitation of this study is that data transfer strategies are used in isolation from the task scheduling algorithm. We plan to address this issue in future work by investigating the use of these strategies inside the workflow scheduling algorithms to implement coscheduling of computations and data transfers. It is also planned to study the choice of optimal strategy depending on the workflow properties, develop advanced strategies that take into account overall network utilization, and incorporate optimizations such as data caching.

Acknowledgments. This work is supported by the Russian Science Foundation (project 16-11-10352).

References

1. Abrishami, S., Naghibzadeh, M., Epema, D.H.: Deadline-constrained workflow scheduling algorithms for infrastructure as a service clouds. Future Gener. Comput. Sys. **29**(1), 158–169 (2013)
2. Bharathi, S., Chervenak, A.: Data staging strategies and their impact on the execution of scientific workflows. In: Proceedings of the Second International Workshop on Data-Aware Distributed Computing, p. 5. ACM (2009)
3. Bharathi S., Chervenak A., Deelman E., Mehta G., Su M.H., Vahi K.: Characterization of scientific workflows. In: 2008 Third Workshop on Workflows in Support of Large-Scale Science, pp. 1–10, November 2008
4. Bryk, P., Malawski, M., Juve, G., Deelman, E.: Storage-aware algorithms for scheduling of workflow ensembles in clouds. J. Grid Comput. **14**(2), 359–378 (2016)

5. Byun, E.K., Kee, Y.S., Kim, J.S., Maeng, S.: Cost optimized provisioning of elastic resources for application workflows. Future Gener. Comput. Syst. **27**(8), 1011–1026 (2011)
6. Casanoya, H., Giersch, A., Legrand, A., Quinson, M., Suter, F.: Versatile, scalable, and accurate simulation of distributed applications and platforms. J. Parallel Distrib. Comput. **74**(10), 2899–2917 (2014)
7. Çatalyürek, Ü.V., Kaya, K., Uçar, B.: Integrated data placement and task assignment for scientific workflows in clouds. In: Proceedings of the Fourth International Workshop on Data-Intensive Distributed Computing, pp. 45–54. ACM (2011)
8. Deelman, E., et al.: Pegasus, a workflow management system for science automation. Future Gener. Comput. Syst. **46**, 17–35 (2015)
9. Juve, G., Chervenak, A., Deelman, E., Bharathi, S., Mehta, G., Vahi, K.: Characterizing and profiling scientific workflows. Future Gener. Comput. Syst. **29**(3), 682–692 (2013)
10. Liu, J., Pacitti, E., Valduriez, P., Mattoso, M.: A survey of data-intensive scientific workflow management. J. Grid Comput. **13**(4), 457–493 (2015)
11. Liu, Z., et al.: A data placement strategy for scientific workflow in hybrid cloud. In: 2018 IEEE 11th International Conference on Cloud Computing (CLOUD), pp. 556–563. IEEE (2018)
12. Nazarenko, A., Sukhoroslov, O.: An experimental study of workflow scheduling algorithms for heterogeneous systems. In: Malyshkin, V. (ed.) PaCT 2017. LNCS, vol. 10421, pp. 327–341. Springer, Cham (2017). https://doi.org/10.1007/978-3-319-62932-2_32
13. Pandey, S., Wu, L., Guru, S.M., Buyya, R.: A particle swarm optimization-based heuristic for scheduling workflow applications in cloud computing environments. In: 2010 24th IEEE International Conference on Advanced Information Networking and Applications, pp. 400–407. IEEE (2010)
14. da Silva, R.F., Filgueira, R., Deelman, E., Pairo-Castineira, E., Overton, I.M., Atkinson, M.P.: Using simple PID controllers to prevent and mitigate faults in scientific workflows. In: WORKS@ SC, pp. 15–24 (2016)
15. Szabo, C., Sheng, Q.Z., Kroeger, T., Zhang, Y., Yu, J.: Science in the cloud: allocation and execution of data-intensive scientific workflows. J. Grid Comput. **12**(2), 245–264 (2014)
16. Taylor, I.J., Deelman, E., Gannon, D.B., Shields, M.: Workflows for e-Science: Scientific Workflows for Grids. Springer, London (2014). https://doi.org/10.1007/978-1-84628-757-2
17. Teylo, L., de Paula, U., Frota, Y., de Oliveira, D., Drummond, L.M.: A hybrid evolutionary algorithm for task scheduling and data assignment of data-intensive scientific workflows on clouds. Future Gener. Comput. Syst. **76**, 1–17 (2017)
18. Topcuoglu, H., Hariri, S., Wu, M.Y.: Performance-effective and low-complexity task scheduling for heterogeneous computing. IEEE Trans. Parallel Distrib. Syst. **13**(3), 260–274 (2002)
19. Velho, P., Schnorr, L.M., Casanova, H., Legrand, A.: On the validity of flow-level TCP network models for grid and cloud simulations. ACM Trans. Model. Comput. Simul. (TOMACS) **23**(4), 23 (2013)
20. Wang, M., Zhang, J., Dong, F., Luo, J.: Data placement and task scheduling optimization for data intensive scientific workflow in multiple data centers environment. In: 2014 Second International Conference on Advanced Cloud and Big Data, pp. 77–84. IEEE (2014)
21. Wu, F., Wu, Q., Tan, Y.: Workflow scheduling in cloud: a survey. J. Supercomput. **71**(9), 3373–3418 (2015)

22. Yu, J., Buyya, R., Ramamohanarao, K.: Workflow scheduling algorithms for grid computing. In: Xhafa, F., Abraham, A. (eds.) Metaheuristics for Scheduling in Distributed Computing Environments. Studies in Computational Intelligence, vol. 146, pp. 173–214. Springer, Heidelberg (2008). https://doi.org/10.1007/978-3-540-69277-5_7
23. Yuan, D., Yang, Y., Liu, X., Chen, J.: A data placement strategy in scientific cloud workflows. Future Gener. Comput. Syst. **26**(8), 1200–1214 (2010)

Preference Based and Fair Resources Selection in Grid VOs

Victor Toporkov[1]([✉]), Dmitry Yemelyanov[1], and Anna Toporkova[2]

[1] National Research University "MPEI",
Ul. Krasnokazarmennaya, 14, Moscow 111250, Russia
{ToporkovVV, YemelyanovDM}@mpei.ru
[2] National Research University Higher School of Economics,
Ul. Myasnitskaya, 20, Moscow 101000, Russia
atoporkova@hse.ru

Abstract. In this work, a preference-based resources allocation algorithm for a job-flow scheduling in Grid virtual organizations (VOs) is proposed and studied. Users' and resource providers' preferences, VOs internal policies, resources geographical distribution along with local private utilization impose specific requirements for efficient scheduling according to different, usually contradictive, criteria. The algorithm performs resources selection optimization according to a specified general criterion and may be used in a variety of scheduling procedures, such as Backfilling or First Fit. Fair scheduling policies in VOs assume resources distribution according to VO stakeholders individual preferences. For this purpose, we consider a target optimization criterion as a linear combination of global (group) and private (user) job scheduling criteria. The mutual importance factor between the private and the global criteria is introduced to achieve a balanced scheduling solution.

Keywords: Scheduling · Grid · Resources selection · Utilization ·
Virtual organization · Preferences · Private · Global

1 Introduction and Related Works

In Grids with non-dedicated resources the computational nodes are usually partly utilized by local high-priority jobs coming from resource owners. Thus, the resources available for use are represented with a set of time intervals (slots) during which the individual computational nodes are capable to execute parts of independent users' parallel jobs. These slots generally have different start and finish times and a performance difference. The presence of a set of slots impedes the problem of resources allocation necessary to execute the job flow from VOs users. Resource fragmentation also results in a decrease of the total computing environment utilization level [1, 2].

Application level scheduling [3] is based on the available resources utilization and, as a rule, does not imply any global resource sharing or allocation policy. Job flow scheduling in VOs [4, 5] suppose uniform rules of resource sharing and consumption, in particular based on economic models [2, 3, 6]. This approach allows improving the job-flow level scheduling and resource distribution efficiency. VO policy may offer

V. Malyshkin (Ed.): PaCT 2019, LNCS 11657, pp. 80–92, 2019.
https://doi.org/10.1007/978-3-030-25636-4_7

optimized scheduling to satisfy both users' and VO global preferences. The VO scheduling problems may be formulated as follows: to optimize users' criteria or utility function for selected jobs [2, 7], to keep resource overall load balance [8, 9], to have job run in strict order or maintain job priorities [10, 11], to optimize overall scheduling performance by some custom criteria [12, 13], etc.

Computing system services support interfaces between users and providers of computing resources and data storages, for instance, in datacenters. Personal preferences of VO stakeholders are usually contradictive. Users are interested in total expenses minimization while obtaining the best service conditions: low response times, high hardware specifications, 24/7/365 service, etc. Service providers and administrators, on the contrary, are interested in profits maximization based on resources load efficiency, energy consumption, and system management costs. The challenges of system management can lead to inefficient resources usage in some commercial and corporate cloud systems.

Thus, VO policies in general should respect all members to function properly and the most important aspect of rules suggested by VO is their fairness. A number of works understand fairness as it is defined in the theory of cooperative games [7], such as fair job flow distribution [9], fair quotas [14, 15], fair user jobs prioritization [11], and non-monetary distribution [16]. In many studies VO stakeholders' preferences are usually ensured only partially: either owners are competing for jobs optimizing users' criteria [3, 17], or the main purpose is the efficient resources utilization not considering users' preferences [10]. Sometimes multi-agent economic models are established [3, 18]. Usually they do not allow optimizing the whole job flow processing.

The goal of the current study is to design a general resources selection procedure and criteria to find a trade-off between VO stakeholders' contradictory preferences. Fair resource sharing assume that every VO stakeholder has mechanisms to influence scheduling results providing own preferences. So, resources selection step may be used for additional job scheduling optimization according to both global (VO) and private (user) criteria. An important feature of the proposed approach is an independence from the particular job-flow scheduling procedure, i.e. First Fit, backfilling or a cycle scheduling scheme [12].

Main contribution of this paper is a general resources selection algorithm combining VO stakeholders preferences in a single target optimization criterion. The algorithm takes into account the system resources configuration and individual jobs features: size, runtime, cost, etc. When used in high-performance distributed computing systems and Grid metaschedulers during the resources allocation step it may improve resources distribution according to fair share policies.

The rest of the paper is organized as follows. Section 2 presents the problem statement and a general job-flow scheduling optimization approach based on conservative backfilling. Section 3 contains experiment setup and the simulation results obtained with different importance ratio of private and global criteria. Finally, Sect. 4 summarizes the paper.

2 Job-Flow Scheduling Optimization

2.1 Problem Statement

We consider a set R of heterogeneous computing nodes with different performance p_i and price c_i characteristics. Each node has a local utilization schedule known in advance for a considered scheduling horizon time L. A node may be turned off or on by the provider, transferred to a maintenance state, reserved to perform computational jobs. Thus, it's convenient to represent all available resources as a set of slots. Each slot corresponds to one computing node on which it's allocated and may be characterized by its performance and price.

In order to execute a parallel job one needs to allocate the specified number of simultaneously idle nodes ensuring user requirements from the resource request. The resource request specifies number n of nodes required simultaneously, their minimum applicable performance p, job's computational volume V and a maximum available resources allocation budget C. The required window length is defined based on a slot with the minimum performance. For example, if a window consists of slots with performances $p \in \{p_i, p_j\}$ and $p_i < p_j$, then we need to allocate all the slots for a time $T = \frac{V}{p_i}$. In this way V really defines a computational volume for each single job subtask with no dynamic load redistribution possible: the worst scenario for the job runtime estimation. Common start and finish times ensure the possibility of inter-node communications during the whole job execution. The total cost of a window allocation is then calculated as $C_W = \sum_{i=1}^{n} T * c_i$.

These parameters constitute a formal generalization for resource requests common among distributed computing systems and simulators.

Additionally we introduce a criterion f as a user preference for the particular job execution during the scheduling horizon L. The criterion f can take a form of any additive function and as an example, one may want to allocate suitable resources with the maximum possible total data storage available before the specified deadline.

2.2 Job-Flow Scheduling with Backfilling

The simplest way to schedule a job-flow execution is to use the First-Come-FirstServed (FCFS) policy. However this approach is inefficient in terms of resources utilization and backfilling [10] was proposed to improve system utilization.

Backfilling procedure makes use of advanced resources reservations which is an important mechanism preventing starvation of jobs requiring large number of computing nodes. Resources reservations in FCFS may create idle slots in the nodes' local schedules thus decreasing system performance. So the main idea behind backfilling is to backfill jobs into those idle slots to improve the overall system utilization. And the backfilling procedure implements this by placing smaller jobs from the back of the queue to these idle slots ahead of the priority order.

There are two common variations to backfilling - conservative and aggressive (EASY). Conservative backfilling enforces jobs' priority fairness by making sure that jobs submitted later can't delay the start of jobs arrived earlier. EASY backfilling aggressively fills jobs as long as they do not delay the start of a *leading* pending job. Conservative backfilling considers jobs in the order of their arrival and either immediately starts a job or makes an appropriate reservation upon the arrival. The jobs priority in the queue may be additionally modified in order to improve system-wide job-flow execution efficiency metrics. Under default FCFS policy the jobs are arranged by their arrival time. Other priority reordering-based policies like Shortest job First or eXpansion Factor may be used to improve overall resources utilization level [10, 19, 20].

Multiple Queues backfilling separates jobs into different queues based on metadata, such as jobs resource requirements: small, medium, large, etc. The idea behind this metaheuristic is that earlier arriving jobs and smaller-sized jobs should have higher execution priority. The number of queues and the strategy for dividing tasks among them can be set by the system administrators. Sometimes different queues may be assigned to a dedicated resource domain segments and function independently. In a single domain the metaheuristic cycles through the different queues in a round-robin fashion and may consider more jobs from the queues with smaller-sized tasks [19].

The look-ahead optimizing scheduler [20] implements dynamic programming scheme to examine all the jobs in the queue in order to maximize the current system utilization. So, instead of scanning queue for single jobs suitable for the backfilling, look-ahead scheduler attempts to find a combination of jobs that together will maximize the resources utilization.

2.3 General Window Search Procedure

Backfilling as well as many other job-flow scheduling algorithms in fact describe a general procedure determining high level policies for jobs prioritization and advanced resources reservations. However, the resources selection and allocation step remains sidelined since its more system specific nature. On the other hand, applying different resources allocation policies based on system or user preferences may affect scheduling results not only for individual jobs but for a whole job-flow.

For a general window search procedure for the problem statement presented in Sect. 2.1, we combine core ideas and solutions from algorithm AEP [21] and system [22]. Both related algorithms perform window search procedure based on a list of slots retrieved from a heterogeneous computing environment.

Following is the general square window search algorithm. It allocates a set of n simultaneously available slots with performance $p_i > p$, for a time, required to compute V instructions on each node, with a restriction C on a total allocation cost and performs optimization according to the criterion f. It takes a list of available slots ordered by their non-decreasing start time as input.

1. Initializing variables for the best criterion value and corresponding best window: $f_{max} = 0$, $w_{max} = \{\}$.
2. From the slots available we select different groups by node performance p_i. For example, group P_k contains resources allocated on nodes with performance $p_i \geq P_k$. Thus, one slot may be included in several groups.
3. Next is a cycle for all retrieved groups P_i starting from the max performance P_{max}. All the sub-items represent a cycle body.

(a) The resources reservation time required to compute V instructions on a node within group P_i is $T_i = \frac{V}{p_i}$.

(b) Initializing variable for a window candidates list $S_W = \{\}$.

(c) Next is a cycle for all slots s_i in group P_i starting from the slot with the minimum start time. The slots of group P_i should be ordered by their non-decreasing start time. All the sub-items represent a cycle body.

  (i)   If slot s_i doesn't satisfy any additional specific user requirements (hardware, software, etc.) then continue to the next slot (3c).

  (ii)   If slot length $l(s_i) < T_i$ then continue to the next slot (3c).

  (iii) Set the new window start time $W_i.start = s_i.start$.

  (iv)   Add slot s_i to the current window slot list S_W

  (v)   Next a cycle to check all slots s_j inside S_W

    (1)   If there are no slots in S_W with performance $p(s_j) = p_i$ then continue to the next slot (3c), as current slots combination in S_W was already considered for previous group P_{i-1}.

    (2)   If $W_i.start + T_i > s_j.end$ then remove slot s_j from S_W as it cannot be part of a window with the new start time $W_i.start$.

  (vi)   If S_W size is greater or equal to n, then allocate from S_W a window W_i (a subset of n slots with start time $W_i.start$ and length T_i) with a maximum criterion value f_i and a total cost $C_i < C$. If $f_i > f_{max}$ then reassign $f_{max} = f_i$ and $W_{max} = W_i$.

4. End of algorithm. At the output variable W_{max} contains the resulting window with the maximum criterion value f_{max}.

2.4   Optimal Slot Subset Allocation

Let us discuss in more details the procedure which allocates an optimal (according to the criterion f) subset of n slots out of S_W list (algorithm step 3c (vi)).

For some particular criterion functions f a straightforward subset allocation solution may be offered. For example for a window finish time minimization it is reasonable to return at step 3c(6) the first n cheapest slots of S_W provided that they satisfy the restriction on the total cost. These n slots (as any other n slots from S_W at the current step) will provide $W_i.finish = W_i.start + T_i$, so we need to set $f_i = -(W_i.start + T_i)$ to *minimize* the finish time at the end of the algorithm.

The same logic applies for a number of other important criteria, including window start time, runtime and a total cost minimization.

However in a general case we should consider a subset allocation problem with some additive criterion: $Z = \sum_{i=1}^{n} c_z(s_i)$, where $c_z(s_i) = z_i$ is a target optimization characteristic value provided by a single slot s_i of W_i.

In this way we can state the following problem of an optimal n-size window subset allocation out of m slots stored in S_W:

$$Z = x_1 z_1 + x_2 z_2 + \cdots + x_m z_m, \tag{1}$$

with the following restrictions:

$$x_1c_1 + x_2c_2 + \cdots + x_mc_m \leq C,$$

$$x_1 + x_2 + \cdots + x_m = n,$$

$$x_i \in \{0,1\}, i = 1..m,$$

where z_i is a target characteristic value provided by slot s_i, c_i is total cost required to allocate slot s_i for a time T_i, x_i - is a decision variable determining whether to allocate slot s_i ($x_i = 1$) or not ($x_i = 0$) for the current window.

This problem relates to the class of integer linear programming problems, which imposes obvious limitations on the practical methods to solve it. However we used 0-1 knapsack problem as a base for our implementation. Indeed, the classical 0-1 knapsack problem with a total weight C and items-slots with weights c_i and values z_i have the same formal model (1) except for extra restriction on the number of items required:$x_1 + x_2 + \cdots + x_m = n$. To take this into account we implemented the following dynamic programming recurrent scheme:

$$f_i(C_j, n_k) = \max\{f_{i-1}(C_j, n_k), f_{i-1}(C_j - c_i, n_k - 1) + z_i\}, \qquad (2)$$

$$i = 1,..,m, j = 1,..,C, k = 1,..,n,$$

where $f_i(C_j, n_k)$ defines the maximum Z criterion value for n_k-size window allocated out of first i slots from S_W for a budget C_j. After the forward induction procedure (2) is finished the maximum value $Z_{max} = f_m(C,n)$. x_i values are then obtained by a backward induction procedure.

An estimated computational complexity of the presented recurrent scheme is $O(m * n * C)$, which is n times harder compared to the original knapsack problem ($O(m * C)$). On the one hand, in practical job resources allocation cases this overhead doesn't look very large as we may assume that $n \ll m$ and $n \ll C$. On the other hand, this subset allocation procedure (2) may be called multiple times during the general square window search algorithm (step 3c(vi)).

2.5   Preference Based Resources Allocation

The proposed Slots Subset Algorithm (SSA) performs window search optimization by a general additive criterion $Z = \sum_{i=1}^{n} c_z(s_i)$, where $c_z(s_i) = z_i$ is a target optimization characteristic value provided by a single slot s_i of window W. These criterion values z_i may represent different slot characteristics: time, cost, power, hardware and software features, etc.

Introducing fair scheduling in VO requires mechanisms to influence scheduling results for VO stakeholders according to their private, group or common integral preferences. Individual users may have special requirements for the allocated resources, for example, total cost minimization or performance maximization. From the other hand,

VO policies usually assume optimization of a joint resources usage according to accepted efficiency criteria. One straightforward example is a maximization of the resources load.

In order to support both private and integral job-flow scheduling criteria we consider the following target criterion function in SSA for a single slot i:

$$z_i^* = z_i^I + \alpha z_i^U. \tag{3}$$

Here z_i^I and z_i^U represent criteria for integral and private jobs execution optimization correspondingly. z_i^I usually represents the same function for every job in the queue, while z_i^U reflects user requirements for a particular job optimization. $\alpha \in [0; +\infty]$ coefficient determines relative importance between private and integral optimization criteria.

By using SSA with z_i^* criterion and different α values it is possible to achieve a balance between private and integral job-flow scheduling preferences and policies. This approach has two important differences from *Anticipation* scheduling scheme [12].

1. SSA may be used as a resources selection algorithm in a variety of scheduling procedures, such as Backfilling or First Fit, and, thus, maintains job's original scheduling priorities and order. Anticipation works as a CSS extension, which schedules jobs in order to optimize the integral job-flow execution criterion. Besides that, Anticipation may require a reference solution for the additional specific jobs execution optimization.
2. General window search scheme in SSA implements optimal resources selection according to the specified criterion z_i^* with regards to the restrictions in problem (1). Anticipation performs heuristic-based resources selection using almost unconfigurable *execution similarity* criterion.

3 Simulation Study

3.1 Implementation and Simulation Details

The experiment was prepared as follows using a custom distributed environment simulator [2, 12, 21]. For our purpose, it implements a heterogeneous resource domain model: nodes have different usage costs and performance levels. A space-shared resources allocation policy simulates a local queuing system (like in CloudSim [6]) and, thus, each node can process only one task at any given simulation time. The execution cost of each task depends on its execution time, which is proportional to the dedicated node's performance level. The execution of a single job requires parallel execution of all its tasks. Some details regarding the computing model were provided in Sect. 2.1.

VO and computing environment were generated automatically during each scheduling simulation with the following properties:

- The resource pool includes 32 heterogeneous computational nodes.
- Node performance level is given as a uniformly distributed random value in the interval [2, 16]. This configuration provides a sufficient resources diversity level while the difference between the highest and the lowest resource performance levels will not exceed one order.

- A specific cost of a node is an exponential function of its performance value (base cost) with an added variable margin distributed normally as ± 0.6 of a base cost.
- The scheduling interval length is 800 time quanta. The initial resource load with owner jobs is distributed hyper-geometrically resulting in 5% to 10% time quanta excluded in total.

Job queue properties:

- Jobs number in the queue is 64.
- Nodes quantity needed for a job is a whole number distributed evenly on $n \in [2, 5]$.
- Node reservation time is a whole number distributed evenly on $V \in [60; 600]$.
- Job budget varies in the way that some of jobs can pay as much as 160% of base cost whereas some may require a discount.

The present configuration of the computing environment and the job-flow allows us to evaluate and compare considered scheduling approaches on a few *cycles*. I.e. during the job-flow execution each node is sequentially loaded with several jobs. Correspondingly, cost and budget parameters were selected so that each job could be executed at least on the cheapest resources. This determines steady market state appropriate for the consistent scheduling results comparison. Resources exponential price function and their initial load are designed to bring the model closer to the real commercial and corporate computing systems.

For the integral job-flow scheduling criterion we used jobs finish time minimization ($z_i^I = -s_i.finishTime$) as a metric for the overall resources load maximization.

For the SSA preference-based resources allocation efficiency study we implemented the following scheduling algorithms.

1. Firstly, we consider two conservative backfilling variations. BFs successively implements start time minimization for each job during the resources selection step. So, BFs criterion for slot i has the following form: $z_i = -s_i.startTime$.

By analogy BFf implements a more solid backfilling strategy of a finish time minimization which is different from BFs in heterogeneous computing environments. BFf target criterion for each job is $z_i = -s_i.finishTime$. BFf configuration represents extreme SSA scenario with $\alpha = 0$.

2. Secondly, we implement a preference-based conservative backfilling (BP) with SSA criterion of the following form: $z_i^* = -s_i.finishTime + \alpha z_i^U$ (3), where z_i^U depends on a private user criterion uniformly distributed between resources performance maximization ($z_i^U = s_i.nodePerformance$) and overall execution cost minimization ($z_i^U = -s_i.usageCost$). So in average half of jobs in the queue should be executed with performance maximization, while another half are interested in the total cost minimization.

Considered α values covered different importance configurations of private and integral optimization criteria: $\alpha \in [0.01; 0.1; 1; 10; 100]$.

3. As a special extreme scheduling scenario with $\alpha \to \infty$ we implemented pure conservative backfilling with SSA criterion $z_i^* = z_i^U$, i.e. without any global parameters optimization.

3.2 Simulation Results

The results of 1000 scheduling simulation scenarios are presented in Figs. 1, 2, 3, and 4. Each simulation experiment includes computing environment and job queue generation, followed by a scheduling simulation independently performed using considered algorithms. The main scheduling results are then collected and contribute to the average values over all experiments.

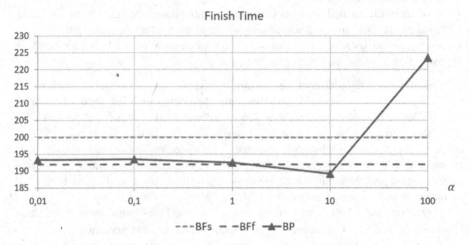

Fig. 1. Simulation results: average jobs finish time.

Figure 1 shows average jobs finish time for BFs, BFf and BP depending on α values on a logarithmic scale. BFs and BFf plots are represented by horizontal lines as the algorithms are independent of α.

As expected BFf provides 5% earlier jobs finish times compared to BFs. BFf with a job finish time minimization considers both job start time and runtime. In computing environments with heterogeneous resources job runtime may vary and depends on the selected resources performance. Thus, BFf implements more accurate strategy for the resources load optimization and a job-flow scheduling efficiency.

Similar results may be observed on Fig. 2 presenting average job queue execution makespan. This time the advantage of BFf by makespan criterion exceeds 10%.

BP approach with $\alpha \leq 10$ and considerable integral z_i^I criterion importance provides average finish time and makespan nearly the same as BFf. Average finish time is 1% later compared to BFf, while makespan is only 0,25% larger. However with increasing α these values are growing rapidly as the importance of the private scheduling preferences is increasing.

Interestingly, with $\alpha = 10$ BP provides even earlier average jobs finish time compared to BFf. In such configuration finish time minimization remains an important factor, while private performance and cost optimization lead to a more efficient resources sharing. At the same time BFf increases advantage by makespan criterion (Fig. 2) as some jobs in BP require more specific resources combinations generally available later in time.

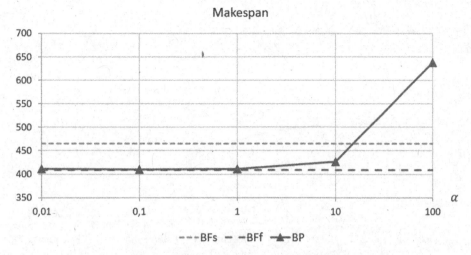

Fig. 2. Simulation results: average jobs queue execution makespan.

Figures 3 and 4 show scheduling results for considered private criteria: average job execution cost and allocated resources performance. BPc and BPp in Figs. 3 and 4 represent BP scheduling results for jobs subsets with cost and performance private optimization correspondingly. Dashed lines show limits for BP, BPc and BPp, obtained in a pure private optimization scenario ($\alpha \to \infty$) without the integral finish time minimization.

The figures show that even with relatively small α values BP implements considerable resource share between BPc and BPp jobs according to the private preferences. The difference reaches 7% in cost and 5% in performance for $\alpha = 0.01$.

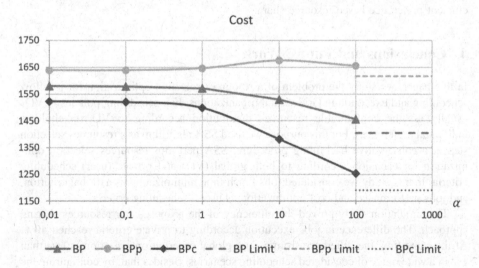

Fig. 3. Simulation results: average jobs execution cost.

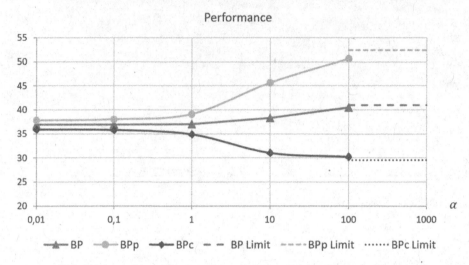

Fig. 4. Simulation results: average performance of the allocated resources.

More noticeable separation up to 30–40% is observed with $\alpha > 1$. With higher importance of the private criteria, BP selects more specific resources and increasingly diverges from the backfilling finish time procedure and corresponding jobs execution order. The values obtained by BP with $\alpha = 100$ are close to the practical limits provided by the pure private criteria optimizations.

We may conclude from Figs. 1, 2, 3, and 4 that by changing a mutual importance of private and integral scheduling criteria it is possible to find a trade-off solution. Even the smallest α values are able to provide a considerable resources distribution according to VO users private preferences. At the same time BP with $\alpha < 10$ maintains an adequate resources utilization efficiency comparable with BFf and provides even more efficient preference-based resource share.

4   Conclusions and Future Work

In this paper, we study the problem of a resources selection optimization for job-flow scheduling and execution in Grid virtual organizations. Fair scheduling policies in VOs usually assume configurable resources distribution according to VO stakeholders individual preferences. For this purpose we used SSA algorithm as a resources selection step in a conservative backfilling procedure. SSA performs resources selection optimization for each job according to both global (VO) and private (user) scheduling criteria. In this study we considered jobs finish time minimization as a global criterion, and jobs performance and cost optimization as users' scheduling criteria.

The simulation study proved the efficiency of the proposed fair resources sharing approach. The difference in jobs execution according to private criteria reached 40%. At the same time the difference from a pure global criterion optimization is less than 1% in a wide range of considered scheduling scenarios. Besides that, by configuring the importance factor between private and integral scheduling criteria it is possible to influence the fair scheduling outcome and propose a balanced solution.

Future work will be focused on a more detailed private and global criteria study, their mutual consistency and possible scheduling strategies to improve resources usage efficiency and the quality of service.

Acknowledgments. This work was partially supported by the Council on Grants of the President of the Russian Federation for State Support of Young Scientists (grant YPhD-2979.2019.9), RFBR (grants 18-07-00456 and 18-07-00534), and by the Ministry on Education and Science of the Russian Federation (project no. 2.9606.2017/8.9).

References

1. Dimitriadou, S.K., Karatza, H.D.: Job scheduling in a distributed system using backfilling with inaccurate runtime computations. In: Proceedings of the 2010 International Conference on Complex, Intelligent and Software Intensive Systems, pp. 329–336 (2010)
2. Toporkov, V., Toporkova, A., Tselishchev, A., Yemelyanov, D., Potekhin, P.: Heuristic strategies for preference-based scheduling in virtual organizations of utility grids. J. Ambient Intell. Humanized Comput. **6**(6), 733–740 (2015)
3. Buyya, R., Abramson, D., Giddy, J.: Economic models for resource management and scheduling in grid computing. J. Concurrency Comput. **14**(5), 1507–1542 (2002)
4. Kurowski, K., Nabrzyski, J., Oleksiak, A., Weglarz, J.: Multicriteria aspects of grid resource management. In: Nabrzyski, J., Schopf, J.M., Weglarz, J. (eds.) Grid Resource Management State of the Art and Future Trends, pp. 271–293. Kluwer Acad. Publ., Dordrecht (2003)
5. Rodero, I., Villegas, D., Bobroff, N., Liu, Y., Fong, L., Sadjadi, S.M.: Enabling interoperability among grid meta-schedulers. J. Grid Comput. **11**(2), 311–336 (2013)
6. Calheiros, R.N., Ranjan, R., Beloglazov, A., De Rose, C.A.F., Buyya, R.: CloudSim: a toolkit for modeling and simulation of cloud computing environments and evaluation of resource provisioning algorithms. J. Softw. Pract. Experience **41**(1), 23–50 (2011)
7. Rzadca, K., Trystram, D., Wierzbicki, A.: Fair game-theoretic resource management in dedicated grids. In: IEEE International Symposium on Cluster Computing and the Grid (CCGRID 2007), Rio De Janeiro, Brazil, pp. 343–350. IEEE Computer Society (2007)
8. Vasile, M., Pop, F., Tutueanu, R., Cristea, V., Kolodziej, J.: Resource-aware hybrid scheduling algorithm in heterogeneous distributed computing. J. Future Gener. Comput. Syst. **51**, 61–71 (2015)
9. Penmatsa, S., Chronopoulos, A.T.: Cost minimization in utility computing systems. Concurrency Comput. Pract. Experience **16**(1), 287–307 (2014)
10. Jackson, D., Snell, Q., Clement, M.: Core algorithms of the Maui scheduler. In: Feitelson, D. G., Rudolph, L. (eds.) JSSPP 2001. LNCS, vol. 2221, pp. 87–102. Springer, Heidelberg (2001). https://doi.org/10.1007/3-540-45540-X_6
11. Mutz, A., Wolski, R., Brevik, J.: Eliciting honest value information in a batch-queue environment. In: 8th IEEE/ACM International Conference on Grid Computing, New York, USA, pp. 291–297 (2007)
12. Toporkov, V., Yemelyanov, D., Toporkova, A., Potekhin, P.: Cyclic anticipation scheduling in grid VOs with stakeholders preferences. In: Malyshkin, V. (ed.) PaCT 2017. LNCS, vol. 10421, pp. 372–383. Springer, Cham (2017). https://doi.org/10.1007/978-3-319-62932-2_36

13. Takefusa, A., Nakada, H., Kudoh, T., Tanaka, Y.: An advance reservation-based co-allocation algorithm for distributed computers and network bandwidth on qos-guaranteed grids. In: Frachtenberg, E., Schwiegelshohn, U. (eds.) JSSPP 2010. LNCS, vol. 6253, pp. 16–34. Springer, Heidelberg (2010). https://doi.org/10.1007/978-3-642-16505-4_2

14. Carroll, T., Grosu, D.: Divisible load scheduling: an approach using coalitional games. In: Proceedings of the Sixth International Symposium on Parallel and Distributed Computing, ISPDC 2007, p. 36 (2007)

15. Kim, K., Buyya, R.: Fair resource sharing in hierarchical virtual organizations for global grids. In: Proceedings of the 8th IEEE/ACM International Conference on Grid Computing, Austin, USA, pp. 50–57. IEEE Computer Society (2007)

16. Skowron, P., Rzadca, K.: Non-monetary fair scheduling cooperative game theory approach. In: Proceedings of the Twenty-fifth Annual ACM Symposium on Parallelism in Algorithms and Architectures, pp. 288–297. ACM, New York (2013)

17. Dalheimer, M., Pfreundt, F.-J., Merz, P.: Agent-based grid scheduling with Calana. In: Wyrzykowski, R., Dongarra, J., Meyer, N., Waśniewski, J. (eds.) PPAM 2005. LNCS, vol. 3911, pp. 741–750. Springer, Heidelberg (2006). https://doi.org/10.1007/11752578_89

18. Thain, T., Livny, M.: Distributed computing in practice: the condor experience. Concurrency Comput. Pract. Experience **17**, 323–356 (2005)

19. Khemka, B., et al.: Resource management in heterogeneous parallel computing environments with soft and hard deadlines. In: Proceedings of 11th Metaheuristics International Conference (MIC 2015) (2015)

20. Shmueli, E., Feitelson, D.G.: Backfilling with lookahead to optimize the packing of parallel jobs. J. Parallel Distrib. Comput. **65**(9), 1090–1107 (2005)

21. Toporkov, V., Toporkova, A., Tselishchev, A., Yemelyanov, D.: Slot selection algorithms in distributed computing. J. Supercomput. **69**(1), 53–60 (2014)

22. Netto, M.A.S., Buyya, R.: A flexible resource co-allocation model based on advance reservations with rescheduling support. In: Technical Report, GRIDSTR-2007-17, Grid Computing and Distributed Systems Laboratory, The University of Melbourne, Australia, 9 October 2007

CAPE: A Checkpointing-Based Solution for OpenMP on Distributed-Memory Architectures

Van Long Tran[1,2(✉)], Éric Renault[2], and Viet Hai Ha[3]

[1] Hue Industrial College, 70 Nguyen Hue Street, Hue City, Vietnam
tvlong@hueic.edu.vn
[2] SAMOVAR, Télécom SudParis, CNRS, Université Paris-Saclay,
9 rue Charles Fourier, 91011 Evry Cedex, France
eric.renault@telecom-sudparis.eu
[3] College of Education, Hue University, Hue, Vietnam
haviethai@gmail.com

Abstract. CAPE, which stands for Checkpointing-Aided Parallel Execution, is a framework that automatically translates and provides runtime functions to execute OpenMP programs on distributed-memory architectures based on checkpointing techniques. In order to execute an OpenMP program on distributed-memory systems, CAPE uses a set of templates to translate an OpenMP source code into a CAPE source code which is then compiled using a regular C/C++ compiler. This code can be executed on distributed-memory systems under the support of the CAPE framework.

This paper aims at presenting the design and implementation of a new execution model based on Time-stamp Incremental Checkpoints. The new execution model allows CAPE to use resources efficiently, avoid the risk of bottlenecks, overcome the requirement of matching the Bernstein's conditions. As a result, these approaches make CAPE improving the performance, ability as well as reliability.

Keywords: CAPE · Checkpointing aided parallel execution ·
OpenMP on cluster · Parallel programming · Distributed computing ·
HPC

1 Introduction

OpenMP and MPI have become the standard tools to develop parallel programs on shared-memory and distributed-memory architectures respectively. As compared to MPI, OpenMP is easier to use. This is due to its ability to automatically execute code in parallel and synchronize results using its directives, clauses, and runtime functions while MPI requires programmers to do all this manually. Therefore, some efforts have been made to port OpenMP on distributed-memory architectures. However, excluding CAPE [7,9,18], no solution has successfully

© Springer Nature Switzerland AG 2019
V. Malyshkin (Ed.): PaCT 2019, LNCS 11657, pp. 93–106, 2019.
https://doi.org/10.1007/978-3-030-25636-4_8

met these two requirements: (1) to be fully compliant with the OpenMP standard and (2) high performance. Most prominent approaches include the use of an SSI [15], SCASH [19], the use of the RC model [13], performing a source-to-source translation to a tool like MPI [1,5] or Global Array [12], or Cluster OpenMP [11].

Among all these solutions, the use of a Single System Image (SSI) is the most straightforward approach. An SSI includes a Distributed Shared Memory (DSM) to provide an abstracted shared-memory view over a physical distributed-memory architecture. The main advantage of this approach is its ability to easily provide a fully-compliant version of OpenMP. Thanks to their shared-memory nature, OpenMP programs can easily be compiled and run as processes on different computers in an SSI. However, as the shared memory is accessed through the network, the synchronization between the memories involves an important overhead which makes this approach hardly scalable. Some experiments [15] have shown that the larger the number of threads, the lower the performance. As a result, in order to reduce the execution time overhead involved by the use of an SSI, other approaches have been proposed. For example, SCASH maps only the shared variables of the processes onto a shared-memory area attached to each process, the other variables being stored in a private memory, and the RC model that uses the relaxed consistency memory model. However, these approaches have difficulties to identify the shared variables automatically. As a result, no fully-compliant implementation of OpenMP based on these approaches has been released so far. Some other approaches aim at performing a source-to-source translation of the OpenMP code into an MPI code. This approach allows the generation of high-performance codes on distributed-memory architectures. However, not all OpenMP directives and constructs can be implemented. As yet another alternative, Cluster OpenMP, proposed by Intel, also requires the use of additional directives of its own (ie. not included in the OpenMP standard). Thus, this one cannot be considered as a fully-compliant implementation of the OpenMP standard either.

CAPE used the Discontinuous Incremental Checkpointing (DICKPT) [8] to implement the OpenMP fork-join model. The jobs of OpenMP work-sharing constructs are divided and distributed to slave nodes using checkpoints. At each slave node, these checkpoints are used to resume execution. In addition, the results after executing the divided jobs on each slave node are also extracted using checkpoints and sent back to the master. It has been demonstrated that this solution is fully compliant with OpenMP and provides high performance. However, there are some limitations:

– to run on top of CAPE, an OpenMP program must fulfill the Bernstein's conditions. This is the reason why the matrix-matrix product has been extensively used in the previous experiments.
– The implementation of CAPE wastes the resources. In the implementation of OpenMP work-sharing constructs on CAPE, the master does not perform a part of the computation. It waits for checkpoint results from the slave nodes and merges them together.

– The risk of bottlenecks and low communication performance at the implementation of the join phase. After executing the divided jobs, each slave node extracts a result checkpoint and sends it back to the master. The master receives, merges checkpoints together and sends the result back to the slave nodes in order to synchronize data.

This paper presents the design and implementation of a new model for CAPE based on Time-stamp Incremental Checkpointing (TICKPT) [24] to bypass the drawbacks mentioned above. The new implementation based on TICKPT improves the performance, capability, and reliability of this solution.

2   Checkpoint Techniques

2.1   Checkpointing

Checkpointing is the technique that saves the image of a process at a point during its lifetime, and allows it to be resumed from the saving's time if necessary [4,17]. Using checkpointing, processes can resume their execution from a checkpoint state when a failure occurs. So, there is no need to take time to initialize and execute it from the begin. These techniques have been introduced for more than two decades. Nowadays, they are used widely for fault-tolerance, applications trace/debugging, roll-back/animated playback, and process migration. To be able to save and resume the state of a process, the checkpoint saves all necessary information at the checkpoint's time. It can include register values, process's address space, open files/pipes/sockets status, current working directory, signal handlers, process identities, etc. The process's address space consists of text, data, *mmap* memory area, shared libraries, heap, and stack segments. Depending on the kind of checkpoints and its application, the checkpoint takes all or some of these information.

Based on the structure and contents of the checkpoint file, checkpointings are categorized into two groups: complete and incremental checkpointing.

– Complete checkpointing [3,4,14] saves all information regarding the process at the points that it generates checkpoints. The advantages of this technique are the reduction of the time of generation and restoration. However, not only a lot of duplicated data are stored each time a checkpoint is taken, there are also duplications in the different generated checkpoints.
– Incremental checkpointing [8,10,17] only saves the modified data. This has to be compared with the previous checkpoint. This technique reduces checkpoint's overhead and checkpoint's size. Therefore, it is widely used in distributed computing.

2.2   Time-Stamp Incremental Checkpointing

Time-stamp Incremental Checkpointing (TICKPT) [24] is an improvement of DICKPT by adding new factor – time-stamp – into incremental checkpoints and by removing unnecessary data based on data-sharing variable attributes of OpenMP programs.

Basically, TICKPT contains three mandatory elements including register's information, modified region in memory of the process, and their time-stamp. As well as DICKPT, in TICKPT, the register's information are extracted from all registers of the process in the system. However, the time-stamp is added to identify the order of the checkpoints in the program. This contributes to reduce the time for merging checkpoints and selecting the right element if located at the same place in memory. In addition, only the modified data of shared variables are detected and saved into checkpoints. It makes checkpoint's size significantly reduced depending on the size of private variables of the OpenMP program.

3   CAPE Based on TICKPT

3.1   Abstract Model

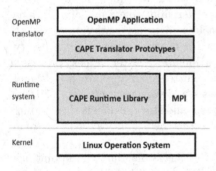

Figure 1 presents the new abstract model for CAPE. It is designed based on TICKPT and uses MPI to transfer data over the network.

As presented in the previous version [21, 22], CAPE provides a set of prototypes to translate OpenMP codes into CAPE codes. An OpenMP CAPE code in C or C++ is replaced by a set of calls to CAPE runtime functions. In this version, the CAPE translator prototypes are

Fig. 1. New abstract model for CAPE.   modified and added to adapt to the new mechanism based on TICKPT. This provides a set of prototypes to translate the common constructs, clauses, and runtime functions of OpenMP.

For the CAPE Runtime library, apart from providing functions to handle OpenMP instructions and to port them on distributed memory systems, some functions have been added to manage the declaration of variables and the allocation of memory on the heap. To transfer data among nodes in the system, instead of using the functions based on sockets like in the previous version, MPI_Send and MPI_Recv functions are called to ensure high reliability.

3.2 RC-Model Based CAPE Memory Model Implementation

OpenMP uses the Relaxed Consistency (RC) memory model. This model allows shared memory allocated in the local memory of a thread to improve memory accesses. When a synchronization point is reached, this local memory is updated in the shared

```
C_id ← generate_checkpoint(flag);
C ← all_reduce (C_id, id, nnodes,
        [operators]);
inject(C) ;
```

Fig. 2. cape_flush() implementation.

memory area that can be assessed by all threads.

CAPE completely implements the RC model of OpenMP on distributed-memory systems. All variables, including private and shared variables, are stored at all nodes of the system, and they can be only accessed locally. At synchronization points, only the modified data of shared variables at each node are extracted and saved into a checkpoint. This checkpoint is sent to the other nodes in the system, and is merged using the **merging** checkpoint operation with the other. Then, the result checkpoint is injected into the application memory to synchronize data.

In the CAPE runtime library, there are two fundamental functions which are called implicitly at synchronization points:

- **cape_flush()** generates a TICKPT, gathers, merges, and injects them into the application memory. This function is described by pseudo code in Fig. 2. Here, the **all_reduce()** function is responsible for gathering and merging the checkpoints generated by the **generate_checkpoint()** function. The gathering and the merging is implemented using both Ring and Recursive Doubling algorithm. The algorithm is automatically selected to be executed by the system depending on the size of the checkpoint.
- **cape_barrier()** sets a barrier and updates shared data between nodes. This function calls **MPI_Barrier()** of the MPI runtime library, and then uses **cape_flush()** to update shared data.

3.3 Execution Model

Figure 3 illustrates the execution model of CAPE. The idea of this model is the use of TICKPT to identify and synchronize the modified data of shared variables of the program among the nodes. OpenMP threads are replaced by processes, and each process runs in a node. At the beginning, the program is initialized and executed at the same time in all nodes of the system. Then, the execution works as the following rules:

- The sequential region or the code inside the **parallel** construct but not belonging to any other constructs is executed in the same way for all nodes.
- When the program reaches a **parallel** region, on each node, CAPE detects and saves the properties of all shared variables that are implicitly declared as sharing. If there are any OpenMP clauses declared in the **parallel** construct,

the relevant runtime functions are called to modify variable properties. Then, the **start** directive of TICKPT is called to save the value of the shared variables.

– At the end of a **parallel** region, the implementation of the **barrier** construct is implicitly called to synchronize data, and the **stop** directive of TICKPT is called to remove all relevant data.

– For the loop construct, each node (including the master node) is responsible for computing a part of the work based on the re-calculation of the range of iterations.

– For the **sections** construct, each node is divided into one or more parts of works that are indicated using **section** construct.

– At the **barrier**, the implementation of the **flush** construct is called to synchronize data.

– When the program reaches the **flush** construct, a TICKPT is generated and synchronized among the nodes to update the modification of shared data. According to [16], a **flush** is implicit at the following locations:
  - At the **barrier**.
  - At the entry to and the exit from **parallel**, **critical**, and **atomic** constructs.
  - At the exit from **for**, **sections**, and **single** constructs unless a **nowait** clause is present.

In this execution model, instead of using the master node to divide jobs and distribute to slave nodes based on incremental checkpoints in order to implement OpenMP work-sharing constructs, each node computes and executes the divided jobs automatically. At synchronization points, a TICKPT is generated at each node. It contains the modified data of shared variables and their time-stamps after executing the divided jobs. These checkpoints are gathered and merged at all nodes in the system using the Ring or Recursive Doubling algorithm [20]. This allows CAPE to void the bottleneck and improve the performance of communication tasks.

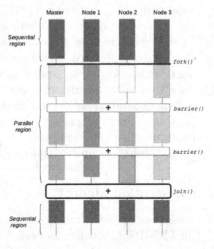

Fig. 3. The new execution model of CAPE.

With the features of TICKPT, checkpoints are able to use checkpoint's operations [23,24]. This allows memory elements to share the same address when computing and makes it simple when merging. Therefore, it allows CAPE to work without the need for the program to match with the Bernstein's conditions. Moreover, the master node takes a part in the computation of the divided jobs. This uses all the resources and improves the system efficiency.

The only missing part of the OpenMP specifications for this implementation is that **dynamic** and **guided** scheduling directives of the work-sharing construct have not been implemented yet. However, one can demonstrate that they can be easily translated into a `static` scheduling.

3.4 Prototypes

To be executed on a distributed-memory system with the support of the CAPE runtime library, the OpenMP source code is translated into a CAPE source code. There, each construct, clause, and runtime function of the OpenMP source code is translated into the relevant runtime function of CAPE. This translation works under the provision of a set of CAPE prototypes.

Based on the general syntax of OpenMP directives, a general template for CAPE prototypes was designed and is illustrated in Fig. 4. They are as follows:

```
cape_begin(directive-name, param-1, param-2);
    [cape_clause_functions]
    ckpt_start();
    //code blocks
cape_end(directive-name, reduction-flag );
```

Fig. 4. General template for CAPE prototypes in `C/C++`.

- `cape_begin()` and `cape_end()` are CAPE runtime functions which perform the actions for entering and exiting OpenMP directives. The `directive-name` is a label declared by CAPE which corresponds to the relevant CAPE runtime function. Depending on this label, the `cape_barrier()` function is called to update the shared data of the system. `param-1` and `param-2` are used to store the range of iterations for `for` loops, otherwise they both are set to zero. The `reduction-flag` is set to `TRUE` if there is a declaration of OpenMP reduction clause, otherwise it is set to `FALSE`.
- `cape_clause_functions` is a set of CAPE runtime functions which is used to implement OpenMP clauses. This implementation is presented in [23].
- `ckpt_start()` marks the location where to start the checkpointing. When reaching the `ckpt_start()` function, the value of shared variables is copied.

4 Experiments

In order to evaluate the performance of this new approach, we designed a set of micro benchmarks and tested them on a Desktop Cluster. The designed programs are based on the Microbenchmark for OpenMP 2.0 [2,6]. These programs have been translated to CAPE and executed on a Cluster to compare the performance.

4.1 Benchmarks

(1) MAMULT2D: This program computes the multiplication of two matrices. Originally, it was written in C/C++ and used the OpenMP `parallel for` construct. It matches Bernstein's conditions. Therefore, it has been used extensively to test CAPE in the previous works.

```
int vector(float A[], float B[], float C[], float D[], int n){
      int i, nthreads, tid;
      #pragma omp parallel shared(C,D,nthreads) private(A, B, i,tid)
    {
            tid = omp_get_thread_num();
            if (tid == 0)
            {
                  nthreads = omp_get_num_threads();
                  printf("Number of threads = %d\n", nthreads);
            }
            printf("Thread %d starting...\n",tid);
            #pragma omp sections nowait
            {
                  #pragma omp section
            printf("Thread %d doing section 1\n",tid);
                  for (i=0; i<N; i++)
                  {
                        for (j= 0 ; j< N; j+=25)
                              A[j] = A[j] * 0.15 ;
                        C[i] = A[i] + B[i];
                        printf("Thread %d: C[%d]= %f\n",tid,i,C[i]);
                  }
                  #pragma omp section
                  printf("Thread %d doing section 2\n",tid);
                  for (i=0; i<N; i++)
                  {
                        for (j= 0 ; j< N; j+=25)
                              B[j] = B[j] + 10.25 ;
                        D[i] = A[i] * B[i];
                        printf("Thread %d: D[%d]= %f\n",tid,i,D[i]);
                  }
            } /* end of sections */
      } /* end of parallel section */
      return 0;
}
```

Fig. 5. OpenMP function to compute vectors using **sections** construct.

(2) *PRIME:* This program counts the number of prime numbers in the range from 1 to N. The OpenMP code uses the **parallel for** construct with data-sharing clauses.

(3) *PI:* This program computes the value of PI by mean of the numeric integration method using Eq. (1).

$$\pi = \int_0^1 \frac{4}{1+x^2} dx \tag{1}$$

(4) VECTOR-1: This program performs operations on vectors. It contains OpenMP runtime functions, data-sharing clauses, a `nowait` clause, and `parallel` and `sections` constructs. The OpenMP code is presented in Fig. 5.

(5) VECTOR-2: This program performs some operations on vectors. It contains OpenMP `parallel` and `for` constructs with a `nowait` clause. The OpenMP code is shown in Fig. 6.

```
int vector2(int A[], int B[], int Y[], int Z[], int n, int m)
{
        int i,j;
        #pragma omp parallel private(A,Z) shared(B, Y)
        {
                #pragma omp for nowait
                for (i=1; i<n; i++){
                        for(j=0; j<n ; j+=20)
                                A[j] = A[j] + 10.25
                        B[i] = (A[i] + A[i-1]) / 2;
                }
                #pragma omp for nowait
                for (i=0; i<m; i++){
                        for(j=0; j<m ; j+=20)
                                Z[j] = Z[j] * 0.025 ;
                        Y[i] = Z[i] * i;
                }
        }
        return 0;
}
```

Fig. 6. OpenMP function to compute vectors using `for` construct.

4.2 Experimental Environment

The experiments have been performed on a 16-node cluster with different computer's configurations. There are two computers with Intel(R) Pentium(R) Dual CPU E2160 at 1.80 GHz, 2 GB of RAM, 5 GB of free HDD; seven computers with Intel(R) Core(TM)2 Duo CPU E7300 at 2.66 GHz, 3 GB of RAM, 6 GB of free HDD; five computers with Intel(R) Core(TM) i3-2120 CPU at 3.30 GHz, 8 GB of RAM, 6 GB of free HDD; and two computers including an AMD Phenom(TM) II X4 925 Processor at 2.80 GHz, 2 GB of RAM, 6 GB of free HDD. All machines are operated by the Ubuntu 14.03 LTS operating system with OpenSSH-Server and MPICH-2. They are interconnected by a 100 Mbps LAN network.

4.3 Experimental Results

Figures 7 and 8 present the execution time in milliseconds for the MAMULT2D program for various size of matrices and different sizes of cluster respectively.

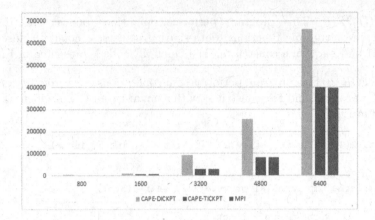

Fig. 7. Execution time (in milliseconds) of MAMULT2D with different size of matrix on a 16-node cluster.

Note that, there are many kinds of processors in different nodes. Some of them include many cores, but a single core was used for each node during the experiments. Three measures are presented at each time: the left one (yellow) for CAPE-DICKPT (the previous version), the middle one (blue) for CAPE-TICKPT (the current version), and the right one (red) for MPI.

Figure 7 presents the execution time for various matrix sizes on a 16-node cluster. The size increases from 800x800 to 6400x6400. The figure shows that the execution times of all methods are proportional to the matrix size. It also shows that the execution time of CAPE-TICKPT is much lower than the one of CAPE-DICKPT and MPI (around 35%) while the execution time of CAPE-TICKPT and MPI are roughly equal.

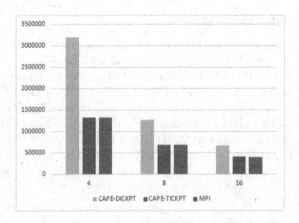

Fig. 8. Execution time (in milliseconds) of MAMULT2D for different cluster sizes.

Figure 8 presents the execution time for a matrix size of 6400x6400 on different cluster size. The number of nodes is successively 4, 8, and 16. The result presented in this figure also shows the similar trend for different matrix size. The execution time of CAPE is significantly reduced so that it is now much closer to an optimized human-written program using MPI.

To demonstrate that the new version of CAPE can run OpenMP programs that do not match with the Bernstein's conditions while achieving high performance, other experiments were conducted and performance were compared with MPI. All of the four other programs presented in Sect. 4.1 have been used to measure the execution time.

Figure 9 presents the execution time in milliseconds of PRIME with $N = 10^6$ on different cluster sizes for CAPE-TICKPT and MPI. It shows that the execution time of MPI is only around 1% smaller than the one of CAPE-TICKPT. In this experiment, the OpenMP `parallel for` directive with the `shared`, `private` and `reduction` clauses are translated and tested for both methods. Table 1 describes the steps executed by the program for both methods. The main different step is the join phase. It gathers the results from all nodes and computes their sum. For the MPI program, the user needs to clearly specify the values that need to be gathered, and then call the `MPI_Reduce()` function after to compute the sum. CAPE-TICKPT automatically identifies the modified value of the shared variables, extracts them into a TICKPT, and then gathers all checkpoints from all the nodes with the `merging` checkpoint operator. However, as the execution time of CAPE-TICKPT is nearly equal to the one of MPI, we consider that we successfully obtained high performance with CAPE.

Table 1. Comparison of the executed steps for the PRIME code for both CAPE-TICKPT and MPI.

Step	CAPE-TICKPT	MPI
Fork	Updates the properties of variables, saves data of shared variables, and re-computes the iterators	Re-computes the iterators
Computation	Computes the divided jobs	Computes the divided jobs
Join	Generates checkpoints, and calls the `merging` checkpoint operator with the `sum` operator	Calls `MPI_Reduce` to gather and sum the results

Figure 10 presents the execution time in milliseconds of PI with a number of steps equal to 10^8 for different cluster sizes using CAPE-TICKPT and MPI. In this experiment, the OpenMP `for` directive with `reduction` clause placed inside the `omp parallel` construct with some clauses are tested. As well as the previous experiments, this figure also shows that CAPE-TICKPT achieves similar performance as MPI.

Figure 11 shows the execution time in milliseconds for the VECTOR-1 program with $N = 10^6$ for different cluster sizes using CAPE-TICKPT and MPI.

In this experiment, OpenMP functions and the `sections` construct with two `section` directives are tested. The figure shows that the larger the number of nodes, the longer the execution time for both methods. The execution time with MPI is smaller than the one of CAPE-TICKPT, but the difference is not significant. Note that there are only two `section` directives in this program, so that both CAPE-TICKPT and MPI distribute the execution to two nodes only. Each node receives and executes the code of a `section`. However, the result has to be synchronized to all nodes on the system. Therefore, the execution time increases when increasing the number of nodes.

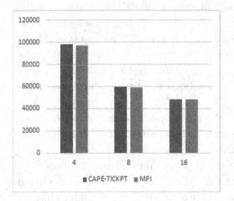

Fig. 9. Execution time (in milliseconds) of PRIME on different cluster sizes.

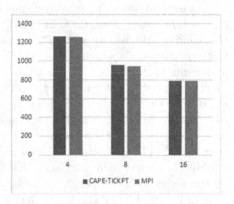

Fig. 10. Execution time (in milliseconds) of PI on different cluster sizes.

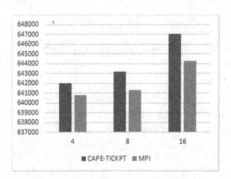

Fig. 11. Execution time (in milliseconds) of VECTOR-1 on different cluster sizes.

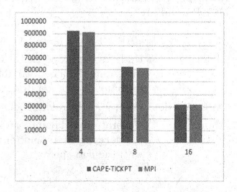

Fig. 12. Execution time (in milliseconds) of VECTOR-2 on different cluster sizes.

Figure 12 shows the execution time in milliseconds for VECTOR-2 with $N = 10^6$ and $M = 1.6 \times 10^6$ on different cluster sizes for both CAPE-TICKPT and MPI. This experiment aims at testing two `omp for` directives with `nowait` clause. The size of the two vectors are different from each other to ensure the nodes take different time to execute the divided jobs. The execution on each node is marked `nowait` until reaching the end block of the `parallel` region. The figure shows the same trend as the previous experiments. The execution time for CAPE-TICKPT is very close to MPI, the difference being negligible.

5   Conclusion and Future Works

This paper presented the design and implementation of a new execution model and prototypes for CAPE based on TICKPT. With this new capability included, CAPE improves the reliability and can run OpenMP programs that do not require to match the Bernstein's conditions. In addition, the analysis and evaluation of performance of this paper demonstrated that CAPE-TICKPT achieves performance very close to a comparable human-optimized hand-written MPI program. This is mainly due to the fact that CAPE-TICKPT takes benefits of the advantages of TICKPT such as checkpoint operators and can use resources more efficiently. The synchronization phase of the new execution model also avoids the risk of bottlenecks that may have occurred in the previous version.

In the near future, base on this mechanism, we will keep on developing the CAPE framework in order to support other OpenMP constructs. Furthermore, we expect to develop CAPE for GPUs.

References

1. Basumallik, A., Eigenmann, R.: Towards automatic translation of OpenMP to MPI. In: Proceedings of the 19th Annual International Conference on Supercomputing, pp. 189–198. ACM (2005)
2. Bull, J.M., O'Neill, D.: A microbenchmark suite for OpenMP 2.0. ACM SIGARCH Comput. Archit. News **29**(5), 41–48 (2001)
3. Chen, Z., Sun, J., Chen, H.: Optimizing checkpoint restart with data deduplication. Sci. Program. **2016**, 11 (2016)
4. Cores, I., Rodríguez, M., González, P., Martín, M.J.: Reducing the overhead of an MPI application-level migration approach. Parallel Comput. **54**, 72–82 (2016)
5. Dorta, A.J., Badía, J.M., Quintana, E.S., de Sande, F.: Implementing OpenMP for clusters on top of MPI. In: Di Martino, B., Kranzlmüller, D., Dongarra, J. (eds.) EuroPVM/MPI 2005. LNCS, vol. 3666, pp. 148–155. Springer, Heidelberg (2005). https://doi.org/10.1007/11557265_22
6. EPCC: EPCC OpenMP micro-benchmark suite. https://www.epcc.ed.ac.uk/research/computing/performance-characterisation-and-benchmarking/epcc-openmp-micro-benchmark-suite
7. Ha, V.H., Renault, E.: Design and performance analysis of CAPE based on discontinuous incremental checkpoints. In: 2011 IEEE Pacific Rim Conference on Communications, Computers and Signal Processing (2011)
8. Ha, V.H., Renault, É.: Discontinuous incremental: a new approach towards extremely lightweight checkpoints. In: 2011 International Symposium on Computer Networks and Distributed Systems (CNDS), pp. 227–232. IEEE (2011)

9. Ha, V.H., Renault, E.: Improving performance of CAPE using discontinuous incremental checkpointing. In: 2011 IEEE 13th International Conference on High Performance Computing and Communications (HPCC), pp. 802–807. IEEE (2011)

10. Heo, J., Yi, S., Cho, Y., Hong, J., Shin, S.Y.: Space-efficient page-level incremental checkpointing. In: Proceedings of the 2005 ACM symposium on Applied computing, pp. 1558–1562. ACM (2005)

11. Hoeflinger, J.P.: Extending OpenMP to clusters. White Paper, Intel Corporation (2006)

12. Huang, L., Chapman, B., Liu, Z.: Towards a more efficient implementation of OpenMP for clusters via translation to global arrays. Parallel Comput. $31(10)$, 1114–1139 (2005)

13. Karlsson, S., Lee, S.-W., Brorsson, M.: A fully compliant OpenMP implementation on software distributed shared memory. In: Sahni, S., Prasanna, V.K., Shukla, U. (eds.) HiPC 2002. LNCS, vol. 2552, pp. 195–206. Springer, Heidelberg (2002). https://doi.org/10.1007/3-540-36265-7_19

14. Li, C.C., Fuchs, W.K.: Catch-compiler-assisted techniques for checkpointing. In: 20th International Symposium Fault-Tolerant Computing. FTCS-20. Digest of Papers, pp. 74–81. IEEE (1990)

15. Morin, C., Lottiaux, R., Vallée, G., Gallard, P., Utard, G., Badrinath, R., Rilling, L.: Kerrighed: a single system image cluster operating system for high performance computing. In: Kosch, H., Böszörményi, L., Hellwagner, H. (eds.) Euro-Par 2003. LNCS, vol. 2790, pp. 1291–1294. Springer, Heidelberg (2003). https://doi.org/10.1007/978-3-540-45209-6_175

16. OpenMP ARB: OpenMP application program interface version 4.0 (2013)

17. Plank, J.S., Beck, M., Kingsley, G., Li, K.: Libckpt: Transparent checkpointing under unix. Computer Science Department (1994)

18. Renault, É.: Distributed implementation of OpenMP based on checkpointing aided parallel execution. In: Chapman, B., Zheng, W., Gao, G.R., Sato, M., Ayguadé, E., Wang, D. (eds.) IWOMP 2007. LNCS, vol. 4935, pp. 195–206. Springer, Heidelberg (2008). https://doi.org/10.1007/978-3-540-69303-1_22

19. Sato, M., Harada, H., Hasegawa, A., Ishikawa, Y.: Cluster-enabled OpenMP: an OpenMP compiler for the SCASH software distributed shared memory system. Sci. Program. $9(2–3)$, 123–130 (2001)

20. Thakur, R., Rabenseifner, R., Gropp, W.: Optimization of collective communication operations in MPICH. Int. J. High Perform. Comput. Appl. $19(1)$, 49–66 (2005)

21. Tran, V.L., Renault, É., Ha, V.H.: Improving the reliability and the performance of CAPE by using MPI for data exchange on network. In: Boumerdassi, S., Bouzefrane, S., Renault, É. (eds.) MSPN 2015. LNCS, vol. 9395, pp. 90–100. Springer, Cham (2015). https://doi.org/10.1007/978-3-319-25744-0_8

22. Tran, V.L., Renault, E., Ha, V.H.: Analysis and evaluation of the performance of CAPE. In: IEEE International Symposium on IEEE Conferences on Ubiquitous Intelligence & Computing, Advanced and Trusted Computing, Scalable Computing and Communications, Cloud and Big Data Computing, Internet of People, and Smart World Congress, pp. 620–627. IEEE (2016)

23. Tran, V.L., Renault, É., Ha, V.H., Do, X.H.: Implementation of OpenMP data-sharing on cape. In: 9th International Symposium on Information and Communication Technology SoICT 2018, pp. 359–366. ACM (2018)

24. Tran, V.L., Renault, É., Ha, V.H., Do, X.H.: Time-stamp incremental checkpointing and its application for an optimization of execution model to improve performance of cape. Informatica $42(3)$ (2018)

Compiler Generated Progress Estimation for OpenMP Programs

Peter Zangerl$^{(\boxtimes)}$, Peter Thoman, and Thomas Fahringer

University of Innsbruck, 6020 Innsbruck, Austria
{peterz,petert,tf}@dps.uibk.ac.at

Abstract. Task-parallel runtime systems have to tune several parameters and take scheduling decisions during program execution to achieve the best performance. In order to decide whether a change was beneficial to the program performance, the runtime needs some kind of feedback mechanism on the progress of the program after such a parameter change was performed. Traditionally, this feedback is derived from metrics only indirectly related to the progress of the program.

To mitigate this drawback, we propose a fully automatic compiler analysis and transformation which generates progress estimates for sequential and OpenMP programs. Combined with a runtime system interface for progress reporting this enables the runtime system to get direct feedback on the progress of the executed program.

We based our implementation on the Insieme compiler and runtime system and evaluated it on a set of eight benchmarks representing a variety of different types of algorithms. Our evaluation results show a significant improvement in estimation accuracy over traditional estimation methods, with an increasing advantage for larger degrees of parallelism.

1 Introduction

A modern runtime system needs to tune several operational parameters to better utilize the underlying hardware and achieve high performance. Examples for this kind of decisions are where to best apply dynamic voltage and frequency scaling (DVFS) [9], how to adjust the granularity of tasks or controling the amount of parallelism [5], and scheduling decisions in case a runtime system is responsible for the co-scheduling of multiple programs [6,10,13].

In order for the runtime system to measure the effectiveness of the decisions it took and reach the most effective combination of parameters, it requires some kind of feedback mechanism which provides information about the performance consequences of parameter changes – a *progress metric.* The system can then monitor the progress development and judge whether or not a particular parameter change was beneficial – enabling it to steer towards optimal settings.

In practice, there are several ways for a runtime system to estimate an application's current progress. An obvious candidate for this kind of progress information are CPU counters. A runtime system can monitor the development of

© Springer Nature Switzerland AG 2019
V. Malyshkin (Ed.): PaCT 2019, LNCS 11657, pp. 107–121, 2019.
https://doi.org/10.1007/978-3-030-25636-4_9

certain counters and thus reason about the amount of work the application has carried out in a given timeframe. However, there are several drawbacks to this approach: (i) the CPU counters do not have a direct relationship to the application's progress; (ii) counter values will also be influenced during the time spent within the runtime system itself, thus skewing the obtained results; and (iii) the use of CPU counters is not portable and the desired counters might not be available on the given target hardware.

A runtime system can also take advantage of its internal state to estimate an application's progress. The runtime's task throughput is a measure of how many tasks the system finished within a given timeframe and thus is also related to the progress of the executed program. This approach has the advantage that the required values are already available in the runtime system or can be added easily without any application code modifications or special permission requirements. On the other hand, this approach is often coarse-grained and not very accurate.

Another popular alternative to the use of counters is manual instrumentation of the input code to inform the runtime system of an application's progress. This eliminates the platform dependent implementation and also is not influenced by time spent within the runtime system itself. However, this method requires a very good understanding of the input program as well as the runtime system, needs to be done manually for each program, and, due to these factors, is often either quite coarse-grained and inaccurate or labor-intensive.

To mitigate these drawbacks, we propose a novel, fully automatic compiler-based analysis and transformation to achieve accurate progress estimations in parallel applications. This enables a low-overhead and platform-independent way for parallel runtime systems to obtain direct feedback on the program's progress upon parameter changes. Our concrete contributions are as follows:

- A compiler based progress estimation analysis and transformation supporting sequential as well as parallel OpenMP input programs.
- An application programming interface for progress information collection and reporting in the runtime system.
- An implementation of the compiler analysis and runtime system facilities based on the Insieme compiler and runtime system [7].
- An evaluation of the achieved progress estimation accuracy of eight benchmark applications on a shared memory system running in different configurations, along with a comparison with the use of CPU counters, task throughput metrics and manual code instrumentation.

2   Motivation and Related Work

Any dynamic optimizing runtime system can take advantage of obtaining a *progress estimation* directly from the scheduled entities. This way, the system can evaluate the choices and parameter tuning it applied and thus steer the scheduling towards optimum settings.

Deriving an *absolute* progress completion rate towards application termination is unattainable for most non-trivial programs. Thus, one form of a good progress estimation would be a value which increases linearly and monotonously with the relative progress of an application. As long as the scheduled program can perform the same amount of useful work towards its goal in two observational timeframes, it should also report the same relative progress estimate.

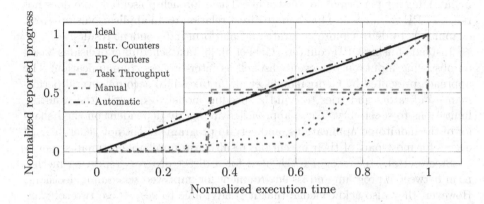

Fig. 1. Comparison of different progress reporting methods in *NPB FT*.

Figure 1 shows an illustration of different progress reporting methods during the runtime of an application by plotting the relative reported progress against the normalized time. The figure shows the results for all the progress estimation methods we evaluated in this paper for the *NPB FT* benchmark running with 64 threads (cf. Sect. 4). A good reporting method would report a value very close to the shown ideal line at any point during the execution. As we can see in this example, some progress estimation methods fail to achieve this criterion. This is the case for both counter approaches, which behave differently in the sequential and parallel phase of this benchmark's execution. The task throughput as well as the manual estimation approach both suffer from their coarse-grained accuracy, essentially rendering them useless for any kind of decision feedback. Our automatic approach on the other hand is able to estimate the progress quite well for most parts of the program execution.

The importance for direct feedback on the progress of scheduled applications has already been well established in the past. There is a wide body of work which tries to base scheduling decisions on the progress made by the scheduled applications to achieve optimal throughput. An example for this is the work by Wu et al. [13], where the authors introduce the concept of an application's progress based on the number of CPU cycles executed during a scheduling period. The goal of this work is a fair scheduling between equally-weighted processes where each of the applications can progress the same amount. Feliu et al. [4] follow a very similar approach. The progress of a process gets estimated by co-scheduling it in a low-contention scenario and thereby determining the maximum possible executed instructions for a given timeframe. By comparing the actual CPU counters with the maximum achievable value they determine the relative progress and use this value to create a fair co-scheduling between different processes.

The same approach has also been applied to scheduling kernels on GPUs by Anantpur et al. [1]. Lee et al. [9] additionally use counter measurements to decide when and where to best apply DVFS for reduced energy consumption while still maintaining set performance constraints. The approach presented here does not rely on CPU counters and has a more direct relation to an application's progress, enabling us to deliver more accurate and platform-independent results.

Instead of using CPU counters, Goel et al. [6] take scheduling decisions based on observing input/output events as well as inter-process communication. The approach presented by Georgakoudis et al. [5] takes into account several performance indicators and tries to build a speedup model to quantify the resulting limitations to scalability. These approaches are highly dependent on the behavior of the monitored applications, and certain programs might not generate such events for most part of their execution, resulting in unreliable estimates.

Steere et al. [10] recognize the need for a direct progress reporting mechanism between a program and its environment for improved scheduling decisions. However, they also acknowledge that it is advisable to keep these two software domains not too tightly interlocked. As a solution, they propose a *symbiotic interface* where e.g. the application notifies the operating system about data buffers and their fill-levels, enabling the latter to reason about the application's progress. The runtime interface proposed by our approach offers a way of directly reporting progress to the surrounding runtime system without burdening application developers with this task, as the invocations of this interface are created automatically with the help of a compiler component.

3   Method

Our approach combines a compiler analysis component with a task-parallel runtime system. Figure 2 provides an overview of our proposed method. As a first step, the input program is translated into a parallelism-aware intermediate representation (IR) by the compiler frontend ①. This IR is then analyzed by our progress estimation component, which will insert reporting nodes at the appropriate locations ②. The compiler backend ③ then creates the output code to be compiled against the runtime system, resulting in the final program binary ④. The full implementation presented in this paper is publicly available[1].

3.1   Compiler Component

As we wanted our analysis to distinguish between sequential and parallel progress of an application, we decided to base it on a compiler with support for parallelism awareness in its intermediate representation. For this reason, we chose the

---

[1] Full implementation along with instructions and evaluation script available at https://github.com/insieme/insieme/tree/progress_estimation.

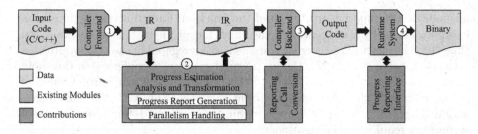

Fig. 2. Method overview for our automatic progress estimation.

Insieme research compiler system with its INSPIRE IR [8], as it allows us to capture the parallel semantics of a variety of input languages. While some parts of our analysis are currently tailored for OpenMP-specific semantics, it is easily extensible to other input languages supported by the Insieme compiler system.

Compiler Analysis. The foundation for our progress estimation is a modified and extended variant of the *effort estimation* component presented by Thoman et al. [11,12]. This analysis allows us to generate effort estimations for arbitrary code parts. In this work, we define the *progress* of an application as the *accumulated effort* of its statements. In the analyses and transformations presented here, we use the same notations as Thoman et al. with the following extensions:

- *is_compound*(n) Checks whether the node n is a compound statement.
- *is_exit_point*(c, n) Checks whether node n is an exit point in compound c.
- *all_child_statements*(c) Returns all child nodes of the given compound c.
- *get_effort*(n) Returns the effort estimation for node n.
- *replace_child*(c, o, n) Replaces child o of node c with n.
- *insert_reporting_call*(c, s, p) Inserts a reporting node with progress p above node s in the compound statement c, returning zero.
- *insert_reporting_call_at_end*(c, p) Inserts a reporting node with progress p at the end of the compound statement c, returning zero.
- *conditional*$(cond, then, else)$ Refers to a conditional statement with its condition *cond* and the branch compound statements *then* and *else*.
- *loop*$(cond, body)$ Refers to a loop with its condition *cond* and the body *body*.
- *all_reporting_addresses*(n) Returns a set of all addresses rooted at node n to progress reporting nodes in any child node of n, at arbitrary depth.
- *is_openmp_{parallel/single/master}*(n) Checks whether node n represents the respective OpenMP construct in INSPIRE.
- *mark_reporting_{parallel/sequential}*(n) Replaces reporting node n with a specialized parallel or sequential version.

Progress Report Generation: A simplified version of the algorithm used to generate progress reportings is depicted by Algorithm 1. In a first phase, the analysis traverses all functions of the program. The *handle_compound* function gets passed the body, the current *progress p* and a flag r indicating whether or not to unconditionally report p at the end of the function. Each function body is analyzed statement by statement, and the effort for all the statements is evaluated (line 19). The effort of the current statement is aggregated in the current *progress estimation value* p (line 27). Before p would overflow a configurable threshold value l, we insert a new IR node into the body reporting the current value of p, and reset p to the effort of the current statement (lines 23–25). This aggregation is applied to every statement within the compound, but several types of statements require special treatment:

- Nested compound statements are handled by recursion (lines 3–5).
- For conditional statements, we accumulate the effort for evaluating the condition (line 7) and then continue to evaluate for each branch individually, reporting at the end of each branch (lines 8–11).
- Before a loop is entered, the current value of p is always reported (line 14). Within the loop, the progress is reported before any exit-point of the loop, as well as at the end of each iteration.

Also, before each exit-point of a function, the current value of p is reported unconditionally (lines 20–21). However, in order to reduce the number of reporting instances and thus the program execution overhead, we remove instances reporting only very small values in single exit-point functions and annotate the functions with the reported value as *unreported progress*. Whenever a statement calls such a function, we then add the unreported progress to the current accumulation and thus effectively *inline* the progress reporting in this case. This optimization is not shown in Algorithm 1 for brevity.

Parallelism: This phase of the analysis is responsible for differentiating between reports in sequential and parallel code. In a second pass through the whole program we traverse all reporting nodes which have been created by the first phase. A simplified version of this transformation pass is outlined in Algorithm 2. The context of each reporting within the program is analyzed for the parallelism at its code location. This is achieved by traversing the path from each reporting location backwards up to the root of the program (line 3). We then decide on the parallel context based on what kind of OpenMP construct we meet first (lines 4 and 7). The reporting nodes are then transformed into specialized versions representing sequential or parallel progress respectively (lines 9–12). Note that, if the same function or set of functions is called in both sequential and parallel contexts, this will generate two distinct versions of these functions in the output program – this is an aspect of our automatic compiler-based system which is particularly cumbersome to replicate in a manual approach.

Tunable Parameters. Our compiler component has a small set of tunable parameters influencing its behavior:

- The most important one is the *progress reporting threshold* l. This is the value above which the aggregated progress will lead to a new progress reporting node being generated within the code.

Algorithm 1. Handle Program Flow

l	the progress reporting threshold

1: **function** HANDLE_COMPOUND(c, p, r)
2:     **for all** $s \in$ all_child_statements(c) **do**
3:         **if** is_compound(s) **then**
4:             $(s', p) \leftarrow$ HANDLE_COMPOUND$(s, p, \bot)$
5:             REPLACE_CHILD(c, s, s')
6:         **else if** $\exists cond, then, else \mid s = $ conditional$(cond, then, else)$ **then**
7:             $p \leftarrow p +$ GET_EFFORT$(cond)$
8:             $(then', \_) \leftarrow$ HANDLE_COMPOUND$(then, p, \top)$
9:             REPLACE_CHILD$(s, then, then')$
10:             $(else', \_) \leftarrow$ HANDLE_COMPOUND$(else, p, \top)$
11:             REPLACE_CHILD$(s, else, else')$
12:             $p \leftarrow 0$
13:         **else if** $\exists cond, body \mid s = $ loop$(cond, body)$ **then**
14:             $p \leftarrow$ INSERT_REPORTING_CALL(c, s, p)
15:             $eCond \leftarrow$ GET_EFFORT$(cond)$
16:             $(body', \_) \leftarrow$ HANDLE_COMPOUND$(body, eCond, \top)$
17:             REPLACE_CHILD$(s, body, body')$
18:         **else**
19:             $p' \leftarrow$ GET_EFFORT(s)
20:             **if** is_exit_point(c, s) **then**
21:                 $p \leftarrow$ INSERT_REPORTING_CALL$(c, s, p + p')$
22:             **else**
23:                 **if** $p + p' > l$ **then**
24:                     INSERT_REPORTING_CALL(c, s, p)
25:                     $p \leftarrow p'$
26:                 **else**
27:                     $p \leftarrow p + p'$
28:     **if** $r \wedge p > 0$ **then**
29:         $p \leftarrow$ INSERT_REPORTING_CALL_AT_END(c, p)
30:     **return** (c, p)

Algorithm 2. Handle Parallelism

m	the *main* program node

1: **for all** $r \in$ all_reporting_addresses(m) **do**
2:     $par \leftarrow \bot$
3:     **for all** $n \in$ reverse_sequence(r) **do**
4:         **if** is_openmp_parallel(n) **then**
5:             $par \leftarrow \top$
6:             **break**
7:         **else if** is_openmp_single$(n) \vee$ is_openmp_master(n) **then**
8:             **break**
9:     **if** par **then**
10:         MAKE_REPORTING_PARALLEL(r)
11:     **else**
12:         MAKE_REPORTING_SEQUENTIAL(r)

- We implemented an optimization which can be beneficial for programs which contain many very fine grained conditional statements. This optimization will – after the normal handling of conditional statements – compare the reported progress of both branches. If the reported values differ only by an amount less than a user-provided threshold, the reportings will be removed from the conditional branches and the analysis will continue after the conditional with the average of the removed values.
- As a last pass of the transformation, we optionally remove reportings of very small values. This is useful for programs with very intricate and tightly nested control flow, where the normal algorithm would lead to a large number of reporting nodes, each of them reporting only tiny amounts of progress.

Listing 1. Runtime system API for progress reporting

```
// report sequential/global progress
void irt_report_progress(uint64_t progress);

// report parallel/per-worker progress
void irt_report_progress_thread(uint64_t progress);
```

All of these parameters can be tuned for a given use case, either to reduce the runtime overhead of our progress reporting method at the cost of slightly reduced accuracy, or alternatively to increase accuracy while potentially introducing more overhead. The default values for these parameters are set to result in reasonable compromise between low overheads and good prediction accuracy for sequential and parallel code parts alike, as shown in Sect. 4.

3.2   Compiler Backend

In the backend of the compiler, the reporting IR nodes need to be translated into calls which will use the runtime system's reporting facilities. The sequential and parallel version of our reporting nodes are translated into distinct runtime function calls, with the reported progress estimate being an argument of the call.

3.3   Runtime System

We extended the Insieme runtime system to support reporting of sequential as well as parallel (per-worker) progress. The runtime interface (cf. Listing 1) consists of two functions which can be used to report progress. For our prototype implementation, a periodic maintenance task within the runtime system is responsible for collecting and combining the reported progress. This thread then prints the combined application progress, allowing us to evaluate the accuracy of our approach. Additionally, these reporting facilities can also be used to implement task throughput estimation as well as manual progress reporting which we used for comparison purposes in our evaluation.

4   Evaluation

Each progress estimation method we investigated comes with a set of requirements and in return offers some features. Table 1 summarizes these properties.

Tracking an application's progress using CPU counters might require certain special permissions on some hardware platforms. More crucially, not every platform will provide all counters which we might be interested in, and different programs might be best measured by distinct counters. On the other hand, we get a very fine grained estimation with minimal overhead. However, by relying on CPU counters we work with estimates which are inherently influenced by work spent within the runtime system itself and can not get per-worker estimates.

Table 1. Requirements and feature set of different progress estimation methods

		CPU counters	Task throughput	Manual	Automatic
Requirement	Source code access	✗	✗	✓	✓
	Program understanding	✗	✗	✓	✗
	Special permissions	(✓)	✗	✗	✗
Feature	Platform independence	✗	✓	✓	✓
	Program independence	✓	✗	✓	✓
	Fine granularity	✓	(✗)	(✗)	✓
	Constant accuracy	✓	(✗)	(✗)	✓
	Low runtime overhead	✓	✓	(✓)	✓/✗
	Unskewed estimate	✗	✓	✓	✓
	Per-worker estimate	✗	✗	(✗)	✓

Using the runtime's task throughput does not impose any additional requirements on the execution, as this value is readily available or easily added to an existing runtime system. However, this method does not allow per-worker performance estimates and also might work poorly with certain kinds of programs which do not produce many tasks. This also implies that its accuracy is often very fluctuating and also rather coarse-grained.

Manual and automatic compiler generated progress estimations both require the application source code in order for the necessary reporting calls to be inserted. Granularity, accuracy as well as the runtime overhead for manual estimation highly depends on how well the programmer understands the program and places the reporting calls. Most often, the result has low estimation overhead with coarse granularity and varying accuracy. Per-worker estimations are rather hard to achieve with manual progress estimation, as any code parts used in both sequential and parallel contexts have to be duplicated.

By generating the reporting calls automatically with the help of a compiler, we can mitigate most of the disadvantages of manual progress reporting, while leveraging its advantages. What remains is a certain overhead at runtime, due to the high number of reporting calls generated for high accuracy. In some programs, these overheads can be quite large and thus render a naive implementation of this approach infeasible. However, these overheads can be minimized by adjusting the tunable parameters of the compiler component (cf. Sect. 3.1).

Table 2. Benchmark overview

Benchmark	Alignment	Strassen	BT	CG	EP	FT	IS	UA
Origin	AKM	Cilk	NPB					
Parameters/Class	prot.100.aa	-n 4096	B	B	B	B	C	A

4.1 Evaluation Setup

The hardware platform we are using for our evaluation is a quad-socket system with four Intel Xeon E5-4650 processors. The 8 cores (or 16 hardware threads) of each CPU are clocked at 2.7 GHz. On the software side, the system is based on CentOS 7.4 running kernel version 3.10.0-693.2.2.el7. All binaries are compiled with GCC 6.3.0 using -O2 optimizations. The thread affinity for all the executions has been fixed with a fill-socket-first policy. Each experiment has been executed ten times and we are always reporting the average values achieved.

We evaluated five different progress estimation methods in this paper, namely (i) CPU counters for executed instructions; (ii) CPU counters for executed floating point instructions; (iii) task throughput statistics gathered in the runtime system; (iv) manual progress estimation; and (v) automatic compiler generated progress estimation as proposed in this paper.

4.2 Benchmarks

To evaluate the approach presented in this paper we chose a set of benchmark applications representing real-world application kernels. Table 2 lists the benchmarks used along with their origin. Most of the benchmarks originate from NASA's parallel benchmark suite [2], with the remainder being derived from the Barcelona OpenMP tasks suite [3].

4.3 Estimation Overhead

The measured overheads averaged by benchmark are shown by Fig. 3. The overhead values reported are relative to the execution of the unmodified benchmarks. Measuring the overheads did produce rather unreliable results for some benchmarks, as they showed some jitter in their execution times between successive runs at higher levels of parallelism. This is caused mainly by the non-deterministic task scheduling and work-distribution of these benchmarks.

As expected, we can observe that the overheads for both CPU counting approaches are negligible in all cases, as reading out these values during program execution should not cause significant overheads. The rather large negative over-head for the floating point counter estimate for the *UA* benchmark is a result of the execution time jitter described above, indicating that an uncertainty range of around 1% has to be considered for overhead evaluation in this benchmark.

Estimating the progress with the help of the runtime's task throughput should also not have a lot of influence on the program execution time. Still, we can observe some small negative overheads for *FT* as well as *IS*, but especially a relatively significant negative overhead for the *EP* benchmark. Also interestingly, on average, the manual estimation method seems to actually speed up the execution of several evaluated benchmarks.

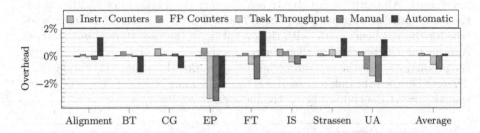

Fig. 3. Overheads for the evaluated progress estimation methods by benchmark

We investigated this behavior in detail, and determined that the reduction in runtime in these benchmarks is related to changes in the binary layout which occur due to the inclusion of additional functions related to progress reporting. These layout changes affect L1 instruction cache effectiveness, particularly for *EP*, and are not specific to the methods we are investigating – even adding or removing unrelated functions in the same translation units causes similar effects.

Regarding our automatic progress estimation, we can note that it shows some minor performance overhead for certain benchmarks, while it seems to improve the performance for others. The latter behavior is caused by similar effects related to the binary layout of functions in GCC as observed for the other progress metrics. Crucially, the performance overhead for our automatic progress estimation approach is less than 2% in all benchmarks.

4.4   Estimation Accuracy

For assessing the quality of the reported progress of our evaluated estimation methods we chose to employ the *mean squared error* (MSE) calculation:

$$\text{MSE} = \frac{1}{n} \sum_{i=1}^{n} (Y_i - \hat{Y}_i)^2 \tag{1}$$

We average the squared difference between every normalized progress report Y_i and the expected value $\hat{Y}_i$ during program execution. The latter is derived as an ideal progress estimation based on constructing a perfectly linear metric after program completion. The smaller the reported MSE, the better the estimation.

For averaging MSE values, we average over the magnitude of the error rather than the absolute value to avoid a single bad pulling the final average to non-representative high values:

$$\text{AVG}_{\text{mag}} = \frac{1}{m} \sum_{j=1}^{m} log_{10}(\text{MSE}_j) \qquad \text{AVG\_MSE} = 10^{\text{AVG}_{\text{mag}}} \qquad (2)$$

Accuracy by Benchmark. Figure 4 shows the accuracy achieved for each benchmark. All methods are able to achieve very good estimations within about the same order of magnitude for the *BT* and *UA* benchmarks. The same holds true for the *EP* benchmark, with the exception of the task throughput method. Also for the *CG* benchmark all evaluated methods result in a similar accuracy of the predictions. For the remaining four benchmarks, the achieved accuracy often diverges between the different methods by one or more orders of magnitude, with the *Alignment*-Benchmark being the most extreme example.

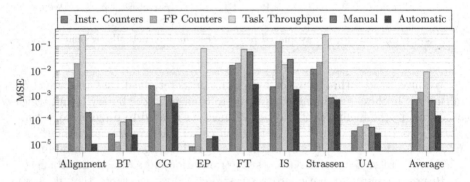

Fig. 4. Accuracy of the evaluated progress estimation methods by benchmark.

Accuracy by Threads. The accuracy achieved by averaging our results across thread counts instead of by benchmark is shown in Fig. 5. We can observe that the estimation accuracy seems to be best for a low number of threads. The accuracy then decreases until we reach the worst results with the maximum number of threads evaluated. Moving from using all available cores to also running on all hardware threads does not have much influence on the estimation accuracy.

Regarding the specific estimation methods, we can observe that:

- The use of instruction CPU counters results in better estimates than the use of floating point CPU counters, regardless of the number of threads.
- Both counter-related estimates have a rather large drop in accuracy when moving from running on a single CPU socket to multiple sockets (16+ threads), with not too much change further on.

- The accuracy achieved by relying on the task throughput estimation is always very bad.
- Results for the manual progress estimation often fall between the accuracy achieved by the use of instruction counters and floating point counters for lower thread counts, but still are always worse than the results for our automatic estimation method.
- The automatic estimation yields the best results overall for any number of cores used, with the advantage over counter based methods increasing with higher numbers of threads.

The final point regarding parallelism is particularly encouraging for our approach: with hardware architectures continuously increasing in the number of cores and hardware threads per socket, it indicates that a parallelism-aware compiler-supported approach such as ours is more suitable for progress estimation on such highly parallel hardware than any of the established alternatives.

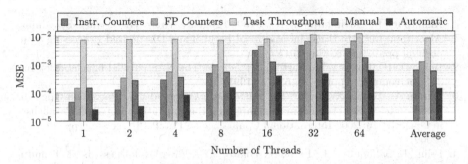

Fig. 5. Accuracy of the evaluated progress estimation methods by number of threads

5   Conclusion

In this work, we presented a novel and fully automatic compiler analysis and transformation to generate progress estimations for OpenMP programs. Our approach provides the runtime system with direct feedback on the progress of an application, without having to resort to metrics only indirectly related to the application's progress or requiring a manual per-application implementation effort. This feedback can be used by the runtime system to measure the effectiveness of parameter changes and thus steer the execution towards optimal settings.

We evaluated our implementation on a set of eight benchmark applications implementing a wide variety of different types of algorithms. The achieved results show a good accuracy of our progress estimation, out-performing any other evaluated progress estimation method for any degree of parallelism evaluated. Crucially, the accuracy advantage of our automatic approach is increasing with a higher degree of parallelism, indicating it to be a valid approach for highly parallel future computing systems.

The work presented here offers several extension opportunities for future research. The compiler analysis itself can be further optimized to generate less reporting calls and thus runtime overhead for code parts which can be fully statically analyzed (e.g. loops with statically constant boundaries). Additionally, the set of tunable parameters of our transformation could be extended to enable a more fine-grained tradeoff between accuracy and runtime overheads. Orthogonally to the improvements of the compiler parts, future research also includes taking advantage of the generated progress estimations in the runtime system. The good accuracy of the provided estimations enables further runtime optimizations ranging from improved scheduling decisions to energy optimizations.

Acknowledgement. This work is supported by the D-A-CH project CELERITY, funded by DFG project CO1544/1-1 and FWF project 13388.

References

1. Anantpur, J., Govindarajan, R.: PRO: Progress Aware GPU Warp Scheduling Algorithm. In: 2015 IEEE International Parallel and Distributed Processing Symposium, pp. 979–988, May 2015
2. Bailey, D.H., Barszcz, E., Barton, J.T., et al.: The NAS parallel benchmarks. Int. J. Supercomput. Appl. **5**(3), 63–73 (1991)
3. Duran, A., Teruel, X., Ferrer, R., Martorell, X., Ayguade, E.: Barcelona OpenMP Tasks Suite: A Set of Benchmarks Targeting the Exploitation of Task Parallelism in OpenMP. In: 2009 International Conference on Parallel Processing, pp. 124–131 (2009)
4. Feliu, J., Sahuquillo, J., Petit, S., Duato, J.: Addressing fairness in SMT multicores with a progress-aware scheduler. In: 2015 IEEE International on Parallel and Distributed Processing Symposium (IPDPS), pp. 187–196. IEEE (2015)
5. Georgakoudis, G., Vandierendonck, H., Thoman, P., Supinski, B.R.D., Fahringer, T., Nikolopoulos, D.S.: SCALO: Scalability-Aware Parallelism Orchestration for Multi-Threaded Workloads. ACM Trans. Archit. Code Optim. **14**(4), 54:1–54:25 (2017)
6. Goel, A., Walpole, J., Shor, M.: Real-rate scheduling. In: 10th IEEE Real-Time and Embedded Technology and Applications Symposium, Proceedings, RTAS 2004, pp. 434–441, May 2004
7. Jordan, H., et al.: A Multi-Objective Auto-Tuning Framework for Parallel Codes. In: 2012 International Conference for High Performance Computing, Networking, Storage and Analysis (SC), pp. 1–12, November 2012
8. Jordan, H., Pellegrini, S., Thoman, P., Kofler, K., Fahringer, T.: INSPIRE: The Insieme Parallel Intermediate Representation. In: Proceedings of the 22nd International Conference on Parallel Architectures and Compilation Techniques, PACT 2013, pp. 7–18. IEEE Press, Piscataway (2013)
9. Lee, S.-J., Lee, H.-K., Yew, P.-C.: Runtime Performance Projection Model for Dynamic Power Management. In: Choi, L., Paek, Y., Cho, S. (eds.) ACSAC 2007. LNCS, vol. 4697, pp. 186–197. Springer, Heidelberg (2007). https://doi.org/10. 1007/978-3-540-74309-5_19
10. Steere, D.C., Goel, A., Gruenberg, J., McNamee, D., Pu, C., Walpole, J.: A feedback-driven proportion allocator for real-rate scheduling. In: OSDI, vol. 99, pp. 145–158 (1999)

11. Thoman, P., Zangerl, P., Fahringer, T.: Task-parallel Runtime System Optimization Using Static Compiler Analysis. In: Proceedings of the Computing Frontiers Conference, pp. 201–210. ACM (2017)
12. Thoman, P., Zangerl, P., Fahringer, T.: Static Compiler Analyses for Application-specific Optimization of Task-Parallel Runtime Systems. J. Sig. Process. Syst., 1–18 (2018)
13. Wu, C., Li, J., Xu, D., Yew, P.C., Li, J., Wang, Z.: FPS: a fair-progress process scheduling policy on shared-memory multiprocessors. IEEE Trans. Parallel Distrib. Syst. **26**(2), 444–454 (2015)

Methods and Tools for Parallel Solution of Large-Scale Problems

Analysis of Relationship Between SIMD-Processing Features Used in NVIDIA GPUs and NEC SX-Aurora TSUBASA Vector Processors

Ilya V. Afanasyev[1]([✉]) [iD], Vadim V. Voevodin[1] [iD], Vladimir V. Voevodin[1] [iD], Kazuhiko Komatsu[2] [iD], and Hiroaki Kobayashi[2] [iD]

[1] Research Computing Center of Moscow State University, Moscow 119234, Russia
afanasiev_ilya@icloud.com
[2] Tohoku University, Sendai, Miyagi 980-8579, Japan

Abstract. This paper presents comprehensive analysis of main SIMD-processing features and computational characteristics of three high performance architectures: two NVIDIA GPU architectures (of Pascal and Volta generations) and NEC SX-Aurora TSUBASA vector processor. Since both these types of architectures strongly rely on using SIMD-processing features, certain similarities of data-processing principles can be found between them. However, despite having vectorised data-processing included in both NVIDIA GPU and NEC SX-Aurora TSUB-ASA architectures, vectorisation features of both architectures are implemented in completely different ways. These differences lead to several fundamental restrictions on classes of algorithms which can be efficiently implemented on corresponding platforms. This paper is devoted to the research of the possibility of porting various classes of programs and algorithms among the discussed architectures with a focus on utilising all vectorisation features available. However, without a detailed analysis of similar and different SIMD-processing features in these architectures, it is impossible to approach this problem. The performed analysis allowed us to identify several important examples of typical applications and algorithms. Some of them demonstrated comparable and the others showed different efficiency on NVIDIA GPUs and NEC SX-Aurora TSUBASA vector processors, including reduction operations, programs relying on frequent indirect memory accesses and data-transfers through co-processor interconnect. Moreover, the conducted analysis allows to easily extend this set of examples to approach the problem of automated porting of programs between the reviewed architectures, what we consider as an important direction of our future research.

Keywords: NEC SX-Aurora TSUBASA · NVIDIA GPU · Vector processing · SIMD

© Springer Nature Switzerland AG 2019
V. Malyshkin (Ed.): PaCT 2019, LNCS 11657, pp. 125–139, 2019.
https://doi.org/10.1007/978-3-030-25636-4_10

1   Introduction

There is a great variety of computational platforms widely used in modern super-computing, that support some form of vectorisation. Vectorised data-processing follows ideas of Single Instruction Multiple Data (SIMD) model [4] – an important computational principle, which allows efficient utilisation of data-level parallelism. At the same time, vectorised data-processing is used both in traditional modern CPUs, such as Intel or IBM Power processors, and dedicated vector systems, such as supercomputers produced by NEC company. Vectorisation in both types of processors can be implemented with two different approaches. Traditional central processors usually contain specialised vector instruction extensions, such as AVX-512 or AltiVec, which allow them to execute a set of predefined vector operations over vector registers of a fixed length. In the meantime, NEC vector processors work according to a combination of parallel and pipelined data-processing principles, utilising several computational pipelines integrated into functional units.

NVIDIA GPU is another important supercomputing architecture. GPUs tend to provide high performance and energy efficient computations, together with high-bandwidth memory. Data-processing model in modern GPUs is called SIMT (Single Instruction Multiple Threads) and is based on a combination of SIMD-processing principles and multithreading. This model operates with thousands of light-weighted computational threads, grouped into so-called warp, each of which working according to the SIMD model. Similarity between GPU executional model and vector-processing allows GPUs to potentially share many computational properties and features with modern high-performance vector processors, such as SX-Aurora TSUBASA.

Thus, many modern architectures include certain features of vectorised data-processing: AVX-512, AltiVec, GPU warps, NEC vector-pipelined units and many others. Despite all of them being referred as "vector processing", these vectorisation features are implemented differently, and thus can potentially impose significant restrictions on classes of algorithms, which can be efficiently implemented on corresponding architectures. At the same time, since all mentioned features are related to a class of architectures operating according to SIMD executional model, it implies possible similarity of data-processing principles between them. This fact makes it very interesting to study a possibility of transferring various program classes between the reviewed platforms with a focus on fully utilising available vectorisation support.

2   Description of Target Architectures

2.1   NEC SX-Aurora TSUBASA

NEC SX-Aurora TSUBASA is the latest SX vector supercomputer with dedicated vector processors [7,11]. SX-Aurora TSUBASA inherits the design concepts of the vector supercomputer and enhances its advantages to achieve higher sustained performance and higher usability. Different from its predecessors in the

SX supercomputer series [3,6], the system architecture of SX-Aurora TSUBASA mainly consists of vector engines (VEs), equipped with a vector processor and a vector host (VH) of an x86 node. The VE is used as a primary processor for executing applications while the VH is used as a secondary processor for executing basic operating system (OS) functions that are offloaded from the VE. The VE has eight powerful vector cores. As each core provides 537.6 GFlop/s of single-precision performance with 1.40 GHz frequency, the peak performance of the VE reaches 4.3 TFlop/s.

Each SX-Aurora vector core consists of three components: scalar processing unit (SPU), vector processing unit (VPU), and memory subsystem. Most computations are performed by VPUs, while SPUs provide functionality of typical CPU. Since SX-Aurora is not just a typical accelerator, but rather a self-sufficient processor, SPUs are designed to provide relatively high performance on scalar computations. VPU of each vector core has its own relatively simple instruction pipeline aimed for decoding and reordering vector instructions incoming from SPU. Decoded instructions are executed on vector-parallel pipelines (VPP). In order to store the results of intermediate calculations, each vector core is equipped with 64 vector registers with a total register capacity equal to 128 KB. Each register is designed to store a vector of 256 double precision elements (DP).

Vector processing on VE core is based on utilising 32 identical vector parallel pipelines, which process vectors located on registers in portions of 32 DP elements according to SIMD model. Thus, one command operating on vectors of 256 elements will be executed in 8 processor cycles, provided that there are no stalls caused by high-latency instructions. In addition, each VPP contains pipelined processing units that operate over scalar elements of input vectors. Each VPP has 3 FMA (Fused Multiply-Add) units, 2 ALUs (Arithmetic and Logic Unit) and 1 unit dedicated for processing high-latency commands (sqrt, division and others), as well as communicating with memory subsystem. Depending on the program structure, required data is redirected between computational units, forming a vector pipeline.

On the memory subsystem side six HBM modules in the vector processor can deliver the 1.22 TB/s world's highest memory bandwidth [3]. This high memory bandwidth contributes to achieve higher sustained performance, especially in memory-bound applications.

2.2 NVIDIA Pascal

Pascal [8] is the codename for modern GPU architecture developed by NVIDIA, a successor of Maxwell and Kepler architectures. NVIDIA Tesla P100 GPU, which is one of the most well-known representatives of Pascal architecture, is used for the performance evaluation in this paper. The P100 GPU is equipped with 3584 light-weighted cores with 1.1 GHz frequency, which allows a single P100 GPU to achieve 9.3 TFlop/s performance on single-precision computations. Cores in Pascal architecture are grouped together into streaming multiprocessors (SM), each one consisting of 64 CUDA-cores.

Program execution model is similar for all recent generations of GPUs, including Pascal and Volta architectures. When computations are launched on a GPU, a significant number of threads is spawned (usually at least several hundreds of thousands). Up to 1024 threads are grouped together into a single computational block. All threads of the computational block are scheduled via Giga Thread Engine into a single streaming multiprocessor. SM processes threads of each block by grouping 32 of them into warps in round-robin order. GPU hardware processes threads from the same warp using SIMD instructions of length 32, assigning each thread to a separate computational CUDA-core. Those instructions have several important limitations, which include processing branch conditions and handling memory accesses.

Each SM of Pascal architecture includes 64 computational CUDA-cores, each one including both FMA and Integer units. In addition Pascal SM is quipped with 32 double-precision cores, 16 load/store cores, and 16 special function cores, aimed to process high-latency instructions. SM also includes registers with a total capacity of 256 KB, shared between all mentioned types of cores.

Device memory of Pascal GPU consists of four HBM2 memory stacks, providing together up to 16 GB memory capability and up to 700 GB/s memory bandwidth.

2.3   NVIDIA Volta

Volta [9], being the latest GPU architecture, is also reviewed in the current paper. In several studies the performance of V100 GPUs is compared to the performance of SX-Aurora processors, since they have comparable technical characteristics, including performance on double-precision computations and memory bandwidth. In general, Volta inherits most computational features of Pascal architecture, but also introduces several important innovations. For example, Volta GPUs include specialised tensor cores, specially designed to speed-up deep learning applications. Each tensor core is dedicated for efficient computation of 4×4 half-precision matrix products, calculating a single matrix product on each GPU cycle using a special pipeline. Volta architecture also supports NVLINK interconnect of version 2.0, which doubles the bandwidth compared to the previous generation of NVLINK. In addition, Volta accelerators are equipped with HBM2 memory modules, providing up to 900 GB/s bandwidth and up to 32 GB of total memory capacity. Volta architecture also uses slightly modified SMs, divided into four processing blocks, each with 16 FP32 cores, 8 FP64 cores, 16 INT32 cores, 2 tensor cores.

3   Comparison of SIMD-Processing in NVIDIA GPUs and NEC SX-Aurora Architectures

Based on the architecture descriptions from the previous section, it is possible to conduct a comparative analysis of similar and different vector-processing features of the reviewed architectures. In many cases, the discussed similar and different architectural features will be illustrated with sample programs and benchmarks, more clearly demonstrating the principal distinguishing features of target platforms.

3.1   Overall System Structure Based on Using Co-processors

Both NVIDIA GPUs and SX-Aurora vector engines are installed into a system as co-processors. Connection to the host is usually implemented via PCI 3.0 bus, although for GPUs NVLINK interconnect is available, capable of providing significantly higher bandwidth. However, despite being installed as accelerators, these architectures have several crucial differences in program execution model, demonstrated on Fig. 1.

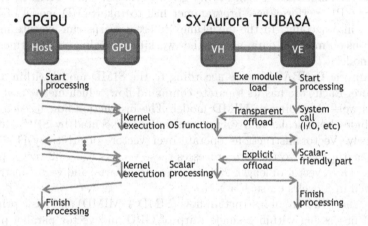

Fig. 1. Execution models of NVIDIA GPU (left) and SX-Aurora TSUBASA (right) architectures.

The first difference affects a process of launching a GPU or vector program. In the case of NVIDIA GPU, the program initially runs on a system host, but uses a special API in order to allocate required data structures inside GPU memory or launch CUDA-kernels, aimed to directly perform parallel computations on GPU. If the program has to perform a sequential region in between different CUDA-kernels, GPU is forced to copy the required data back to the host and perform sequential calculations on the CPU.

The situation on SX-Aurora TSUBASA is different. A vectorised program initially launches on scalar processing units of vector engine. While control flow logic is processed on SPUs, all vector instructions are redirected to vector processing units. Usually, SPUs provide high-enough performance required for processing sequential parts of typical algorithms. In addition, all of system calls are transparently offloaded to VH. For this reason no data copies between VH and VE are required. However, if there is a fundamental necessity for high-performance execution of a sequential program region, the required calculations can be explicitly redirected to vector host with corresponding data copies. In this case execution models of SX-Aurora and GPU are similar.

3.2 Warp-Based GPU SIMD and SX-Aurora TSUBASA Vector SIMD

NVIDIA GPUs execute program code on numerous threads, which are grouped into warps in round-robin order. Each warp consists of 32 threads, which means that warp schedulers of SMs process GPU code using SIMD instructions of width 32. From programmer's point of view, each CUDA-thread is executing it's own scalar instructions, but on the hardware side thread execution is organised according to the principles of SIMD-processing, when all threads from the same warp execute single instruction at any given period of time. Thus, in order to maximise GPU performance, programmer has to take SIMD model of thread execution into account. In the meantime, different warps can execute multiple instructions on multiple data, all together working according to the principles of MIMD model.

SX-Aurora TSUBASA works according to the SIMD model within a single vector core. Each core has its separate command flow, which means that different cores work according to MIMD model. The instruction flow of each vector core includes both scalar and vector instructions, executed by SPU and VPU respectively. Vector instructions operate over vectors of arbitrary (1–256 elements) length. Each vector core processes these vectors using 32 vector parallel pipelines. Thus, vector of length 256 is processed in parts, and vector instruction is executed in 8 cycles on such a vector.

Thus, the structure of instruction flow (SIMD + MIMD) and basic principles of SIMD processing within a single warp of GPU and vector parallel pipeline of SX-Aurora are similar for both architectures. This leads to an equivalent behaviour of various programs and algorithms on both architectures in such cases as execution of conditional branch operators, loading data from memory subsystem or performing various types of calculations. At the same time, principles of interaction between GPU warps and SX-Aurora vector cores are very different, which leads to an important source of computational distinctions between the reviewed architectures, which will be described in future sections.

3.3 Control-Flow Divergence

An important feature of SIMD-processing in almost any computational platform is a sequence of hardware actions performed in the case of divergence (i.e. different behaviour) inside single SIMD instruction. Divergence, which may significantly bottleneck the performance of vectorised code, usually occurs in two cases, discussed in this and next subsections. The first type of divergence is usually referred as control-flow divergence. It may occur in the process of executing conditional branch operators, such as if-then-else operator in C/C++ languages. In this case vector instruction is executed depending on some external data, that forces SX-Aurora to generate extra vector instructions and NVIDIA GPU to implement special warp behaviour.

Control-flow divergence on SX-Aurora architecture is handled using masking instructions. A typical program with possible divergence within single vector instruction is shown in Listing 1. SX-Aurora architecture converts the provided conditional operator into 4 separate vector operations: (1) a comparison of A and B vectors, resulting into vector logical mask, (2) a masked vector copy, (3) an inversion of logical vector mask from step 1, and (4) another masked vector copy using an inverted mask. Thus, regardless of branching structure inside if-then-else operator within single SIMD instruction, additional vector instructions are generated and executed even if the whole SIMD instruction follows the same conditional branch.

Listing 1. An example of simple data-driven control-flow divergence on vector processors.

```
#pragma simd
for (i = 0; i < 256; ++i)
    if (a[i] >= b[i])  // (1)
        c[i] = a[i]     // (2)
    else                // (4)
        c[i] = b[i]     // (3)
```

GPU handling of similar if-then-else construct is quite different. On the same program CUDA platform will instruct the warp to execute the "if" part first, and then proceed to the "else" part. While executing the "if" part, all threads falling into "false" branch (i.e. "else" threads) are effectively deactivated. When execution proceeds to the "else" condition, the situation is reversed. Thus, "if" and "else" parts are executed in serial but not in parallel as it could be expected, which is very similar to SX-Aurora behaviour in the same situation. However, different behaviour occurs in the case when all threads inside single GPU warp execute the same conditional branch – then there will be no performance degradation on the GPU, while in the case of SX-Aurora architecture the performance will be always lower since extra vector instructions are always generated. This may result in 2 times slower execution in the case of two conditional branches inside if-then-else-statement, and up to n times slower execution in the case of n independent branches for both architectures.

Listing 2. Program benchmarking different types of data-driven control-flow divergence on vector processors and GPUs.

```
#pragma simd
for (int idx = 0; idx < _size; idx++)
 {
        if(_condition[idx] == 0)
        {
                float t1 = _x[idx];
```

```
          float t2 = _y[idx];
          float t3 = (float)_condition[idx] + _seed1;
          _z[idx] += ((t1 / t2) + (t1 / t3)) * (t2 + t1 + t3);
      }
   else
   {
          float t1 = _x[_size - 1 - idx];
          float t2 = _y[_size - 1 - idx];
          float t3 = (float)_condition[_size - 1 - idx] + _seed2;
          _z[idx] *= ((t2 / t1) + (t2 / t3)) * (t2 - t1 - t3);
      }
}
```

An example of a benchmark program, which is affected by thread divergence on both architectures is shown in listing 2. This program selects executional branch depending on values from a "condition" array. These values can be easily varied in order to provide different divergence structure inside single SIMD instruction. The difference between execution times of various divergence structures is presented in Table 1. "No conditional statement" refers to the program from listing 2, which executes only the first conditional branch without any conditional operator. Other table rows correspond to different distributions of values from "condition" array – a number of consecutive zero values, which lead to execution of "if" conditional branch.

Table 1. Control-flow divergence differences for program from listing 2.

Test type	P100 (Pascal) GPU time (ms)	GPU warp execution efficiency (%)	SX-Aurora time (ms)
No conditional statement	2,71	100%	2,8
Divergence (every 1 vector element)	4,38	55,7%	6,5
Divergence (every 2 vector element)	4,38	55,7%	6,7
Divergence (every 32 vector element)	2,93	96,1%	6,7
Divergence (every 256 vector element)	2,93	96,1%	8,3

3.4 Memory Divergence

The second type of divergence is called memory divergence. It may while loading data from the memory subsystem, if several elements of vector instruction or warp threads fail to load data from the cache, and thus have to wait until a transaction to the main memory is finished. This situation results into a stall of entire vector instruction or warp, since hardware has to process the whole instruction in the same way. Both GPU and SX-Aurora architectures process cases of memory divergence in a similar way: when a cache-miss occurs for at least one element of SIMD instruction, an entire vector instruction or warp is stalled until memory request is complete, which significantly reduces the performance.

3.5   Utilisation of High-Bandwidth Memory

Both of the reviewed architectures utilise High Bandwidth Memory 2 (HBM2) technology. NEC SX-Aurora TSUBASA processors are equipped with 6 HBM2 memory stacks, operated by two memory controllers; these stacks provide up to 48 GB of total memory. NVIDIA P100 GPUs of Pascal architecture use 4 HBM2 memory stacks operated by 8 controllers, providing a total capacity of up to 16 GB. NVIDIA V100 GPUs of Volta architecture also use 4 HBM2 memory stacks managed by 8 controllers, but with larger memory capacity (up to 32 GB). Both architectures provide the same access speed for all computational vector and CUDA cores no matter which memory stack stores the requested data.

Table 2. Memory bandwidths and capacity characteristics of the reviewed architectures.

Architecture	Memory type	Memory capacity	Theoretical peak bandwidth (GB/s)	Bandwidth achieved on STREAM benchmark (GB/s)	The ratio of bandwidth achieved on STREAM to theoretical
SX-Aurora TSUBASA	HBM2	48 GB	1200	995	82%
NVIDIA Pascal P100	HBM2	16 GB	732	628	85%
NVIDIA Volta V100	HBM2	32 GB	900	809	89%

Table 2 provides technical characteristics of memory subsystems of the studied architectures. Theoretical peak memory bandwidth values from this table are provided by hardware vendors. Achievable bandwidth values have been obtained based on the standard STREAM TRIAD benchmark [1]. The proposed comparison allows to conclude that SX-Aurora architecture has significantly larger memory capacity, as well as slightly better memory bandwidth on serial memory access pattern (22%), and thus can perform better on memory-intensive workloads with sequential memory access pattern.

Memory latency, which is another important characteristic for each level of memory hierarchy is listed in Table 3. Comparison from this table shows that memory access latency for GPU is significantly higher compared to SX-Aurora processor. However, in order to efficiently hide this high memory latency, GPUs

Table 3. Memory latency of the reviewed architectures

Level of memory hierarchy	NVIDIA V100 and P100 GPUs	NEC SX-Aurora
L1 cache/shared memory	~1–2 cycles	- (memory transactions from VPUs go directly through L3 cache)
L2 cache	~70 cycles	- (memory transactions from VPUs go directly through L3 cache)
L3 cache	- (L3 cache is not used in GPUs)	~25 cycles
Main memory	~200–300 cycles	~45 cycles

use light-weighted context switches of computational threads: each streaming multiprocessor can have up to 7 idle threads per 1 active. Idle threads are usually waiting for memory transfers to be complete, while active threads execute necessary computations on warp schedulers. Thus, GPUs have comparable average latency with SX-Aurora architecture in practice ($300/8 \sim 37$ cycles).

3.6   Available Computational Parallelism

An important characteristic of any modern processor is the maximum number of computational operations which can be executed on each cycle by all available computational units across all cores. On the one hand, this value can be viewed just as another definition of widely-used peak performance metric. However, on the other hand this metric can be viewed in a different context – as a fundamental restriction on algorithms which can be efficiently implemented on this architecture. If an algorithm does not include sufficient amount of parallelism required to fully utilise all computational resources of target processor, this algorithm will be processed inefficiently. The amount of required parallelism may also be affected by other architecture properties. For example, GPU architecture requires a much larger amount of computational threads actively running in order to effectively hide memory access latency.

Table 4 compares the amount of parallel resources provided by each of the reviewed architectures, depending on the data types used during computations. Since both SX-Aurora and Pascal GPU architectures do not explicitly support half-precision computations, we assume that these processors have exactly the same half-precision and single-precision performance. The values in the table are calculated as follows. For GPU architectures, the amount of parallelism on single(double) precision floating-point computations is equal to $2 * SMs\_per\_GPU * FP(DP)\_cores\_per\_SM$. The amount of parallelism over integer arithmetics is equal to $SMs\_per\_GPU * Integer\_cores\_per\_SM$. For SX-Aurora the formulas are different: $2 * 2 * vector\_cores * VPPs\_per\_core * FMA\_units\_per\_VPP$ corresponds to single-precision computational parallelism, while $2 * vector\_cores \times$

$VPPs\_per\_core*FMA\_units\_per\_VPP$ corresponds to double precision. Finally, the amount of integer arithmetics parallelism for SX-Aurora is equal to $vector\_cores *VPPs\_per\_core*(FMA\_units\_per\_VPP+ALU\_units\_per\_VPP)$.

Table 4. The amount of computational parallelism (in required multiply and add operations) available for each architecture on each cycle.

Datatype	NVIDIA Pascal (P100)	NVIDIA Volta (V100)	NEC SX Aurora-TSUBASA
Floating-point (FP) single precision	7168	10752	3072
FP double precision	3584	5376	1536
FP half precision	7168	86016	3072
Integer	3584	5376	1280

Table 4 demonstrates that both generations of GPU architecture require algorithms with significantly bigger parallelism available than SX-Aurora architecture. For example, GPU V100 requires the program to include in average 7 times more parallel operations than SX Aurora, while GPU P100 – 4.6 times.

Another significant difference between architectures is the support of context switching. If an algorithm requires frequent memory accesses, both GPU architectures use context switches in order to hide memory latency. To achieve this, GPU is required to have even more (approximately 8 times) parallel operations running, which are executed by temporarily idle warps, waiting for memory transfers to complete. On the other side, SX-Aurora vector engine doesn't support thread context switches at all, so optimal number of threads is equal to the number of VE cores. This is a significant computational difference on its own, but in the context of comparing the resource of parallelism between different platforms this means that Volta architecture requires in average 8 times more parallel operations from programs, than values listed in table 4. This fact even further narrows the class of algorithms which can be efficiently implemented on GPU architectures, compared to SX-Aurora.

3.7   Communication Principles in SX-Aurora Vector Cores and GPU Warps

NVIDIA GPUs and SX-Aurora architectures have several fundamental differences in how they perform thread synchronisations and communications. GPUs provide tools required for synchronising computations only within a single computational CUDA block – thus only among several (up to 32) adjacent warps. In order to synchronise computations between different blocks and different SMs, GPU has to explicitly launch a new CUDA-kernel, which can cause significant overhead. SX-Aurora allows to synchronise computations between each pair of vector instructions using standard OpenMP barrier synchronisation. Moreover, SX-Aurora provides functionality for efficient data-sharing between different vector cores, implemented based on storing the required data inside shared between all vector cores LLC cache, while GPUs can only share data inside single computational block, using shared memory. Thus, a lack of efficient GPU-wide synchronisation and data-sharing mechanisms imposes significant restrictions on algorithms which could be efficiently implemented on GPUs.

Reduction is an important example of computational operation that requires frequent synchronisations between different computing units. Reduction, together with other operations which operate over long vectors with a low computational intensity (e.g. SAXPY), is a memory-bound problem, and thus its overall efficiency can be measured in terms of utilised memory bandwidth (in GB/s). Utilised memory bandwidth is calculated as a ratio of the amount of bytes loaded from memory during the calculations to the actual computation time. The percentage ratio of utilised to theoretical peak memory bandwidths is another important efficiency metric of memory-bound applications, which can also be used in the reduction case.

Reduction implementation for SX-Aurora architecture is very straightforward. Since NEC compiler parallelisation is based on OpenMP directives together with automatic multithread parallelisation, it is possible to use OpenMP reduction clause to allow the compiler automatically parallelise and vectorise sequential reduction. Implementation principles of parallel reduction for NVIDIA GPU architectures are fundamentally different. Due to the lack of effective synchronisation mechanisms among different warps and the requirement to utilise all available computational units, an efficient implementation of parallel reduction operation is much more complicated for GPU. Possible optimisations of GPU parallel reduction are described in [5]. Efficient reduction implementations are also available in Thrust library [2], which will be used for the comparative performance analysis later in this section.

Table 5. The performance of parallel reduction operation for different architectures.

Architecture	Reduction type	Vector size	Execution time	Achieved bandwidth	Efficiency (achieved and theoretical peak bandwidths ratio)
P100 Pascal GPU	Sum	512 MB	1.46 ms	365 GB/s	52%
SX-Aurora	Sum	512 MB	0.53 ms	1009 GB/s	84%

Table 5 compares performance characteristics of parallel reduction implementations for Pascal GPU and SX-Aurora architectures. The initial time required to copy input data arrays into GPU memory is not included in time measurements. The provided metrics demonstrate a clear advantage of SX-Aurora architecture on parallel reduction operation both in execution time (2.7 times faster) and bandwidth efficiency (1.6 times better).

3.8   Processing Indirect Memory Accesses

Both classes of the reviewed architectures have fundamental differences in the technology of processing indirect memory accesses. When accessing data using an irregular pattern, both architectures at first check the presence of the required data inside cache memory. Since most likely this will result in cache-misses, it will be followed by the request to the main memory, causing memory divergence described in the previous section. Memory requests are handled by SX-Aurora and GPU architectures differently. In the case of GPU, threads from a single warp load the required data using several transactions based on memory coalescing approach, which implies combining multiple memory accesses into a single transaction. Coalesced memory access idea is described in [10], but it is important to highlight that on irregular memory accesses GPU will generally load significantly more data than required.

SX-Aurora architecture processes indirect memory accesses using special gather and scatter vector instructions, which involve additional latency caused by placing indexes into gather instruction. Moreover, irregular accesses cause frequent memory port conflicts, which reduce effective bandwidth even further in this code.

Figure 2 presents a comparison of effective bandwidth values achieved on random memory access benchmark implemented for Pascal GPU and SX-Aurora TSUBASA platforms. Different element types and sizes of array (which is accessed with an indirect pattern) were used. The effective bandwidth values achieved on both architectures are comparable in the case when the indirectly accessed array can be entirely placed inside the last level cache. As soon as the size of array exceeds the size of last level cache (16 MB for SX-Aurora and 2 MB for Pascal GPU), the effective bandwidth drops significantly for both platforms.

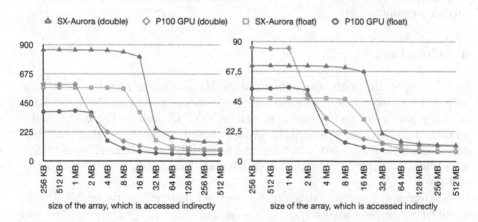

Fig. 2. Bandwidth (in GB/s, left) and ratio percentages (right) achieved on P100 and SX-Aurora platforms on indirect memory accesses benchmarks.

3.9   Processing Small Data Types

Volta architecture provides specialised support for half-precision computations. Not only it demonstrates higher performance due to a significant number of specialised half-precision computational units (tensor-cores), but also because of an ability of loading significantly less data from memory due to an explicit support of half-precision datatypes. On the contrary, SX-Aurora architecture does not support vectorised computations both with half-precision arithmetics and small data types, such as bool, char, and short.

Graph algorithms is an important class of problems demonstrating significantly higher performance due to the possible usage of small data types, which can significantly reduce the amount of data loaded from memory with an irregular access pattern. Another important application is neural networks training, which allows to perform all the required matrix-matrix multiplications in half-precision, thus effectively halving the required bandwidth.

3.10   Computational Scheduling and Execution

Several significant differences are caused by fundamental differences between GPU CUDA-cores and SX-Aurora VPPs. GPU SMs execute active warps using warp schedulers, which are dual-issue capable, as long as there are 2 independent instructions in the instruction flow of the same warp. These 2 SIMD instructions can be executed on any type of CUDA cores – integer, FP, DP, SFU or load/store. Thus, GPU is able to utilise different types of computational cores on each cycle, including simultaneous usage of single-precision and double-precision computations. In the meantime, SX-Aurora uses same FMA units for both single- and double-precision computations (single-precision values are packed and processed in pairs), and thus is incapable of performing simultaneous single- and double-precision computations.

4   Conclusions

In this paper the main computational SIMD-processing features have been reviewed for two types of modern high-performance platforms: NVIDIA GPU of Pascal and Volta architectures, as well as NEC SX-Aurora TSUBASA architecture. A detailed analysis of SIMD-processing features of the reviewed architectures have been conducted, which allowed to identify several examples of typical algorithms, demonstrating similar and different efficiency on the reviewed architectures.

Examples of programs that can be executed more efficiently on SX-Aurora architecture include algorithms, which require frequent synchronisations and data exchanges between different computational units, sequential memory accesses or regular execution of sequential regions. Typical examples of programs more suitable for execution on GPU architectures include algorithms with frequent computations over small data types or half-precision calculations.

The presented study allows to address the problem of efficiently porting programs from GPU architectures to vector platforms and vice versa, based on architecture properties highlighted in the current paper. This is an important direction for future research.

This project was partially supported by JSPS Bilateral Joint Research Projects program, entitled "Theory and Practice of Vector Data Processing at Extreme Scale: Back to the Future". The reported study was supported by the Russian Foundation for Basic Research, project No. 18-57-50005.

References

1. STREAM Benchmark. https://www.cs.virginia.edu/stream/
2. Thrust Library. https://thrust.github.io
3. Egawa, R., et al.: Potential of a modern vector supercomputer for practicalapplications: performance evaluation of SX-ACE. J. Supercomput. **73**(9), 3948–3976 (2017). https://doi.org/10.1007/s11227-017-1993-y
4. Flynn, M.J.: Very high-speed computing systems. Proc. IEEE **54**(12), 1901–1909 (1966)
5. Harris, M., et al.: Optimizing parallel reduction in CUDA. Nvidia Dev. Technol. **2**(4), 70 (2007)
6. Komatsu, K., Egawa, R., Isobe, Y., Ogata, R., Takizawa, H., Kobayashi, H.: An approach to the highest efficiency of the HPCG benchmark on the SX-ACE supercomputer. In: Proceedings of the Conference on High Performance Computing Networking, Storage and Analysis (SC15), Poster, pp. 1–2, November 2015
7. Komatsu, K., et al.: Performance evaluation of a vector supercomputer SX-aurora TSUBASA. In: Proceedings of the International Conference for High Performance Computing, Networking, Storage, and Analysis, SC 2018, pp. 54:1–54:12. IEEE Press, Piscataway (2018). http://dl.acm.org/citation.cfm?id=3291656.3291728
8. NVIDIA: Nvidia Tesla P100: The most advanced datacenter accelerator ever built featuring Pascal GP100, the world's fastest GPU. Whitepaper (2016)
9. NVIDIA Tesla: V100 GPU architecture (2017)
10. Wu, B., Zhao, Z., Zhang, E.Z., Jiang, Y., Shen, X.: Complexity analysis and algorithm design for reorganizing data to minimize non-coalesced memory accesses on GPU. In: ACM SIGPLAN Notices, vol. 48, pp. 57–68. ACM (2013)
11. Yamada, Y., Momose, S.: Vector engine processor of NECs brand-new supercomputer SX-aurora TSUBASA. In: Intenational Symposium on High Performance Chips (Hot Chips 2018) (2018)

Efficient Parallel Solvers
for the FireStar3D Wildfire Numerical
Simulation Model

Oleg Bessonov[1（✉）] and Sofiane Meradji[2]

[1] Ishlinsky Institute for Problems in Mechanics RAS, 101, Vernadsky ave.,
119526 Moscow, Russia
`bess@ipmnet.ru`
[2] IMATH, EA 2134, University of Toulon, Avenue de l'Université,
83957 La Garde, France
`sofiane.meradji@univ-tln.fr`

Abstract. This paper presents efficient parallel methods for solving ill-conditioned linear systems arising in fluid dynamics problems. The first method is based on the Modified LU decomposition, applied as a preconditioner to the Conjugate gradient algorithm. Parallelization of this method is based on the use of nested twisted factorization. Another method is based on a highly parallel Algebraic multigrid algorithm with a new smoother developed for anisotropic grids. Performance comparisons demonstrate superiority of new methods over commonly used variants of the Conjugate gradient method.

Keywords: Ill-conditioned linear systems · Conjugate gradient · Preconditioners · Multigrid · Smoothers · Parallelization

1   Introduction

The multi-physical FireStar3D numerical simulation model was developed in order to predict the behavior of wildfires at local scales (up to 500 m) [1,2]. This model consists of solving the conservation equations of a coupled system composed of vegetation and the surrounding gaseous medium. The model is able to account explicitly for all mechanisms of degradation of vegetation and various interactions between the gas mixture and the vegetation cover such as drag force, heat transfer by convection and radiation, and mass transfer.

Solving a three-dimensional nonstationary multi-physical problem requires significant computational resources. An appreciable part of the computational time is spent on solving large sparse linear systems arising from the discretization of partial differential equations in the above model [3].

The most popular iterative methods used to solve large linear systems are the Conjugate Gradient for symmetric matrices and its non-symmetric variants (BiCGStab, GMRES etc.) [4]. To accelerate convergence, these methods require

© Springer Nature Switzerland AG 2019
V. Malyshkin (Ed.): PaCT 2019, LNCS 11657, pp. 140–150, 2019.
https://doi.org/10.1007/978-3-030-25636-4_11

preconditioning [5]. There exists also a family of multigrid methods which possess very good convergence and parallelization properties [6,7].

The applicability of solvers depends on the nature of the underlying physical processes and on the speed of propagation of physical information. In particular, incompressible viscous fluid flows can be driven by three basic mechanisms with different propagation speeds:

- convection: slow propagation, Courant condition can be applied (one or few grid distances per time-step); using an iterative solver with few iterations;
- diffusion: faster propagation (tens grid distances per time-step), well-conditioned linear system; using an iterative solver with more iterations;
- pressure: instant propagation, ill-conditioned linear system; using an iterative solver with a robust preconditioner or a multigrid or a direct solver.

The choice of the solution method is determined by the above property. In the FireStar3D code, robust and efficient methods are used to solve the most time-consuming Poisson equation for pressure – the preconditioned Conjugate Gradient and the Algebraic multigrid. To solve the coupled system of convection-diffusion equations, for which a robust solver is not required, the BiCGStab method is applied.

In the previous papers [5,7,8], we analyzed various properties of iterative methods from the point of view of mathematics, convergence, efficiency and parallelization. In this paper we will consider the application of these methods for wildfire modeling, taking into account specific properties and requirements of the corresponding numerical simulation model.

The remaining part of the paper is organized as follows. Section 2 briefly presents the mathematical and geometric formulation of the FireStar3D model. Section 3 discusses the preconditioned Conjugate gradient method and describes the parallelization approach for the implicit MILU preconditioner. Section 4 introduces the multigrid method and describes a new smoother for anisotropic grids. Section 5 presents and analyzes the performance comparison results.

2   Mathematical Model

The mathematical model is based on a multiphase formulation [1]. It consists of two. parts, that are solved on two distinct grids. The first part is described by the equations of the reacting turbulent flow in the gaseous phase, consisting of a mixture of fresh air with gaseous products resulting from the degradation of the solid phase and homogeneous combustion in the flaming zone. The second part consists of the equations governing the state and composition of the solid phase subjected to an intense heat flux coming from the flaming zone.

Solving the gaseous phase model consists in the resolution of conservation equations of mass, momentum, energy (in enthalpy formulation), and chemical species filtered using an unsteady RANS approach. Degradation of the vegetation is governed by three temperature-dependent mechanisms: drying, pyrolysis, and charcoal combustion.

The balance equations in the gaseous phase are solved numerically using the fully implicit finite volume method in a segregated formulation [9,10]. The Finite Volume discretization is applied to the non-uniform Cartesian staggered grid. The transport equations are solved by a fully implicit segregated method based on the PISO algorithm [11].

Figure 1 shows the computational domain of the wildfire numerical simulation model with two distinct grids [2].

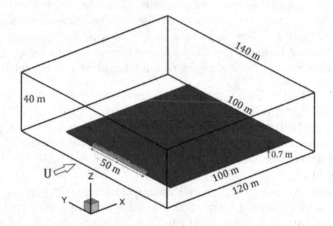

Fig. 1. Perspective view showing the computational domain and vegetation cover. The ignition line is shown on the left side of the vegetation cover

3   Preconditioned Conjugate Gradient Method

3.1   Explicit and Implicit Preconditioners

The original non-preconditioned Conjugate Gradient method (CG) [4] for solving a linear system $A\boldsymbol{x} = \boldsymbol{b}$ is simple to implement and can be easily parallelized. However, due to the explicit nature, it has a low rate of convergence and requires about $O(N)$ iterations, where N is the dimension of the problem in one spatial direction.

Because of this, the CG method is usually applied to the preconditioned linear system $(M^{-1}A)\boldsymbol{x} = M^{-1}\boldsymbol{b}$ where M is a symmetric positive-definite matrix that is "close" to the main matrix A (also symmetric and positive-definite). In practice, the system to be solved looks like $(L^{-1}AL^{-T})\boldsymbol{x}^* = L^{-1}\boldsymbol{b}$ where $LL^{-1} = M$ (Incomplete LU decomposition), but in the preconditioned CG algorithm, only computations of the form $\boldsymbol{x} = M^{-1}\boldsymbol{z}$ or $M\boldsymbol{x} = \boldsymbol{z}$ are required [4].

Preconditioning works well if the condition number of the matrix $L^{-1}AL^{-T}$ is much less than that of the original matrix A. The easiest way to reduce this condition number and speed up the convergence is to apply an "explicit" preconditioner $(B = M^{-1})$ than does not require the inversion of M (i.e. $\boldsymbol{x} = B\boldsymbol{z}$ is to be computed).

A good example of this kind is the polynomial Jacobi preconditioner [8], based on the truncated approximation series $1/(1-a) = 1 + a + a^2 + \ldots$

$$B = M^{-1} = \sum_{k=0}^{n} (H^k) P^{-1} \text{ where } P = \text{diag}(A), \ H = P^{-1}(P - A) = I - P^{-1}A$$

For $n = 0$, this expression degenerates into a diagonal preconditioner $B = P^{-1}$, which, due to its simplicity, is usually not considered as a true preconditioner. For $n = 1$, the Jacobi preconditioner looks like $B = (I+(I-P^{-1}A))P^{-1}$ and improves the acceleration rate twice (with some increase in computational complexity). This exactly corresponds to the expansion of the computational stencil in one iteration of the algorithm. Therefore, it can be easily applied and parallelized.

Unfortunately, neither kind of the simple explicit preconditioner can drastically improve convergence. The reason is that the explicit preconditioner acts locally using a stencil of limited size and propagates information through the domain with low speed. On the other hand, the implicit preconditioner, based on solving auxiliary linear systems, operates globally and propagates information almost instantly. Due to this, the implicit preconditioner works much faster and has a better than linear dependence of convergence on the geometric size of the problem. For this reason, to solve an ill-conditioned linear system, it is necessary to apply a preconditioner of the implicit kind.

In the FireStar3D code, the explicit Jacobi preconditioner is used to solve well-conditioned linear systems resulting from the discretization of a coupled system of convection-diffusion equations. Due to the non-symmetric nature of these linear system, the BiCGStab method is used.

3.2   Parallelization of the Implicit Preconditioner

The parallel properties of preconditioners are strongly dependent on how information is propagated in the algorithm. For this reason, it can be difficult to parallelize an implicit preconditioner, and a lot of effort is required to find the geometric and algebraic approach to parallelization. In particular, this applies to Incomplete LU-decomposition (ILU).

For the Cartesian computational domain, the geometric potential of parallelization can be revealed. The initial idea of the method is taken from the twisted parallelization of the tridiagonal linear system, when Gauss elimination is performed from both sides simultaneously. This idea can be naturally generalized to three dimensions. The resulting method is called "nested twisted factorization" [8,12].

In this method, the rectangular parallelepipedic domain is divided into 8 octants by separator planes (Fig. 2). In each octant, Gauss elimination is performed from the corner in the direction inwards independently in different threads (Fig. 2, left).

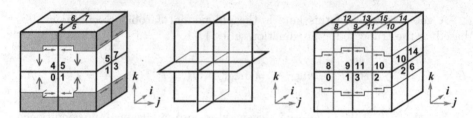

Fig. 2. Parallelization of the nested twisted factorization: illustration of the method (left); separator planes (center). Parallelization for 16 threads, staircase method (right)

After doing eliminations at internal octant points, they are performed in quadrants of separator planes in the same way (Fig. 2, center). Then, the intersection lines of the separator planes are processed and, finally, the solution is calculated at the central point. The following backsubstitution is performed in the reverse order, from the central point outwards.

Parallelization for 16 threads can be achieved by applying the staircase method shown on Fig. 2 (right). Here, each octant is divided into two halves in the direction j (see bottom left octant, divided between threads 0 and 1). Computations in the plane (i,j) for a certain k cannot be performed by thread 1 until they are completed by thread 0. However, they can be performed by a pipelined fashion: thread 1 computes the layer for some k at the same time when thread 0 computes the next layer for k+1 (this looks like a step on the stairs). At the backsubstitution stage of the algorithm, the computations are performed in the reverse order.

Additional parallelization of the method for more threads seems to be impractical due to synchronization overhead. Nevertheless, this method can be used on a computer with more cores, since the performance of the algorithm is mainly limited by the memory bandwidth (i.e. the method belongs to the memory-bound class). Because of this, it is possible to implement a procedure for any reasonable number of threads, and not just for 8 or 16. To achieve this, it is necessary to distribute the active threads of the method (8 or 16) among all cores of the computing system, thus ensuring load balance. As a result of this modification, the method works well on up to 32 cores of a bi-processor computer.

The convergence of the ILU preconditioner depends on how the decomposition is calculated. The most accurate variant of the method, Modified ILU (MILU), requires about $O(N^{\frac{1}{2}})$ iterations, where N is the dimension of the problem in one spatial direction [8,13]. As a result, this algorithm becomes 5 to 6 times faster than the Conjugate gradient method with explicit Jacobi polynomial preconditioner.

3.3   Modified ILU Preconditioner for Periodic Boundary Conditions

The Modified ILU preconditioner can be mathematically strictly implemented and parallelized only for a rectangular parallelepipedic domain with non-periodic boundary conditions. In the case of periodic conditions, the algorithm becomes not strict, and its convergence properties deteriorate. In particular, while the convergence estimate for a strict MILU is $O(N^{\frac{1}{2}})$ iterations, the loss of these properties leads to an estimate of $O(N)$ iterations.

However, in the problem under consideration, the flow properties in the periodic transverse direction are almost uniform with some fluctuations. For this reason, it becomes possible to use the original MILU preconditioner, which does not care on the periodic boundaries. This preconditioner is applied on the top of the Conjugate gradient algorithm with accurate treatment of the periodicity. This algorithm smooths the solution around periodic boundaries and maintains relatively fast convergence.

This new algorithm was implemented and tested. Its convergence with an accuracy 10^{-10} for a problem size $100 \times 200 \times 224$ is 68 iterations, compared with about 50 iterations of the original algorithms applied to a non-periodical problem of a similar size. This is much less than 350 iteration of the Conjugate gradient method with explicit Jacobi polynomial preconditioner. In term of the computational time, the new algorithm is about 4 times faster.

4   Algebraic Multigrid

The multigrid method is potentially the most efficient one for solving ill-conditioned linear systems because of its ability to suppress error components of all scales. Also, it can be parallelized to a large number of threads. This method solves differential equations using a hierarchy of discretizations.

In one multigrid cycle (V-cycle, Fig. 3), both short-range and long-range components of the error are smoothed out, so information is instantaneously transmitted throughout the domain. As a result, this method becomes very efficient for elliptic problems that spread physical information infinitely fast.

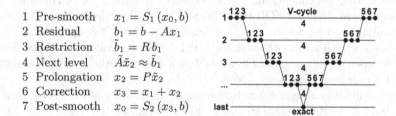

1	Pre-smooth	$x_1 = S_1(x_0, b)$
2	Residual	$b_1 = b - Ax_1$
3	Restriction	$\tilde{b}_1 = R\,b_1$
4	Next level	$\tilde{A}\tilde{x}_2 \approx \tilde{b}_1$
5	Prolongation	$x_2 = P\tilde{x}_2$
6	Correction	$x_3 = x_1 + x_2$
7	Post-smooth	$x_0 = S_2(x_3, b)$

Fig. 3. Scheme of the multigrid algorithm (left); illustration of the V-cycle (right)

In the FireStar3D code, the Algebraic multigrid (AMG) approach [6,7] is applied. This method is based on matrix coefficients rather than on geometric parameters of the domain. The main computational operations in the multigrid cycle are smoothing (usually an iteration of the Gauss-Seidel or SOR method) and, to a lesser extent, restriction (fine-to-coarse grid conversion by averaging) and prolongation (coarse-to-fine conversion by interpolation).

4.1   Smoothers for Anisotropic Grids

The multigrid is a very efficient method, its convergence does not depend on the problem size. However, it does not perfectly work on anisotropic grids (with cells that have a high aspect ratio). The reason is that the typically used smoothing procedure (Gauss-Seidel or SOR) effectively suppresses error components only along the shortest cell dimension. In the considered problem, the cell aspect ratio reaches 15:1. Because of this, the traditional approach leads to extremely slow convergence (up to 300 iterations against a typical value of the order of 10).

There are several approaches to resolve this problem. The most straightforward method is semi-coarsening [6]. However, after applying this procedure, the grid becomes non-structured, and the overall method becomes very complex and numerically less efficient. Another method is based on the use of incomplete matrix factorization as a smoother [14], which improves the performance and convergence of the multigrid. Other approaches originate on building a more robust smoother that is not sensitive to grid anisotropy [15]. They are based on the replacement of point relaxation methods with plane relaxation ones.

If the grid cells are compressed in a single spatial direction, it becomes possible to apply the line Gauss-Seidel (line GS) or the line SOR smoothing procedure in this direction. The idea is to solve the GS or SOR equation for the full line of grid points, rather than separately for each grid point. As a result, the smoothing of error components along the longest cell dimension is not suppressed. The new procedure requires solving a tridiagonal linear system along a compressed direction and, therefore, is slightly more expensive than the standard one.

It was found that the line smoother successfully solves the above problem, but it is not efficient enough to smooth the error components in the remaining part of the domain. To improve the convergence, this procedure was supplemented by standard (point) Gauss-Seidel or SOR smoother, which costs less. The above approach was applied for all levels of the multigrid algorithm.

To achieve good convergence, it is necessary to determine the optimal over-relaxation parameters for SOR procedures. These values depend on the size and configuration of the grid. For the first (finest) grid level, the optimal values are about 1.3–1.4 for line smoothers and about 1.6–1.65 for point smoothers. For the upper (coarser) levels, a plain GS is used as a line smoother, while the optimal values for point smoothers are about 1.6–1.9.

The application of over-relaxation reduces the number of iterations from 40–50 to 10–11 (for grid sizes up to $100 \times 200 \times 504$ and relative accuracy 10^{-10}).

4.2   Parallelization of Smoothers

An iteration of the Gauss-Seidel or SOR method looks like an implicit procedure: $(D + L)\boldsymbol{x}_{k+1} = b - U\boldsymbol{x}_k$ (here D, L and U are diagonal, lower and upper parts of the matrix A in the equation $A\boldsymbol{x} = \boldsymbol{b}$). To avoid dependences that prevent parallelization, a multicolor grid partitioning is required. For the first level of the grid with 7-point stencils, a two-color (red-black) scheme can be used. In this scheme, the procedure is divided into two explicit steps: $D^{(1)}\boldsymbol{x}_{k+1}^{(1)} = \boldsymbol{b}^{(1)} - U\boldsymbol{x}_k^{(2)}$

and $D^{(2)}x^{(2)}_{k+1} = b^{(2)} - Lx^{(1)}_{k+1}$ (superscripts $^{(1)}$ and $^{(2)}$ refer to red and black grid points, respectively). After that, elements with the same color can be processed independently, and, as a consequence, parallel splitting can be applied.

For line smoothers, red-black partitioning is applied to whole lines.

For the upper levels of the grid with 27-point stencils, a 4-color scheme is used, also applied to whole lines.

Multicolor processing of the computational domain can be performed in several passes according to the number of colors. However, each pass needs access to all the elements of the data arrays. Since the performance of the algorithm depends primarily on the memory access rate, this proportionally increases the computational time.

To reduce the number of passes, it is necessary to somehow combine the processing of different colors, while retaining the property of a multicolor scheme. The idea of the combination technique is illustrated in Fig. 4. Shown here are the cross-sections of the computational domain perpendicular to the compressed direction (i.e. the direction where the line GS or SOR is applied). The proposed idea is expressed in terms of rows and columns assuming that, in lexicographic order, rows are processed first.

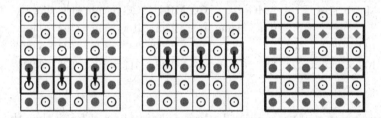

Fig. 4. Illustration of a multicolor smoothing procedure: alternating iterations of the red-black pass (left, center); single pass of the 4-color case (right) (Color figure online)

For the red-black case, processing is performed in a single pass with alternating iterations. At even iterations of the pass (Fig. 4, left), adjacent pairs of red and black elements of even columns are calculated (first red, then black). At odd iterations (Fig. 4, center), similar pairs of elements of odd columns with a row number increased by one are processed (in the same order).

For the 4-color case, two passes are required (Fig. 4, right) – one pass for even rows and another pass for odd ones. Within each pass, two sub-passed are performed – one for each color in a row. The second sub-pass does not require costly memory accesses, since most of the data is cached after the first sub-pass.

Multicolor partitioning allows to implement in the shown cross-section any splitting of the computational domain required for parallelization.

The above technique ensures regular and efficient memory accesses as an important requirement of computational efficiency. It is supplemented by the vectorization of arithmetic operations (in frame of the AVX vector extension) and by other optimizations.

5   Performance Comparison

The convergence and performance of the new solvers were evaluated using matrices and data taken from the typical runs of the FireStar3D code (Table 1). The first matrix corresponds to a larger problem with the periodic boundary conditions for the second spatial dimension. The second matrix was taken from a smaller problem with non-periodic boundary conditions (see Fig. 1 for the geometric illustration of this problem).

The tests were conducted on a cluster node built on two 16-core Xeon Gold 6142 processors at the Mesocentre computer center (Marseille, France). In addition to the new solvers described in the paper (Algebraic multigrid and MILU-preconditioned Conjugate gradient), two variants of the Conjugate gradient method were tested – one with explicit Jacobi preconditioner and another one with simple diagonal scaling. Results are presented for parallel runs on 32 cores of a cluster node with the relative accuracy 10^{-10}, time for solving a linear system is shown in seconds.

Table 1. Comparison of convergence and performance of different solvers

Matrix size	AMG		CG MILU		CG Jacobi		CG diag	
	Iter.	Time	Iter.	Time	Iter.	Time	Iter.	Time
$100 \times 200 \times 504$	10	0.440	48	0.788	267	3.15	541	5.02
$100 \times 248 \times 224$	11	0.293	46	0.398	317	2.06	638	3.22

It can be seen that the multigrid solver is about 11 times faster than the plain Conjugate gradient (CG diag). Using the explicit Jacobi preconditioner makes the CG method 1.5 times faster due to more optimal structure of the algorithm, but does not change its convergence properties, so the number of iterations still depends linearly on the largest dimension of the discretized problem.

Compared to the Conjugate gradient method with the MILU-preconditioner, the multigrid solver is 36% faster for the problem with non-periodic boundary conditions and 79% faster for the problem with a periodicity. The latter can be explained by two properties of the CG MILU method – sensitivity to the size of the problem (as opposed to the multigrid) and some decrease in convergence due to explicit treatment of periodic boundary conditions.

Another advantage of the multigrid method is better scalability. In particular, the speedup for this method for 32 threads ranges from 15 to 17 (depending on the size of the matrix), while for CG MILU it is at the level of 10–11. For both methods, the speedup is limited by the memory bandwidth, but the second method is more memory-bound than the first one. In addition, CG MILU parallelization is limited to 16 threads.

For these reasons, the multigrid method is more preferable for using in the FireStar3D code. On the other hand, variants of the CG MILU method do not require adjustment of the parameters of over-relaxation, as is necessary for the multigrid. Therefore, it may happen to be more robust for some problems. Thus, this method can still be applicable, at least, for running on computer systems with a smaller number of processor cores.

There is another promising approach, Algebraic multigrid as a preconditioner for the Conjugate gradient method. At the moment, it does not benefit when solving linear systems in FireStar3D runs, unlike happens in other fluid dynamics problems [7]. However, in the future this approach will be examined more carefully in order to achieve faster convergence and lower computational costs.

6 Conclusion

In this paper, we presented two parallel methods for solving ill-conditioned linear systems arising from the discretization of partial differential equations as applied to the FireStar3D wildfire numerical simulation model.

The first of the presented methods is based on a parallel MILU-preconditioned Conjugate gradient algorithm. This method has been extended to run on any number of processor cores and to support periodic boundary conditions. The second method is based on an Algebraic multigrid. It uses a new smoothing algorithm that can work with highly compressed grids. This smoother is optimally parallelized using multicolor grid partitioning and a special processing scheme.

New methods were used to build efficient parallel solvers for the FireStar3D code. Both solvers were evaluated using data taken from the production runs of the code. They demonstrate robustness and superiority over the widely used variants of the Conjugate gradient method. In particular, the multigrid solver is more than ten times faster than the diagonally scaled Conjugate gradient solver.

Of these two methods, the multigrid algorithm is faster and more scalable for large number threads. On the user hand, it requires some tuning to achieve faster convergence. For this reason, MILU-based methods remain attractive because of their robustness and therefore can be used for running with fewer threads.

Acknowledgements. This work was supported by the Russian State Assignment under contract No. AAAA-A17-117021310375-7. The work was granted access to the HPC resources of Aix-Marseille Université financed by the project Equip@Meso (ANR-10-EQPX-29-01) of the program Investissements d'Avenir supervised by the Agence Nationale pour la Recherche (France).

References

1. Morvan, D., Accary, G., Meradji, S., Frangieh, N., Bessonov, O.: A 3D physical model to study the behavior of vegetation fires at laboratory scale. Fire Saf. J. **101**, 39–53 (2018). https://doi.org/10.1016/j.firesaf.2018.08.011

2. Frangieh, N., Morvan, D., Meradji, S., Accary, G., Bessonov, O.: Numerical simulation of grassland fires behavior using an implicit physical multiphase model. Fire Saf. J. **102**, 37–47 (2018). https://doi.org/10.1016/j.firesaf.2018.06.004
3. Saad, Y.: Iterative Methods for Sparse Linear Systems. PWS Publishing, Boston (2000)
4. Shewchuk, J.R.: An Introduction to the Conjugate Gradient Method Without the Agonizing Pain. School of Computer Science, Carnegie Mellon University, Pittsburgh (1994)
5. Bessonov, O.: Parallelization properties of preconditioners for the conjugate gradient methods. In: Malyshkin, V. (ed.) PaCT 2013. LNCS, vol. 7979, pp. 26–36. Springer, Heidelberg (2013). https://doi.org/10.1007/978-3-642-39958-9_3
6. Stüben, K.: A review of algebraic multigrid. J. Comput. Appl. Math. **128**, 281–309 (2001). https://doi.org/10.1016/S0377-0427(00)00516-1
7. Bessonov, O.: Highly parallel multigrid solvers for multicore and manycore processors. In: Malyshkin, V. (ed.) PaCT 2015. LNCS, vol. 9251, pp. 10–20. Springer, Cham (2015). https://doi.org/10.1007/978-3-319-21909-7_2
8. Accary, G., Bessonov, O., Fougère, D., Gavrilov, K., Meradji, S., Morvan, D.: Efficient Parallelization of the preconditioned conjugate gradient method. In: Malyshkin, V. (ed.) PaCT 2009. LNCS, vol. 5698, pp. 60–72. Springer, Heidelberg (2009). https://doi.org/10.1007/978-3-642-03275-2_7
9. Patankar, S.V.: Numerical Heat Transfer and Fluid Flow. Hemisphere Publishing, New York (1980)
10. Versteeg, H., Malalasekera, W.: An Introduction to Computational Fluid Dynamics: The Finite Volume Method. Prentice Hall, Harlow (2007)
11. Moukalled, F., Darwish, M.: A unified formulation of the segregated class of algorithms for fluid flow at all speed. Numer. Heat Transf. Part B **37**, 103–139 (2000). https://doi.org/10.1080/104077900275576
12. van der Vorst, H.A.: Large tridiagonal and block tridiagonal linear systems on vector and parallel computers. Parallel Comput. **5**, 45–54 (1987). https://doi.org/10.1016/0167-8191(87)90005-6
13. Gustafsson, I.: A class of first order factorization methods. BIT **18**, 142–156 (1978). https://doi.org/10.1007/BF01931691
14. Axelsson, O.: Analysis of incomplete matrix factorizations as multigrid smoothers for vector and parallel computers. Appl. Math. Comput. **19**, 3–22 (1986). https://doi.org/10.1016/0096-3003(86)90094-9
15. Llorente, I.M., Melson, N.D.: Robust multigrid smoothers for three dimensional elliptic equations with strong anisotropies. Technical report 98-37, ICASE (1998)

Optimizing a GPU-Parallelized Ant Colony Metaheuristic by Parameter Tuning

Andrey Borisenko[1(✉)] and Sergei Gorlatch[2]

[1] Tambov State Technical University, Tambov, Russia
`borisenko@mail.gaps.tstu.ru`
[2] University of Muenster, Münster, Germany
`gorlatch@uni-muenster.de`

Abstract. We address the problem of accelerating the GPU-parallelized Ant Colony Optimization (ACO) metaheuristic used for an important class of optimization problems – design of multiproduct batch plants, with a particular use case of a *Chemical-Engineering System* (CES). We propose and implement a novel approach to ACO's parameter tuning, with the following advantages compared to previous work: we accelerate tuning by using GPU, and we do not require additional constructs like function mapping in fuzzy logic, algorithms for online-tuning, etc. We report our experimental results that confirm the efficiency of parameter tuning and the advantages of our approach.

Keywords: Constraint Satisfaction Problem ·
Ant Colony Optimization · Tuning metaheuristics ·
Parallel metaheuristics · GPU computing ·
Multi-product batch plant design

1 Motivation and Related Work

The Ant Colony Optimization (ACO) metaheuristic is a popular approach to solving optimization problems. It can be viewed as a multi-agent system in which agents (ants) interact with each other in order to reach a global goal [10]. ACO follows the idea of collective intelligence in colonies of ants: the ants cooperatively search for food and bring this food to their nest. While walking between food sources and the nest, ants deposit a chemical substance called *pheromone* on their path. The pheromone is used to find the shortest path from their nest to food. Parameters of ACO determine the probability with which ants follow the pheromone deposited by previous ants, and how fast the pheromone evaporates.

We apply ACO to an important class of real-world optimization problems – optimal design of multiproduct batch plants, with a particular use case of a *Chemical-Engineering System* (CES). Such a system is a set of equipment units (reactors, tanks, filters, dryers etc.) which manufacture products, and the

© Springer Nature Switzerland AG 2019
V. Malyshkin (Ed.): PaCT 2019, LNCS 11657, pp. 151–165, 2019.
https://doi.org/10.1007/978-3-030-25636-4_12

problem is finding the optimal number of units at processing stages and their main sizes for the given input that includes: demand for each product of assortment, production horizon, accessible equipment set, etc. This problem is NP-hard, i.e., the time to solve a problem instance grows exponentially with the instance size. Therefore, metaheuristics are often the only feasible way to obtain good-quality solutions at acceptable computational cost [11]. In our previous work [4], we develop a hybrid parallel algorithm consisting of two metaheuristics: (1) ACO finds an initial soluting of the Constraint Satisfaction Problem (CSP); (2) this initial solution is then optimized using Simulated Annealing (SA). The hybrid ACO+SA algorithm is parallelized for Graphics Processing Units (GPU) and successfully solves the design optimization problem for CES of size up to $12^{16} \approx 10^{17}$ variants; it demonstrates a significant time saving as compared to the traditional branch-and-bound optimization method.

In this paper, we aim at further acceleration of the GPU-parallelized ACO method, in order to apply it to even larger sizes of problems that arise in practice: already starting with the size 12^{16}, the run time of the original algorithm becomes prohibitively high despite the use of a highly parallel GPU. Our approach is to use the additional performance potential offered by the tunable parameters of the ACO algorithm.

A significant amount of work has been devoted to tuning ACO parameters [2] that has proven to be a hard problem [11, 24].

The approaches to parameters tuning can roughly be divided into offline versus online procedures. A tuning framework [25] is based on the sequential optimization of perturbed regression models. Paper [1] presents a methodology combining statistical and artificial intelligence methods in the fine-tuning of metaheuristics. Paper [21] uses a fuzzy system for parameter adaptation in the ACO metaheuristic. In [23], the problem of finding the parameters of a meta-heuristic algorithm is formulated as a meta-optimization problem solved by an evolutionary metaheuristic. An enhanced ACO with dynamic mutation and ad hoc initialization for generating the initial ant solutions to improve the accuracy of fuzzy system design is proposed in [8]. Paper [7] explores a new fuzzy approach for diversity control in ACO. In [12], a parameter tuning methodology for metaheuristics based on the design of experiments is proposed. Paper [13] uses a Particle Swarm Optimization (PSO) algorithm to optimize the ACO parameters.

We propose and implement a novel approach to parameter tuning for an ACO algorithm that solves the CSP (Constraint Satisfaction Problem) part of our global problem. The main differences of the proposed approach to the previous work are as follows. By tuning for CSP, rather than for the optimization problem as in [17, 21], we can apply frequency analysis. For calculating how often values occur within a range of values, we do not need any specific and non-obvious information, like functions mapping in fuzzy logic, an algorithm for online-tuning, etc. [7, 8, 21]. The parallelization of the algorithm and the use of modern GPUs allow us to conduct a large number of computational experiments to accumulate statistical data in a short time. An advantage is also that the found optimal values of parameters for can be used for both parallel and sequential versions of the algorithm. Summarizing, the advantages of our approach as compared to previous work are two-fold: (1) we exploit the computation power of GPU in the tuning process, and (2) we do not rely on any additional information like functions mapping in fuzzy logic, an algorithm for online-tuning, etc.

In the remainder of the paper, Sect. 2 outlines the CES optimization problem and its GPU-implementation. In Sect. 3, we analyze our ACO algorithm for CSP problem and the roles of its parameters, and we describe our novel methodology of ACO parameters tuning. In Sect. 4, we report our experimental results that confirm the advantages of our approach, and Sect. 5 concludes.

2   GPU-Algorithm for Designing Multi-product Plants

Our application use case is designing a *Chemical-Engineering System* (CES) – a set of equipment (reactors, tanks, filters, dryers etc.) for manufacturing diverse products. Assuming that the number of units at every stage of CES is fixed, the problem can be formulated as follows (for a detailed formulation, see [3]). A CES consists of a sequence of I processing stages; i-th stage can be equipped with equipment units from a finite set X_i, with J_i being the number of equipment units variants in X_i. The goal is to find the optimal number of units at stages and their main sizes; the input data are: production horizon, demand for each product, available equipment, etc. Each system's variant Ω_e has to be in an operable condition (*compatibility constraint*) expressed by function S: $S(\Omega_e) = 0$. If T_{max} is the total available time horizon, then an operable variant of a CES must also satisfy a *processing time constraint*: $T(\Omega_e) \leq T_{max}$.

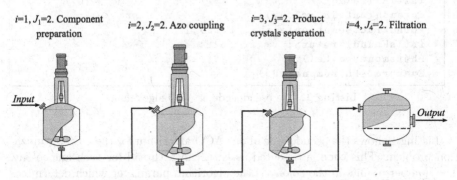

$i=1, J_1=2.$ Component preparation $i=2, J_2=2.$ Azo coupling $i=3, J_3=2.$ Product crystals separation $i=4, J_4=2.$ Filtration

Input   *Output*

Accessible equipment set with the main sizes (volume for reactors and surface for filter) $x_{i,j} \in X_i$:

$$X_1 = \begin{Bmatrix} x_{1,1}=6.3\ \text{m}^3 \\ x_{1,2}=10.0\ \text{m}^3 \end{Bmatrix} \quad X_2 = \begin{Bmatrix} x_{2,1}=20.0\ \text{m}^3 \\ x_{2,2}=25.0\ \text{m}^3 \end{Bmatrix} \quad X_3 = \begin{Bmatrix} x_{3,1}=16.0\ \text{m}^3 \\ x_{3,2}=20.0\ \text{m}^3 \end{Bmatrix} \quad X_4 = \begin{Bmatrix} x_{4,1}=5.0\ \text{m}^2 \\ x_{4,2}=12.5\ \text{m}^2 \end{Bmatrix}$$

Fig. 1. Example: a simple Chemical-Engineering System (CES)

Figure 1 shows an example CES consisting of 4 stages ($I = 4$), where each stage can be equipped with 2 devices ($J_1 = J_2 = J_3 = J_4 = 2$); the number of all possible system variants in this case is $2^4 = 16$. The optimization works on the search tree, in which each path from the root to one of the leaves in this tree corresponds to a candidate solution of the optimization problem.

In [3], we create a hybrid approach to optimizing CES; it combines two meta-heuristics – *Ant Colony Optimization (ACO)* and *Simulated Annealing (SA)*. We parallelize and implement it on a CPU-GPU system using CUDA [19] and we show that it is preferable to the popular Branch-and-Bound (B&B) method [5]. In our approach, the solution of the optimization problem is divided into two stages: (1) construct a *feasible* (i.e., functionally operative CES variant) initial solution using ACO; (2) improve this feasible solution using SA. While for classical optimization problems, e.g., Traveling Salesman Problem (TSP), it is possible to use a random initial solution [17], in our case the random initialization is unacceptable, because the *compatibility* and the *processing time* constraints must be satisfied. Our search for a feasible solution in the first stage is a Constraint Satisfaction Problem (CSP) [26] which consists in finding an operable variant of a CES, where both the compatibility constraint and the processing time constraint are satisfied. For solving this problem, we use ACO.

In our parallel implementation on a GPU [3], the ACO kernel function searches for the first feasible solution using the Multiple Ant Colonies approach [9]: all colonies work as threads in parallel to solve a problem independently.

```
1  AntColonyOptimization()
2  { isFound = false;  /* repeat while solution not found */
3    while(!isFound && iterCounter < maxIterNumber){
4      Initialize();  /* initialize pheromone value */
5      foreach(ant in colony){/* colony has M ants */
6        ConstructSolution(alpha, beta);}
7      if(isFound) return;  /* if solution is found, then end */
8      PheromoneUpdate();  /* update pheromone */
9      EvaporatePheromone(rho);}
```

Listing 1. The pseudocode of ACO algorithm.

Listing 1 shows the pseudocode of our ACO algorithm for the CES optimization problem. This code is executed as kernel in a thread for each ant colony. The number of ants in the colony is the algorithm parameter which determines the trade-off between the number of iterations and the breadth of the search at each iteration: the larger the number of ants per iteration, the fewer iterations are needed in ACO [24]. The local iteration counter is used by each thread as a nonstop operation protection (line 3): if ants in this thread cannot find the solution after `maxIterNumber` iterations (which is in principle possible for stochastic algorithms), then the thread terminates.

Up to now, most improvement work for ACO has concentrated on the tour construction and pheromone update. But there is also a question how to decide the termination condition of ACO algorithms in practice [29]. The possible variants of termination condition include: (1) the algorithm has found a solution within a predefined distance from a lower bound on the optimal solution quality; (2) a maximum number of tour constructions or a maximum number of algorithm iterations has been reached; (3) a maximum CPU time has been spent; (4) the algorithm shows stagnation behaviour [29]. These variants have shortcomings: e.g., we may not know the optimal solution, so (1) will lose the effect in the algorithm, while (2) and (3) are often not economical [29]. We use a combination of termination variants (1) and (2); they are good in our case, because for CSP, it is clear when constraints are fulfilled and when not. According to recommendations in [28,29] and our previous work we use `maxIterCount = 100`.

The first potential candidate ACO parameter for tuning in Listing 1 could be the size M of the ant colony. However, different sizes of colonies would adversely affect the GPU-algorithm, because of divergent branches and memory operations that cause uncoalesced accesses or bank conflicts [5,6]. The NVIDIA Streaming Multiprocessors (SMs) only get one instruction at a time and all CUDA cores execute the same instruction. Threads within a *warp* (a group of 32 threads, that are used in hardware to coalesce memory access and instruction dispatch) must execute the same instruction at each cycle. The most common code construct that can cause thread divergence is branching in an *if-then-else* statement: it can hurt performance due to a lower utilization of the processing elements, which cannot be compensated for via increased amount of parallelism [14]. To reduce this divergence, we use one value of M for all threads.

As confirmed by numerous experiments in previous work [18,22,24] and our own work [4], a good approximation for the number of ants in a colony is $M = 100$, so we use this value as default in all our experiments described in this paper.

We now turn to other tunable parameters of ACO which are the subject in this work. Ants in Listing 1 all behave in a similar way: every ant moves from the top of the tree-structured search space to the bottom. Once the ant selects a node $r = n_{i,j}$ at tree level i, it can pick the next child node $s = n_{i+1,j}$. The tour of an ant ends in the leaves of the tree (level I); each path corresponds to a potential solution of the problem. The ant transition from node r to s is probabilistically biased by two values: pheromone trail τ_{rs} and heuristic information η_{rs} as follows: $p_{rs} = \tau_{rs}^{\alpha} \cdot \eta_{rs}^{\beta} / \sum_{k \in C_r} (\tau_{rk}^{\alpha} \cdot \eta_{rk}^{\beta})$, where C_r is the set of child nodes for r [10,27], and k are indices of these nodes. The evaporation (line 9 in Listing 1) is performed at a constant rate ρ at the end of each iteration. It allows the ant colony to avoid an unlimited increase of the pheromone value and to "forget" poor choices made previously [24]. We implement this by the assignment: $\tau_{rs} = \rho \cdot \tau_{rs}$, where $\rho \in [0,1]$ is the trail persistence parameter. In calculating heuristic information, we make a unit which satisfies the constraint for the beginning part of the CES and larger main size more preferable than a unit with the unsatisfied compatibility constraint and smaller main size.

We observe that parameters α and β influence the pheromone value and heuristic value, respectively. They control the relative importance of the pheromone trails and the heuristic information, as we explain in the following.

We use the following rule for the pheromone update: $\tau_{rs} = \tau_{rs} + Q/\sum_{m=1}^{M} L_m$, where Q is some constant and L_m is the tour length of the m-th ant, M is the swarm size. The smaller is the value of L_m the larger is the value added to the previous pheromone value. We use L_m as a fitness value that indicates how close is a given solution to achieving the required goals.

3 ACO Parameter Tuning

For our target applications, we solve the constraint satisfaction problem (CSP), rather than the optimization problem as in previous work. A specific feature of our CSP is that only the existence of a valid solution is required. The quality criterion is the frequency of feasible solutions for particular parameters values. We use offline tuning in terms of [24] to configure the ACO parameters used to solve the CSP. Our objective function is the algorithm run time. Since ACO is a probability-based algorithm, its results are different if run multiple times on the same instance of a problem, with varying run time. So, in order to achieve reliable results, we run each instance multiple times and take the average value.

3.1 Choosing Parameters for Tuning

In the case of CES, we tune the following three parameters of ACO.

The *Information Elicitation Factor* α reflects the importance of the pheromone accumulation with regard to the ants' path selection. If α is large, the ants tend to choose the same path as the preceding ants, resulting in a stronger cooperation among the ants [16]. Although the convergence speed of ACO in this case increases, it is likely for the algorithm to fall into a locally optimal solution, i.e. large α reduces the global search ability. Conversely, if α is small, the convergence speed of the ACO is slowed down, although of the fact that the global search ability of the algorithm can be improved.

The *Expected Heuristic Factor* β represents the relative importance of the mutual ants' visibility, i.e., it reflects the importance of the heuristic information with regard to the ants' path selection. If the value is very large, the probability of a state transition is close to that of a greedy algorithm. If β is small, the heuristic information has virtually no effect on the path selection, which may lead ACO to fall into stagnation or a local optimum.

The third parameter, *Pheromone Evaporation Rate* $\rho \in [0, 1]$ regulates the degree of the decrease in pheromone level in trails. If ρ is high (near to 1) then pheromone values will persist longer, while low values of ρ (near to 0) allow forgetting quickly of previous choices and, hence, allow faster adaptation to changes [24]. In other words, smaller ρ reduces the global search ability of ACO, is while larger, ρ improves this ability but limits the convergence speed.

3.2 Our Tuning Method: The Idea

For tuning parameters α, β and ρ of parallel ACO, we use a statistical analysis of the experimental data obtained as a result of computational experiments on the CPU-GPU system. The application code for a CPU-GPU systems consists of a sequential code (*host* code executed on the CPU) that invokes hundreds or thousands of parallel threads on the *device* (GPU), where threads execute the *kernel* code shown in Listing 1. If some thread finds a solution of CSP then all threads finish their work. With an increasing number of threads, the probability of finding a solution increases, and, therefore, the search time is typically reduced.

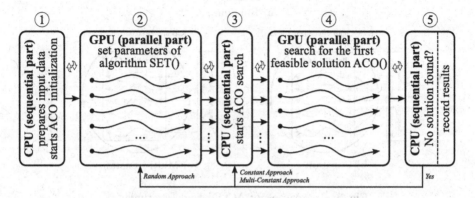

Fig. 2. General method of ACO parameter tuning.

Figure 2 shows the main steps of our tuning method, as follows: (1) CPU reads the input data (number of CES stages I, number of devices J[I], production horizon Tmax etc.) and starts on the GPU the parameter initialization α, β and ρ; (2) GPU initializes ACO parameters by one of approaches described below; (3) GPU starts the kernel function of Listing 1; (4) the ACO kernel searches for the first feasible solution – the initial CES-variant; if some thread finds a solution then all threads finish their work; (5) a if solution is not found, repeat step 2 or 3 depending on the approach; otherwise CPU receives the obtained feasible solution and records results for further processing.

Figure 3 shows that within the general tuning method consisting of steps (1)–(5), we distinguish three particular approaches to parameter tuning *Random* 3(a), *Constant Approach* 3(b) and *Multi-Constant* 3(c), as follows.

Random Approach. In the Random Approach, algorithm parameters are initialized in step 2 by uniformly distributed random values from the intervals $[\alpha_a, \alpha_b]$, $[\beta_a, \beta_b]$, $[\rho_a, \rho_b]$. We set the bounds of these intervals as recommended in literature, e.g. [15]. We use high-performance, GPU-accelerated random number generator from NVIDIA's native cuRAND library (CUDA RAndom Number Generation) [20]. Function curand_uniform() returns a uniformly distributed value in the interval (0.0, 1.0]. We generate random numbers within a specified interval $(a, b]$ as follows: rnd(a,b) = curand_uniform() * (b-a) + a. So, the SET() function for the Random approach reads as in Listing 2.

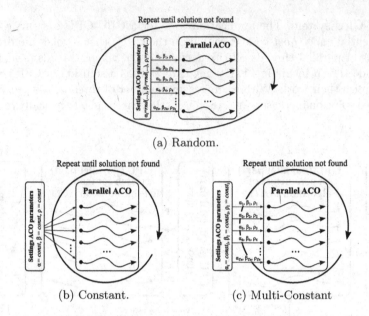

(a) Random.

(b) Constant.   (c) Multi-Constant

Fig. 3. Approaches for ACO-parameters tuning.

```
1  __global__ void SET() {
2  ...  /* obtaining thread identifier */
3  threadID = blockDim.x * blockIdx.x + threadIdx.x;
4  alpha[threadID] = rnd(alpha_a, alpha_b);
5  beta[threadID] = rnd(beta_a, beta_b);
6  rho[threadID] = rnd(rho_a, rho_b); ...}
```

Listing 2. Random Approach: the SET() pseudocode.

If a particular combination of parameter values produces a feasible solution then these values are saved for the further processing. This way, we obtain a set of triples of parameter values α, β and ρ, for which ACO finds feasible solutions. If a feasible solution in step 4 of the Random approach is not found after maxIterNumber iterations then the approach goes to step 2 of Fig. 2.

Constant Approach. The Constant approach, see Fig. 3(b) differs from the Random: all threads use the same values α, β and ρ for all threads (see Listing 3).

We obtain the starting values on the basis of a frequency analysis of the set of α, β and ρ obtained using, for example, the Random approach. In the following iterative process, we obtain a set of triples of parameter values α, β and ρ, for which ACO finds feasible solutions. If a feasible solution is not found in step 2 after maxIterNumber iterations then we proceed to step 3 of Fig. 2.

```
1  __global__ void SET() {
2  ... /* obtaining thread identifier */
3  threadID = blockDim.x * blockIdx.x + threadIdx.x;
4  alpha[threadID] = const_alpha;
5  beta[threadID] = const_beta;
6  rho[threadID] = const_rho; ...}
```

Listing 3. Constant approach: the SET() kernel pseudocode.

```
1  __global__ void SET() {
2  ... /* obtaining thread identifier */
3  threadID = blockDim.x * blockIdx.x + threadIdx.x;
4  alpha[threadID] = const_alpha[threadID];
5  beta[threadID] = const_beta[threadID];
6  rho[threadID] = const_rho[threadID]); ...}
```

Listing 4. Multi-constant approach: the SET() pseudocode.

Multi-constant Approach. In the Multi-Constant Approach (see Fig. 3c), all threads use different initial values of α, β and ρ for all threads.

In this case, parallel ACO algorithm with a Multi-Constant approach can be viewed as "learning" from the random parameter tuning, since only those triples of the parameters α, β and ρ are saved for which the solution was found. If the Multi-const approach does not find a feasible solution in maxIterNumber iterations then it goes to step 3.

4   Experimental Evaluation

Our experiments are conducted on a hybrid system comprising: (1) a CPU: Intel Xeon Gold 5118, 12 cores with Hyper-Threading, 2.3 GHz with 192 GB RAM, and (2) a GPU: NVIDIA Tesla V100-SXM2-16GB with 80 multiprocessors, each with 64 CUDA cores (total 5 120 CUDA cores), GPU max clock rate 1.53 GHz, 16 GB of global memory. We use CentOS Linux release 7.5.1804, NVIDIA Driver version 410.72, CUDA version 10.0 and GNU C++ Compiler version 6.4.0. On the GPU we employ 5 120 threads as the number of CUDA cores for Tesla v100.

As our test case, we evaluate the use of ACO for designing a CES consisting of 16 processing stages with 11 to 20 variants of devices at every stage (in total from $11^{16} \approx 10^{17}$ up to $20^{16} \approx 10^{21}$ CES variants). Note that this size is significantly larger than was possible in our previous work [4] without parameter tuning.

In the experiments, for each size of the problem from 11^{16} to 20^{16} (total 10 series of experiments), the algorithm is launched 100 times. For each launch, the run time for finding a feasible solution is measured, the average run time is calculated, and the corresponding values of α, β and ρ are recorded.

Random Approach. For the first series of experiments, values of α and β are set using a random uniform distribution in range $(0,2]$, and ρ in range $(0,1]$, as recommended in [15]. After the entire series of experiments, we obtain $10 \cdot 100 = 1\,000$ triples of values α, β and ρ with the problem sizes for which solutions were found. The total run time spent by the GPU for the first series of experiments with the Random approach is $50\,575\,\text{s} \approx 14\,\text{h}$.

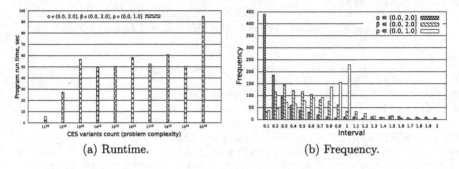

(a) Runtime.   (b) Frequency.

Fig. 4. Random approach for ACO-parameters tuning. First iteration.

Figure 4(a) shows the average run time of solving the optimization problem depending on the problem size. Figure 4(b) shows the frequency of the found feasible solutions represented as histograms for different intervals/ranges of parameter values. We observe from the histograms that $\approx 90\%$ of the solutions are obtained when the ACO parameters are in the intervals: $\alpha \in (0.0, 1.2]$, $\beta \in (0.1, 1.2]$, $\rho \in (0.2, 1.0]$.

Our idea is to stepwise reduce the intervals for α, β, ρ by moving to values where feasible solutions are more frequent. Therefore, we repeat the procedure as above to obtain new $1\,000$ solutions with the reduced parameter ranges, and again analyze the frequency of solutions. We repeat this process (search for $1\,000$ solutions – frequency analysis – correction of ACO-parameters intervals) altogether 7 times. After these 7 repetitions, the parameter ranges narrow to a single point: $\alpha = 0.2$, $\beta = 0.5$, $\rho = 0.9$. We use it as the ACO parameter values for Constant Approach in Subsect. 4.

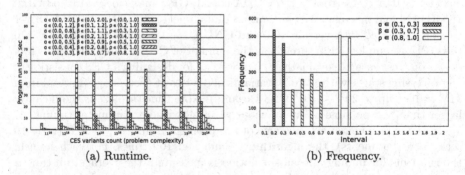

(a) Runtime.   (b) Frequency.

Fig. 5. Random approach for ACO-parameters tuning.

Figure 5(a) shows the change in the average run time of the algorithm for problems of various dimensions 11^{16} to 20^{16}. Figure 6 shows the run time for the parameter triple $\alpha = 0.2$, $\beta = 0.5$, $\rho = 0.9$, together with other results for the Constant approach. Figure 5(b) shows the frequency achieved after the final, 7th iteration of experiments for $\alpha \in (0.1, 0.3]$, $\beta \in (0.3, 0.7]$, $\rho \in (0.8, 1.0]$.

Summarizing the results achieved by the Random approach in our experiments, we can conlclude that, due to the parameter tuning, the average run time of the algorithm decreased by ≈ 29 times (from ≈ 50 s for $\alpha \in (0.0, 2.0]$, $\beta \in (0.0, 2.0]$ and $\rho \in (0.0, 1.0]$ to ≈ 1.7 s for $\alpha \in (0.1, 0.3]$, $\beta \in (0.3, 0.7]$, $\rho \in (0.8, 1.0]$). The total tuning time spent by the GPU for all seven iterations of experiments was $83\,116$ s ≈ 23 h. In the sequel, we use thus obtained values of α, β, ρ for problems of different size as initial values in the Constant and Multi-constant approach.

Constant Approach. In the Constant approach, each GPU thread uses the same triple of values α, β, ρ that was obtained by the Random approach.

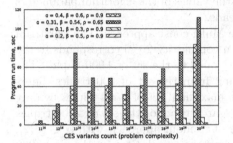
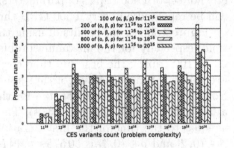

Fig. 6. Constant approach: run time depending on the problem size.

Fig. 7. Multi-constant approach: run time depending on the problem size.

Figure 6 shows the results. The first triple of parameter values $\alpha = 0.4$, $\beta = 0.6$, $\rho = 0.9$ are the same which we empirically used in our previous articles [3, 4] – we present them here for comparison.

As we described in the previous subsection, after the first series of experiments with the Random approach we obtained first 1000 triples of α, β, ρ values. An interesting finding of our experiments is that, although it may seem intuitively apparent that the average values of the found parameter values would serve as good candidates for parameter values, this hypothesis was not confirmed. Indeed, the second triple in Fig. 6 ($\alpha = 0.31$, $\beta = 0.54$, $\rho = 0.65$) is calculated as the average values of the parameter intervals obtained after the first iteration of the Random approach. We observe in the figure that these values seriously worsened the run time of the algorithm by ≈ 1.4 times.

The third triple $\alpha = 0.1$, $\beta = 0.3$, $\rho = 0.9$ in Fig. 6 is set based on the analysis of the data frequency shown in Fig. 4(b) after the first iteration of the Random approach (we take the values that provide the highest frequency). This reduces the run time by ≈ 8.5 times. The best, fourth triple of values $\alpha = 0.2$, $\beta = 0.5$, $\rho = 0.9$ in Fig. 6 is obtained after seven iterations of the Random approach. This triple reduces the run time of the algorithm by ≈ 35 times. The tuning time spent by the GPU is the same 23 h as in the Random approach.

Multi-constant Approach. The Multi-constant approach differs from the previously discussed Constant approach in that each GPU thread uses its own triple of constants α, β, ρ that are obtained as a result of the first series of the Random approach. The approach yields a total of 1 000 triples of ACO-parameters for problems of various sizes from 11^{16} up to 20^{16}, for which feasible solutions are obtained. When running the program on the GPU (for our case on the Tesla v100 we use 5 120 threads, which corresponds to the number of CUDA cores for this GPU-model), the initialization of the algorithm parameters is performed cyclically. If we have the set of 100 triples of algorithm parameters, then on the GPU, after every 100 GPU threads, the values of algorithm parameters will be repeated, for set of 200 triples repetition values will be every 200 GPU threads, etc. For set of 1 000 triples of parameters, the values of the algorithm parameters will be repeated on every 1 000 GPU threads.

Figure 7 shows the results of using the Multi-const approach. The worst run time (especially for the maximum problem complexity 20^{16}) corresponds to the set of 100 triples, the best run time corresponds to the set of 1 000 triples. The average run time of the algorithm starting from the set of 200 triples differs from the runtime of the algorithm for the maximum set of 1 000 triples by only a factor of ≈ 1.16 (2.75 s vs. 2.37 s). The search time for the set of 1 000 triples is equal to the time of the first iteration of random approach with $\alpha \in (0, 2]$, $\beta \in (0, 2]$ and $\rho \in (0, 1]$ is 50 575 s ≈ 14 h, and the search time for the set of 200 triples with the same random approach for a problem complexity $11^{16} + 12^{16}$ is 3 282 sec ≈ 54 min, which is 15.4 times faster.

Comparison of Tuning Approaches. Figure 8 compares the best run time results obtained by each of our three tuning approaches. For the Multi-constant approach, we compare also to the variant with 200 triples that provides still acceptable results achieved in a significantly shorter tuning time.

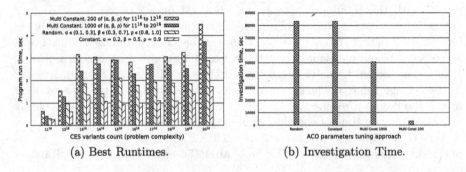

(a) Best Runtimes.   (b) Investigation Time.

Fig. 8. Comparison of approaches.

We observe in Fig. 8(a) that the fastest run time is achieved by using the Constant approach with constant parameter values $\alpha = 0.2$, $\beta = 0.5$, $\rho = 0.9$ (average execution time $\approx 1.1\,\text{s}$). Figure 8(b) shows that the tuning time for the most economical Multi-const approach with 200 samples ($\approx 54\,\text{min}$) is 25 times shorter, while the resulting run time of the optimization process is only ≈ 2.5 times slower.

5   Conclusion

Our contribution is a new set of three approaches to parameter tuning of the GPU-parallelized Ant Colonies Optimization (ACO) metaheuristic. The advantage of our approaches is that they work for different metaheuristics and different optimization problems. As a particular demonstration, this paper describes the use case when ACO is used for solving the Constraint Satisfaction Problem (CSP) in the process of optimizing the multiproduct batch plants design. Our three tuning approaches – Random, Const and Multi-Const – proceed by using a statistical analysis of solution frequences in particular intervals of parameter values. By stepwise narrowing these intervals, we arrive at the intervals or even single parameter values that provide good solutions in short time.

Using modern high-performance CPU-GPU systems, it is possible to conduct a large number of computational experiments (e.g., overnight), and to use their results for a statistical frequency analysis. We demonstrate that the user can choose between longer experiments with a very good quality of solutions and shorter experiments that still provide a acceptable level of quality. This shows that it is possible to use the parameters values obtained for problems of a small complexity for solving problems of a large complexity. It should be noted that despite the relatively short time of the algorithm (minutes) without tuning on high-performance equipment (Tesla v100), the values of the ACO parameters obtained as a result of our approach can be applied for a different equipment (e.g., Tesla k20s), as well as when implementing a sequential version of the algorithm on the CPU, since ACO parameter values are device-independent. ACO is a stochastic algorithm. On the one hand, the α, β, ρ parameters of the algorithm are not associated with a specific implementation, so they are architecture-independent. On the other hand, with an increase in the number of threads, the probability of finding a solution, and, consequently, the speed of the algorithm, increases. Therefore, improving the implementation for a particular target architecture allows to additionally increase the speed of finding the solution.

While the Const Approach achieves eventually the best performance, it requires the most investigation time due to multiple repetitions of Random Approach with narrowing of the parameter value intervals. MultiConst approach can significantly reduce the investigation time, but its results are applicable only for the parallel implementation of the algorithm. Our approach can be used for tuning other metaheuristic algorithms and for other applied problems based on constraint satisfiability.

Acknowledgements. We are grateful to the anonymous reviewers for their very help-ful comments, and to the Nvidia Corp. for the donated hardware used in our experi-ments. This work was supported by the DAAD (German Academic Exchange Service) and by the Ministry of Education and Science of the Russian Federation under the "Mikhail Lomonosov II"-Programme, and by the HPC2SE project of BMBF (Federal Ministry of Education and Research, Germany).

References

1. Barbosa, E., Senne, E.: Improving the fine-tuning of metaheuristics: an approach combining design of experiments and racing algorithms. J. Optim. **2017**, 1–7 (2017). https://doi.org/10.1155/2017/8042436

2. Birattari, M.: Tuning Metaheuristics. Studies in Computational Intelligence, vol. 197. Springer, Heidelberg (2009). https://doi.org/10.1007/978-3-642-00483-4

3. Borisenko, A., Gorlatch, S.: Parallelizing metaheuristics for optimal design of mul-tiproduct batch plants on GPU. In: Malyshkin, V. (ed.) PaCT 2017. LNCS, vol. 10421, pp. 405–417. Springer, Cham (2017). https://doi.org/10.1007/978-3-319-62932-2_39

4. Borisenko, A., Gorlatch, S.: Comparing GPU-parallelized metaheuristics to branch-and-bound for batch plants optimization. J. Supercomput. 1–13 (2018). https://doi.org/10.1007/s11227-018-2472-9

5. Borisenko, A., Haidl, M., Gorlatch, S.: A GPU parallelization ofbranch-and-bound for multiproduct batch plants optimization. J. Supercomput. **73**(2), 639–651 (2017). https://doi.org/10.1007/s11227-016-1784-x

6. Burtscher, M., Nasre, R., Pingali, K.: A quantitative study of irregular programs on GPUs. In: 2012 IEEE International Symposium on Workload Characterization (IISWC), pp. 141–151. IEEE, November 2012. https://doi.org/10.1109/IISWC.2012.6402918. http://ieeexplore.ieee.org/document/6402918/

7. Castillo, O., Neyoy, H., Soria, J., Melin, P., Valdez, F.: A new approach for dynamic fuzzy logic parameter tuning in ant colony optimization and its application in fuzzy control of a mobile robot. Appl. Soft Comput. **28**, 150–159 (2015). https://doi.org/10.1016/j.asoc.2014.12.002

8. Chen, C.C., Liu, Y.T.: Enhanced ant colony optimization with dynamic mutation and ad hoc initialization for improving the design of TSK-type fuzzy system. Com-put. Intell. Neurosci. **2018**, 1–15 (2018). https://doi.org/10.1155/2018/9485478

9. Delévacq, A., Delisle, P., Gravel, M., Krajecki, M.: Parallel ant colony optimization on graphics processing units. J. Parallel Distrib. Comput. **73**(1), 52–61 (2013). https://doi.org/10.1016/j.jpdc.2012.01.003

10. Dorigo, M., Birattari, M.: Ant colony optimization. In: Encyclopedia of Machine Learning, pp. 36–39. Springer, Heidelberg (2011). https://doi.org/10.1007/978-1-4899-7687-1_22

11. Dorigo, M., Stützle, T.: Ant colony optimization: overview and recent advances. In: Gendreau, M., Potvin, J.Y. (eds.) Handbook of Metaheuristics, vol. 272, pp. 311–351. Springer, Cham (2018). https://doi.org/10.1007/978-3-319-91086-4_10

12. Fallahi, M., Amiri, S., Yaghini, M.: A parameter tuning methodology for meta-heuristics based on design of experiments. Int. J. Eng. Technol. Sci. **2**(6), 497–521 (2014)

13. Gómez-Cabrero, D., Ranasinghe, D.N.: Fine-tuning the ant colony system algo-rithm through particle swarm optimization. arXiv preprint arXiv:1803.08353 (2018)

14. Han, T.D., Abdelrahman, T.S.: Reducing branch divergence in GPU programs. In: Proceedings of the Fourth Workshop on General Purpose Processing on Graphics Processing Units - GPGPU-4, pp. 1–3. ACM Press, New York, March 2011. https://doi.org/10.1145/1964179.1964184

15. Khan, S., Bilal, M., Sharif, M., Sajid, M., Baig, R.: Solution of n-Queen problem using ACO. In: 2009 IEEE 13th International Multitopic Conference, pp. 1–5. IEEE, December 2009. https://doi.org/10.1109/INMIC.2009.5383157

16. Li, P., Zhu, H.: Parameter selection for ant colony algorithm based on bacterial foraging algorithm. Math. Probl. Eng. 1–12 (2016). https://doi.org/10.1155/2016/6469721. https://www.hindawi.com/journals/mpe/2016/6469721/

17. Mahi, M., Baykan, Ö.K., Kodaz, H.: A new hybrid method based on particle swarm optimization, ant colony optimization and 3-opt algorithms for traveling salesman problem. Appl. Soft Comput. **30**, 484–490 (2015). https://doi.org/10.1016/j.asoc.2015.01.068

18. Maier, H.R., et al.: Ant colony optimization for design of water distribution systems. J. Water Resour. Plann. Manag. **129**(3), 200–209 (2003)

19. NVIDIA Corporation: CUDA C programming guide 10.0, October 2018. http://docs.nvidia.com/cuda/pdf/CUDA_C_Programming_Guide.pdf

20. NVIDIA Corporation: The NVIDIA CUDA random number generation library (cuRAND), December 2018. https://developer.nvidia.com/curand

21. Olivas, F., Valdez, F., Castillo, O.: Dynamic parameter adaptation in ant colony optimization using a fuzzy system for TSP problems. In: IFSA-EUSFLAT, pp. 765–770 (2015)

22. Simpson, A., Maier, H., Foong, W., Phang, K., Seah, H., Tan, C.: Selection of parameters for ant colony optimization applied to the optimal design of water distribution systems. In: Proceedings of the International Congress on Modeling and Simulation, Canberra, Australia, pp. 1931–1936 (2001)

23. Skakov, E.S., Malysh, V.N.: Parameter meta-optimization of metaheuristics of solving specific NP-hard facility location problem. J. Phys.: Conf. Ser. **973**, 012063 (2018). https://doi.org/10.1088/1742-6596/973/1/012063

24. Stützle, T., et al.: Parameter adaptation in ant colony optimization. In: Hamadi, Y., Monfroy, E., Saubion, F. (eds.) Autonomous Search, pp. 191–215. Springer, Heidelberg (2011). https://doi.org/10.1007/978-3-642-21434-9_8

25. Trindade, Á.R., Campelo, F.: Tuning metaheuristics by sequential optimization of regression models. arXiv preprint arXiv:1809.03646, pp. 1–22, September 2018

26. Tsang, E.: Foundations of Constraint Satisfaction: The Classic Text. BoD-Books on Demand, Norderstedt (2014)

27. Valadi, J., Siarry, P.: Applications of Metaheuristics in Process Engineering. Springer, Cham (2014). https://doi.org/10.1007/978-3-319-06508-3

28. Veluscek, M., Kalganova, T., Broomhead, P.: Improving ant colony optimization performance through prediction of best termination condition. In: 2015 IEEE International Conference on Industrial Technology (ICIT), pp. 2394–2402. IEEE, March 2015. https://doi.org/10.1109/icit.2015.7125451

29. Zhang, Z., Feng, Z., Ren, Z.: Approximate termination condition analysis for ant colony optimization algorithm. In: 2010 8th World Congress on Intelligent Control and Automation, pp. 3211–3215. IEEE, July 2010. https://doi.org/10.1109/wcica.2010.5554984

Parallel Dimensionality Reduction
for Multiextremal Optimization Problems

Victor Gergel[ID], Vladimir Grishagin[✉][ID], and Ruslan Israfilov[ID]

Lobachevsky State University, Gagarin Avenue 23, 603950 Nizhni Novgorod, Russia
gergel@unn.ru, vagris@unn.ru, ruslan@israfilov.com

Abstract. The paper is devoted to consideration of numerical global optimization methods in the framework of the approach of reducing dimensionality based on nested optimization schemes. For the adaptive nested scheme being more efficient in comparison with its classical prototype a new algorithm of parallel implementation is proposed. General descriptions of the parallel techniques both for synchronous and asynchronous versions are given. Results of numerical experiments on a set of complicated multiextremal test problems of high dimension are presented. These results demonstrate essential acceleration of asynchronous parallel algorithm in comparison with the sequential version.

Keywords: Multiextremal optimization · Global optimum ·
Dimensionality reduction · Parallel algorithms

1 Introduction

Global optimization problems aimed at finding the global optimum of multiextremal functions are complicated decision making models and describe many important applications in engineering, economy, scientific researches, etc. (see some examples in [3,9,13,26,30,32,36,43]). The complexity of these problems depends crucially on the dimension (number of model parameters) because in general case the growth of the computational costs measured, for example, in number of objective function evaluations is exponential when increasing the dimension. There exist several approaches to analyzing global optimization problems oriented at different classes of multiextremal functions defined by their specific properties. The wide spectrum of directions in the field of global optimization can be found in the fundamental monographs [23,31,32,35,39,44].

Among the approaches generating efficient algorithms to solving multiextremal optimization problems with objective functions satisfying the Lipschitz condition one can mention the approach based on different partition schemes (component approach) and the class of methods which apply the ideas of reducing multidimensional problems to one or a family of univariate subproblems for solving those by means of well-developed one-dimensional optimization algorithms.

© Springer Nature Switzerland AG 2019
V. Malyshkin (Ed.): PaCT 2019, LNCS 11657, pp. 166–178, 2019.
https://doi.org/10.1007/978-3-030-25636-4_13

In the framework of the component approach the search region is partitioned into several subregions (components), every component are evaluated numerically for the purpose of its efficiency for search continuation, and after that a new iteration is carried out in the most "perspective' subregion. The first class of component methods called characteristical ones was proposed and theoretically investigated in the work [18], and later it was generalized to multidimensional case by many researchers (see, for example, publications [24,25,27,31,32,35]).

As for the approach transforming a multidimensional problem to the univariate case, it includes two different schemes. The first one is based on applying the Peano space-filling curves which are continuous mappings of a multidimensional hypercube onto the unit interval of the real axis [4,14,22,28,29,34,39]. The second scheme reduces a multidimensional problem to a family of univariate subproblems connected recursively (nested optimization) [5,6,11,12,16,37–39]. These schemes can be combined when inside the recursive procedure the subproblems of the less dimensionality are considered and solved by means of Peano mappings [40]. As it has been shown in [19], among algorithms of this type the adaptive scheme of nested optimization has demonstrated the best efficiency.

A promising way to overcome the complexity of the multiextremal optimization problems consists in parallelizing sequential schemes of optimization algorithms. Following this idea, some optimization methods have been proposed (see [2,8,15,17,20,33,39,40]). In this paradigm the usual way consists in performing parallel trials (computations of objective function values) [8,17,20,21,39,40]. The algorithm [2] using multiple Peano mappings performs parallel computations of trial couples corresponding to several Peano evolvents. Very interesting approach is used in parallel branch and bound algorithms which build a hierarchical structure of feasible domain partitions and parallelize the procedure of partitioning. For example, the paper [21] describes a model using threads within one computational node and the publication [1] suggests a parallel strategy of partitioning in distributed memory.

As opposed to above approaches the methods on the base of nested optimization scheme [15,33] implement parallelization by means of parallel performance of internal subtasks. In this paper we consider a parallel algorithm being a generalization of the adaptive scheme of global optimization [11] which belongs to the type of recursive reduction techniques and applies for solving the nested univariate subproblems the information characteristical method [33,39]. The main goal of the work is to describe a new model of parallel computations inside the adaptive scheme realizing "parallelization by subtask" approach and to estimate the effectiveness of parallelizing measured as speedup of the parallel adaptive scheme compared to the sequential one.

The rest of the paper is organized as follows. Section 2 contains the statement of multiextremal optimization problem to be studied and the general algorithm of the nested optimization scheme. Section 3 describes the model of parallelism organization in the framework of the nested adaptive dimensionality reduction. Section 4 presents results of numerical experiments and speedup estimations of the parallel adaptive scheme. The last section concludes the paper.

2   Nested Optimization Scheme

The statement of the optimization problem to be considered is as follows. It is necessary to find in a hyperparallelepiped H of the N-dimensional Euclidean space $\mathbb{R}^N$ the least value (global minimum) F_* of an objective function $F(u)$ and the coordinate $u_* \in H$ of the global minimum (global minimizer). This problem can be written in a symbolical form as

$$F(u) \rightarrow \min, \ u = (u_1, \ldots, u_N) \in H \subseteq \mathbb{R}^N, \tag{1}$$
$$H = \{u \in \mathbb{R}^N : a_i \leq u_i \leq b_i, 1 \leq i \leq N\}, \tag{2}$$

The objective function $F(u)$ is supposed to satisfy in the search domain H the Lipschitz condition

$$|F(u') - F(u'')| \leq L\|u' - u''\|, \ u', u'' \in H, \tag{3}$$

where $L > 0$ is a finite value called Lipschitz constant and $\| \cdot \|$ denotes the Euclidean norm in $\mathbb{R}^N$. Under condition (3) the problem (1)–(2) is, in general case, multiextremal and non-smooth.

The nested scheme of dimensionality reduction served as the source for different global optimization methods [5,6,11,12,16,37–39]. It is based on the known relation [5,39]

$$\min_{u \in H} F(y) = \min_{u_1 \in H_1} \min_{u_2 \in H_2} \cdots \min_{u_N \in H_N} F(u_1, \ldots, u_N), \tag{4}$$

where H_i is a line segment $[a_i, b_i]$, $1 \leq i \leq N$.

Let us give the general description of the nested scheme introducing recursively a family of reduced function $F^i(\tau_i)$, $\tau_i = (u_1, \ldots, u_i)$, $1 \leq i \leq N$, in the following manner.

$$F^N(\tau_N) \equiv F^N(u) \equiv F(u), \tag{5}$$
$$F^{i-1}(\tau_{i-1}) = \min_{u_i \in H_i} F^i(\tau_i), \ 2 \leq i \leq N. \tag{6}$$

Then, instead of minimizing in (1) the N-dimensional function $F(u)$ we can search for the global minimum of the univariate function $F^1(u_1)$ as, in accordance with (4),

$$F_* = \min_{u_1 \in H_1} F^1(u_1). \tag{7}$$

However, any numerical optimization method in the course of solving the problem (7) has to calculate values of the function $F^1(t_1)$. But such a computation at a point $\tilde{t}_1$ requires solving the problem

$$F^2(\tilde{t}_1, t_2) \rightarrow \min, \ t_2 \in H_2, \tag{8}$$

which are one-dimensional again as the argument $\tilde{t}_1$ is fixed, and so on. Following this way, we reach the level N, where the problem

$$F^N(\tilde{\tau}_{N-1}, t_N) \rightarrow \min, \ t_N \in H_N, \tag{9}$$

is one-dimensional as well because the vector $\tilde{\tau}_{N-1} = (\tilde{t}_1, \ldots, \tilde{t}_{N-1})$ is fixed (its coordinates are given at previous levels of recursion). As $F^N(t) \equiv F(t)$ then evaluation of objective function values in the problem (9) consists in calculation of the values $F(\tilde{\tau}_{N-1}, t_N)$ of the given function from (1).

The procedure (7)–(9) described above is recursive and enables to find the solution of the multidimensional problem (1)–(2) via solving the family

$$F^i(\tau_{i-1}, u_i) \to \min, \ u_i \in H_i, 1 \leq i \leq N, \tag{10}$$

of univariate subproblems. Such the scheme is called *the nested scheme of dimensionality reduction*.

The recursive structure of generation of the subproblems in the family (10) can be presented as a tree of connections between generating (parental) and generated (child) subtasks (see Fig. 1).

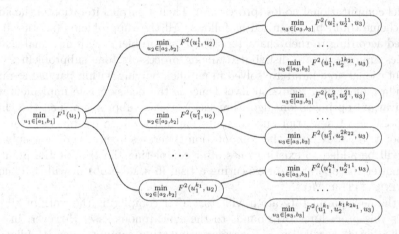

Fig. 1. Tree of subtasks in the nested optimization scheme for dimension 3.

In this tree the problem (7) is the root one and the problems (9) are leaves of the tree. Of course, the tree is built in dynamics, and Fig. 1 shows the full tree obtained after completing multidimensional optimization. It should be noted that conducting one trial (computation of objective function value at a point) in one-dimensional subproblem of minimization of $F^i(\tau_{i-1}, u_i)$, $1 \leq i \leq N-1$, generates a subtree in the tree of subtasks. As a consequence, any subproblem (10) is parental for subproblems in subtrees generated by its trials.

In classical implementation of the nested scheme the subproblems (10) are solved until a stopping rule of applied univariate method holds true for all of them. It means that in the course of optimization only subproblems which belong to a sole path from the root to a leaf can interact inter se. It leads to loss of search information obtained in the course of optimization and worsens the efficiency of classical scheme significantly.

In order to overcome this drawback of the classical nested scheme, its improved version called *adaptive nested scheme of dimensionality reduction* has been proposed in the paper [11]. As opposed to the classical nested scheme, the adaptive extension considers all the currently existing subproblems (10) simultaneously. A numerical value of "significance" called characteristic is assigned to each subproblem of the current family and all the subproblems are decreasingly ordered according to their characteristics. Then, the subproblem with the maximal characteristic is chosen, and in the best subproblem a new iteration of the univariate method connected with this subproblem is executed. The detailed algorithm of the sequential adaptive scheme has been described in [11].

3   Parallel Adaptive Scheme

A natural way of parallelizing the adaptive scheme consists in solving several subproblems in parallel. Let us suppose that at our disposal there are $P > 1$ parallel computational nodes (processors). Then a parallel iteration of the adaptive scheme could be organized as follows. All the subproblems (subtasks) are ordered according to their characteristics, P subproblems with maximal characteristics are chosen, are distributed among processors (one subproblem to one computational node) and are solved in parallel. Solving within parallel iteration one subproblem of a recursion level l means the decision rule implementation of univariate optimization algorithm used in this subproblem, i.e., the choice of a point u_l of new trial and computation of the objective function value at this point. If $l < N$, such the computation generates a subtree of new subtasks that will be added to existing ones after completing the trial at the point u_l. Hereinafter the operation of executing a trial in a subproblem will be denoted as EXECUTEITERATION.

If the next parallel iteration will start after completing the work of all processors this procedure corresponds to the synchronous case. However, in such organization of parallelism a processor completing computations is obliged to wait until the other processors finish and will stand idle. To avoid this drawback one can to use more effective, but more complicated asynchronous organization of parallelism when a processor completing its work take the best subtask from the pool of non- distributed subproblems.

Further we consider more detailed how both synchronous and asynchronous parallelisms can be organized for the nested adaptive scheme. As a detailed code of the parallel implementation is very large we will give a general algorithmic description of parallel adaptive scheme on the base of an abstract one-dimensional optimization method. For this formal description it is necessary to introduce several notions and designations.

Let at a stage of the adaptive scheme implementation all subproblems renumber with integer numbers from 1 to λ, where λ is the number of subtasks (10) generated already and the root subproblem (7) is the first one. A univariate method in the course of minimizing a subproblem $F^l(\tau_{l-1}, u_l)$ generates a sequence of trial points $u_l^1, \ldots, u_l^k$ at which the values $z_l^1, \ldots, z_l^k$ are computed where $z_l^i = F^l(\tau_{l-1}, u_l^i), 1 \leq i \leq k$. These points and values form the set of pairs

$$\omega_k = \{(u_l^1, z_l^1), \ldots, (u_l^k, z_l^k)\}, \tag{11}$$

that can be interpreted as the current state of the search for this subproblem. It should be remembered that any computation of value z_l^i requires solving a one-dimensional subproblem at the next $(l+1)$-th level and, if $l < N$, building a subtree of subproblems (10). Uniting the subtrees of all trials we get the subtree generated by the subproblem on the whole.

Taking these circumstances into account, we identify a subtask $t \in \{1, \ldots, \lambda\}$ as a tuple

$$t = \langle l, \tau_{l-1}, k, \omega_k, h, t^p, T^c, W \rangle. \tag{12}$$

Here l is the number of the recursion level, which the subproblem belongs to, τ_{l-1} is a vector of fixed coordinates obtained from preceding levels, k is the number of trials executed by the univariate algorithm, ω_k from (11). The indicator $h = h(\omega_k)$ shows whether solving this subproblem has been completed, namely, if $h = 0$ then the algorithm solving the described subproblem terminates its execution, if $h = 1$ the optimization has to be continued. The number t^p corresponds to the parental subproblem having generated the current one, and, finally, T^c presents the set of all subtasks (up to the level N) generated by the subproblem considered and, finally, W is a numerical characteristic of the subproblem significance. The set of all subtasks t, $t \in \{1, \ldots, \lambda\}$, we will denote as T.

It should be noted that tuple (12) is not applicable to the root subtask (7) because it has no parents. In order to include the root subproblem into the unified description let us introduce as a parent of the root an "empty" subtask t^0 and define $t_0 = \varnothing$.

For starting the adaptive scheme (both sequential and parallel) it is necessary to create an initial set T. It could be done applying the classical nested scheme with a few trials in one dimensional search. We will consider just a general procedure INITIALIZE implementing this initial stage without its concretization. It is executed only once and it is not important whether it is sequential or parallel.

As for parallel implementation of the main body of the adaptive scheme we will deal with a computational system with distributed memory. The system is supposed to consist of P computational nodes. Each node has just one processor and memory, to which the processor of the node only has the direct access. The remote direct access to this memory (RDMA) is considered to be impossible. It means that recording the data of j-th node in the memory of i-th node $(i \neq j)$ can be carried out by means of operations of data transmission only.

The simplest way of parallelizing the adaptive scheme in a distributed system can consist in employment of the program model MapReduce [7]. A generalized algorithm of the parallel adaptive scheme could be presented as the Algorithm 1.

Algorithm 1. Parallel adaptive scheme on the base of MapReduce

1: $l \leftarrow 1$, $t^1 \leftarrow 1$
2: $T \leftarrow$ INITIALIZE()
3: **while** $h(t^1) = 1$ **do**
4:     $T' \leftarrow \{t_1, \ldots, t_P : W(t) \le W(t_i),\ 1 \le i \le P,\ t \in T \setminus \{t_1, \ldots, t_P\}\}$
5:     $T'' \leftarrow$ MAPREDUCE(T', EXECUTEITERATION)
6:     $T \leftarrow T \cup T''$
7:     $T \leftarrow T \setminus \{t \in T' : h(t) = 0\}$
8: **end while**

In the Algorithm 1, after initialization in the loop until the termination condition in the root subproblem is satisfied parallel iterations of the adaptive scheme are executed. At Stage 4 the set T' containing P subtasks with the best characteristics is formed. Stage 5 distributes the subproblems from T' to processors which in parallel execute one trial in their subtasks with the help of procedure EXECUTEITERATION. After completing all the trials a set T'' of new subproblems obtained in the course of computations is formed. Stage 6 complements the set T with new subproblems and Stage 6 removes from the set T the terminated subproblems.

Practical implementation of the described algorithm can be realized in the framework of such the platforms as Hadoop [41] or Spark [42]. Unfortunately, this algorithm is synchronous and requires significant number of data transmissions. Moreover, implementation of Algorithm 1 implies that one processor (master node) plays the main role and coordinates the work of the other (slave) processors, i.e., the organization of the parallel processes is centralized.

To improve the parallel implementation of the adaptive scheme we propose for the adaptive scheme an asynchronous decentralized model of parallel computations where all processors are equal in rights.

Let, as earlier, a distributed system have P processors. We change Algorithm 1 so that procedure INITIALIZE after creating an initial set T splits this set into P parts and send each part to separate processor. Moreover, i-th processor is supposed to be able to connect independently with any other node and to execute the information interchange with it after completing a trial in its subproblems. Under this assumption the full set T of subtasks can be stored portionwise on different nodes and in order to get the best subproblems, a node can request from other nodes only the best subproblems from their local subsets.

In this situation there exists no integrated iteration implemented by all the processors jointly and we can deal with iterations executed by processors separately. Completing its iteration a node can request immediately the best subproblems from other nodes and begin a new iteration. Two examples of requests are shown in Fig. 2.

Under these assumptions we propose an asynchronous algorithm of the parallel adaptive scheme that is presented below in a pseudo code. This algorithm is supposed to be executed on every node.

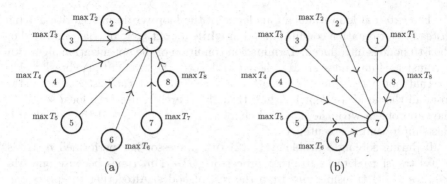

Fig. 2. Gathering information for new iteration to the 1-st node (a) and to the 7-th node (b).

Algorithm 2. Parallel asynchronous adaptive scheme

1: **procedure** RECEIVETASKFROMNODE(j)
2:     $t_j^* \leftarrow \max T_j^{\text{local}}$
3:     $T_j^{\text{local}} \leftarrow T_j^{\text{local}} \setminus \{t_j^*\}$
4:     **return** t_j^*
5: **end procedure**
6:
7: **procedure** RUNONNODE(i)
8:     $T_i^{\text{local}} \leftarrow$ PARTOFINITIALTASKSET()
9:     **for** $n \in \{1, 2, \ldots, n_{\max}^i\}$ **do**
10:         **for** $j \in \{1, 2, \ldots, P\} \setminus \{i\}$ **do**
11:             $t_j^* \leftarrow$ RECEIVETASKFROMNODE(j)
12:             $T_i^{\text{local}} \leftarrow T_i^{\text{local}} \cup \{t_j^*\}$
13:         **end for**
14:         $t^* \leftarrow \max T_i^{\text{local}}$
15:         $t^1, T^1 \leftarrow$ EXECUTEITERATION(t^*)
16:         $T_i^{\text{local}} \leftarrow T_i^{\text{local}} \cup T^1$
17:         **if** $h(t^*) = 0$ **then**
18:             **if** $l(t^*) = 1$ **then**
19:                 BROADCASTSTOPSIGNAL()
20:                 **break**
21:             **end if**
22:             $T_i^{\text{local}} \leftarrow T_i^{\text{local}} \setminus \{t^*\}$
23:         **end if**
24:     **end for**
25: **end procedure**

Let us give some remarks about the Algorithm 2. T_i^{local} is the subset of subproblems stored on the i-th node. Procedure RECEIVETASKFROMNODE provides receiving the best subtask from other node. In line 8 the procedure PARTOFINITIALTASKSET forms the initial set T_i^{local} from subtasks obtained by the procedure INITIALIZE.

The external loop **for** is an analogue of the loop **while** in Algorithm 1 but instead of termination condition used in **while** a limit $n_i^{\max}$ of trials executed on the i-th node is introduced. Termination condition of the optimization algorithm is transferred into lines 17–21, where $l(t)$ denotes the level of the subtask t. The internal loop carries out collecting the subproblems with the best characteristics from all the nodes (except for the i-th node). Further, from the local set T_i^{local} the subproblem with the maximal characteristic is chosen and the new trial in this subproblem is executed.

If during solving a problem (1)–(2) i-th processor has performed n_i trials, it has transferred tasks to other processors $(P - 1)n_i$ times because one trial requires $P - 1$ transmissions from the rest of nodes. Altogether the processors have performed $n = \sum_{i=1}^{P} n_i$ trials and, consequently, executed $(P - 1)n$ transmissions. The estimation of transmissions number for synchronous Algorithm 1 gives the result about $(P^2 + P)n$, which is worse essentially compared with the asynchronous case.

4 Numerical Experiments

To evaluate the effectiveness of the parallelism described in Sect. 3 a computational experiment aimed at comparison of sequential and asynchronous parallel implementations of the adaptive nested optimization schemes has been carried out. The simpler synchronous version did not participated in comparison because it is inferior to the asynchronous one according to theoretical estimations. The experiment consisted in solving a set of functions from the test class GKLS [10] of essentially multiextremal functions (hard subclass). These functions have complicated structure with tens of local minima. Nowadays this class is a classical tool for comparison of global optimization methods.

In experiment 50 functions of dimension 8 have been taken and solved both sequential and parallel adaptive schemes. As one-dimensional method for solving the subproblems 10 in both the schemes the information global search algorithm GSA [38,39] was taken with reliability parameter $r = 6.5$ and accuracy in termination condition $\varepsilon = 0.01$. Computations were executed on the cluster consisting of 64 nodes, where each node is equipped with Intel® Xeon® Gold 6148 processor having 20 physical cores. Mellanox® Infiniband FDR was used as interconnection technology. The parallelism was provided on the base of MPI, version Intel® MPI 2019. Only one MPI rank was assigned to one node.

The global minima have been found with given accuracy in all the test problems. The results of the experiment are reflected in Table 1 and Fig. 3. The table contains average time spent by the parallel scheme per one test problem and speedup in time achieved by the parallel technique in comparison with the sequential one for different number of MPI ranks.

Table 1. Speedup in time on GKLS test class

MPI rank	1	2	4	8	16	32	64
Time (sec.)	7409.39	4919.94	2422.97	1202.00	717.19	380.39	198.88
Speedup	1	1.50	3.05	6.16	10.33	19.47	37.26

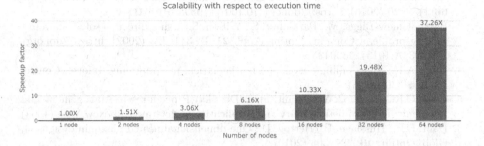

Fig. 3. Speedup in time on GKLS test class

5 Conclusion

The paper proposes general descriptions of new parallel algorithms implementing methods of multiextremal optimization on the base of adaptive nested schemes reducing a multidimensional problem to a family of one-dimensional subproblems. Two parallel versions are presented in synchronous and asynchronous variants for computational distributed systems. Efficiency of the parallelism are investigated experimentally on the test class GKLS of complicated multiextremal multidimensional problems. The results of the experiment have shown essential speedup of the optimization process in case of applying the asynchronous adaptive scheme.

Combining the general parallel procedure of the adaptive scheme with fast univariate optimization methods (like characteristical ones) enables to construct new efficient techniques for solving multiextremal problems of high dimensions. Moreover, it is promising to develop new parallel implementations of the adaptive scheme oriented at other parallel architectures, for example, at supercomputers with mixed types of memory. It would be very interesting as well to compare the proposed algorithm with parallel optimization methods based on the other principles of parallelizing. These problems can be fruitful directions of further researches.

Acknowledgements. The research has been supported by the Russian Science Foundation, project No 16-11-10150 "Novel efficient methods and software tools for time-consuming decision make problems using superior-performance supercomputers".

References

1. Androulakis, I.P., Floudas, C.A.: Distributed branch and bound algorithms for global optimization. In: Pardalos, P.M. (ed.) Parallel Processing of Discrete Problems. The IMA Volumes in Mathematics and its Applications, vol. 106, pp. 1–35. Springer, New York (1999). https://doi.org/10.1007/978-1-4612-1492-2_1

2. Barkalov, K., Gergel, V.: Parallel global optimization on GPU. J. Glob. Optim. **66**(1), 3–20 (2016). https://doi.org/10.1007/s10898-016-0411-y

3. Bartholomew-Biggs, M., Parkhurst, S., Wilson, S.: Using direct to solve anaircraft routing problem. Comput. Optim. Appl. **21**(3), 311–323 (2002). https://doi.org/10.1023/A:1013729320435

4. Butz, A.R.: Space-filling curves and mathematical programming. Inform. Control **12**, 314–330 (1968)

5. Carr, C.R., Howe, C.W.: Quantitative Decision Procedures in Management and Economic: Deterministic Theory and Applications. McGraw-Hill, New York (1964)

6. Dam, E.R., Husslage, B., Hertog, D.: One-dimensional nested maximin designs. J. Glob. Optim. **46**, 287–306 (2010)

7. Dean, J., Ghemawat, S.: MapReduce: simplified data processing on large clusters. In: Sixth Symposium on Operating System Design and Implementation, OSDI 2004, San Francisco, CA, pp. 137–150 (2004)

8. Evtushenko, Y.G., Malkova, V.U., Stanevichyus, A.A.: Parallel globaloptimization of functions of several variables. Comput. Math. Math. Phys. **49**(2), 246–260 (2009). https://doi.org/10.1134/S0965542509020055

9. Famularo, D., Pugliese, P., Sergeyev, Y.: A global optimization technique for checking parametric robustness. Automatica **35**, 1605–1611 (1999)

10. Gaviano, M., Kvasov, D.E., Lera, D., Sergeyev, Y.D.: Software for generation ofclasses of test functions with known local and global minima for globaloptimization. ACM Trans. Math. Softw. **29**(4), 469–480 (2003)

11. Gergel, V.P., Grishagin, V.A., Gergel, A.V.: Adaptive nested optimization scheme for multidimensional global search. J. Glob. Optim. **66**, 35–51 (2016)

12. Gergel, V.P., Grishagin, V.A., Israfilov, R.A.: Local tuning in nested scheme of global optimization. Proc. Comput. Sci. **51**, 865–874 (2015)

13. Gergel, V.P., Kuzmin, M.I., Solovyov, N.A., Grishagin, V.A.: Recognition of surface defects of cold-rolling sheets based on method of localities. Int. Rev. Autom. Control **8**, 51–55 (2015)

14. Goertzel, B.: Global optimization with space-filling curves. Appl. Math. Lett. **12**, 133–135 (1999)

15. Grishagin, V.A., Israfilov, R.A.: Multidimensional constrained global optimization in domains with computable boundaries. In: CEUR Workshop Proceedings, vol. 1513, pp. 75–84 (2015)

16. Grishagin, V.A., Israfilov, R.A.: Global search acceleration in the nested optimization scheme. In: AIP Conference Proceedings, vol. 1738, p. 400010 (2016)

17. Grishagin, V.A., Sergeyev, Y.D., Strongin, R.G.: Parallel characteristical algorithms for solving problems of global optimization. J. Glob. Optim. **10**, 185–206 (1997)

18. Grishagin, V.: On convergence conditions for a class of global search algorithms. In: Proceedings of the 3-rd All-Union Seminar Numerical Methods of Nonlinear Programming, pp. 82–84 (1979, in Russian)

19. Grishagin, V., Israfilov, R., Sergeyev, Y.: Convergence conditions and numerical comparison of global optimization methods based on dimensionality reduction schemes. Appl. Math. Comput. **318**, 270–280 (2018). https://doi.org/10.1016/j.amc.2017.06.036. http://www.sciencedirect.com/science/article/pii/S0096300317304496. Recent Trends in Numerical Computations: Theory and Algorithms

20. He, J., Verstak, A., Watson, L.T., Sosonkina, M.: Design and implementation of a massively parallel version of DIRECT. Comput. Optim. Appl. **40**(2), 217–245 (2008). https://doi.org/10.1007/s10589-007-9092-2

21. Herrera, J.F.R., Salmerón, J.M.G., Hendrix, E.M.T., Asenjo, R., Casado, L.G.: On parallel branch and bound frameworks for global optimization. J. Glob. Optim. **69**(3), 547–560 (2017). https://doi.org/10.1007/s10898-017-0508-y

22. Hime, A., Oliveira Jr., H., Petraglia, A.: Global optimization using space-filling curves and measure-preserving transformations. Soft Comput. Industr. Appl. **96**, 121–130 (2011)

23. Horst, R., Pardalos, P.M.: Handbook of Global Optimization. Kluwer Academic Publishers, Dordrecht (1995)

24. Jones, D.R.: The DIRECT global optimization algorithm. In: Floudas, C., Pardalos, P.M. (eds.) Encyclopedia of Optimization, pp. 431–440. Kluwer Academic Publishers, Dordrecht (2001)

25. Jones, D.R., Perttunen, C.D., Stuckman, B.E.: Lipschitzian optimization without the Lipschitz constant. J. Optim. Theory Appl. **79**, 157–181 (1993)

26. Kvasov, D.E., Menniti, D., Pinnarelli, A., Sergeyev, Y.D., Sorrentino, N.: Tuning fuzzy power-system stabilizers in multi-machine systems by global optimization algorithms based on efficient domain partitions. Electr. Power Syst. Res. **78**, 1217–1229 (2008)

27. Kvasov, D.E., Pizzuti, C., Sergeyev, Y.D.: Local tuning and partition strategies for diagonal GO methods. Numer. Math. **94**, 93–106 (2003)

28. Lera, D., Sergeyev, Y.D.: Lipschitz and Hölder global optimization using space-filling curves. Appl. Numer. Math. **60**, 115–129 (2010)

29. Lera, D., Sergeyev, Y.D.: Deterministic global optimization using space-filling curves and multiple estimates of Lipschitz and holder constants. Commun. Nonlinear Sci. Numer. Simul. **23**, 328–342 (2015)

30. Modorskii, V.Y., Gaynutdinova, D.F., Gergel, V.P., Barkalov, K.A.: Optimization in design of scientific products for purposes of cavitation problems. In: AIP Conference Proceedings, vol. 1738, p. 400013 (2016)

31. Paulavičius, R., Žilinskas, J.: Simplicial Global Optimization. Springer, NewYork (2014). https://doi.org/10.1007/978-1-4614-9093-7

32. Pintér, J.D.: Global Optimization in Action. Kluwer Academic Publishers, Dordrecht (1996)

33. Sergeyev, Y.D., Grishagin, V.A.: Parallel asynchronous global search and the nested optimization scheme. J. Comput. Anal. Appl. **3**, 123–145 (2001)

34. Sergeyev, Y.D., Strongin, R.G., Lera, D.: Introduction to Global Optimization Exploiting Space-Filling Curves. Springer, New York (2013). https://doi.org/10.1007/978-1-4614-8042-6

35. Sergeyev, Y., Kvasov, D.: Deterministic Global Optimization: An Introduction to the Diagonal Approach. Springer, New York (2017). https://doi.org/10.1007/978-1-4939-7199-2

36. Shevtsov, I.Y., Markine, V.L., Esveld, C.: Optimal design of wheel profile for railway vehicles. In: Proceedings of the 6th International Conference on Contact Mechanics and Wear of Rail/Wheel Systems, Gothenburg, Sweden, pp. 231–236 (2003)
37. Shi, L., Ólafsson, S.: Nested partitions method for global optimization. Oper. Res. **48**, 390–407 (2000)
38. Strongin, R.G.: Numerical Methods in Multiextremal Problems (Information-Statistical Algorithms). Nauka, Moscow (1978, in Russian)
39. Strongin, R.G., Sergeyev, Y.D.: Global Optimization with Non-convex Constraints: Sequential and Parallel Algorithms. Kluwer Academic Publishers/Springer, Dordrecht/Heiselberg (2014)
40. Sysoyev, A., Barkalov, K., Sovrasov, V., Lebedev, I., Gergel, V.: Globalizer – a parallel software system for solving global optimization problems. In: Malyshkin, V. (ed.) PaCT 2017. LNCS, vol. 10421, pp. 492–499. Springer, Cham (2017). https://doi.org/10.1007/978-3-319-62932-2_47
41. White, T.: Hadoop: The Definitive Guide. O'Reilly Media, Inc., Newton (2009)
42. Zaharia, M., Chowdhury, M., Franklin, M.J., Shenker, S., Stoica, I.: Spark: cluster computing with working sets. In: Proceedings of the 2Nd USENIX Conference on Hot Topics in Cloud Computing, HotCloud 2010, p. 10. USENIX Association, Berkeley (2010). http://dl.acm.org/citation.cfm?id=1863103.1863113
43. Zhao, Zh., Meza, J.C., Van Hove, M.: Using pattern search methods for surface structure determination of nanomaterials. J. Phys.: Condens. Matter **18**(39), 8693–8706 (2006)
44. Zhigljavsky, A.A., Žilinskas, A.: Stochastic Global Optimization. Springer, NewYork (2008). https://doi.org/10.1007/978-0-387-74740-8

Multiple-Precision Scaled Vector Addition on Graphics Processing Unit

Konstantin Isupov[✉][iD] and Alexander Kuvaev[iD]

Vyatka State University, Kirov 610000, Russia
ks_isupov@vyatsu.ru, kyvaevy@gmail.com

Abstract. Many large problems need linear algebra operations with a precision exceeding the standard floating-point binary64 format. In this paper, we implement a multiple-precision scaled vector addition BLAS routine (WAXPBY) on graphics processing units. We use a residue number system (RNS) to represent significands of floating-point values. In RNS, large numbers replace with their residues and the operations of addition, subtraction and multiplication perform on these residues in parallel and without carry propagation. Our parallel WAXPBY algorithm is divided into a number of steps, and each step is carried out by a separate GPU kernel. Experiments show that the developed routine clearly outperforms parallel CPU-based multiple-precision implementations.

Keywords: High-precision computations · Computer arithmetic · Residue number system · BLAS · CUDA

1 Introduction

For given three floating-point vectors x, y and w of length N and scalars α and β, the scaled vector addition routine (WAXPBY) scales x by α and y by β, adds these two vectors to one another and stores the result in w, that is, $w \leftarrow \alpha x + \beta y$. This operation extends the original AXPY operation and is included in the updated set of Basic Linear Algebra Subprograms (BLAS) [3].

There are several linear algebra libraries that implement WAXPBY, e.g. ATLAS, Arm Allinea Studio, and the Perl Math::BLAS package. Moreover, Intel MKL, OpenBLAS, as well as several other optimized BLAS libraries implement the AXPBY routine ($y \leftarrow \alpha x + \beta y$). Most of these libraries support double precision floating-point format (binary64) with a significand part of 53 bits. However, there are a large number of linear algebra applications that require higher precision, up to hundreds or even thousands of digits [2,7,15].

Since arbitrary length data types are not natively supported by general-purpose hardware, software emulation of multiple-precision arithmetic is used. In this paper, we present a parallel multiple-precision implementation of the WAXPBY operation for graphic processor units (GPUs) compatible with the NVIDIA Compute Unified Device Architecture (CUDA).

© Springer Nature Switzerland AG 2019
V. Malyshkin (Ed.): PaCT 2019, LNCS 11657, pp. 179–186, 2019.
https://doi.org/10.1007/978-3-030-25636-4_14

2   Related Works

There are a number of extended- and multiple-precision linear algebra packages, such as XBLAS [9] and MPACK [12]. XBLAS provides double-double precision (128-bit total, 106-bit significand) reference implementations for the dense and banded BLAS functions. MPACK consists of MBLAS and MLAPACK, multiple-precision versions of BLAS and LAPACK, respectively. For multiple-precision arithmetic, MPACK uses the well-known GMP, QD and MPFR libraries. Both XBLAS and MPACK target the CPU architecture.

There is also some research on multiple-precision computations on GPUs. Mukunoki and Takahashi [11] used double-double arithmetic to implement three BLAS functions, AXPY, GEMV and GEMM on GPUs using CUDA. Lu et al. [10] proposed GARPREC and GQD, which are GPU-based implementations of the well known high-precision packages for CPUs, ARPREC and QD, respectively. GQD consists of double-double and quad-double precision, while GARPREC is suitable for arbitrary precision computation. Nakayama implemented CUMP [13], a library for arbitrary precision arithmetic on CUDA-enabled GPUs. It is based on the GMP library and can be a substitute of GMP if arbitrary precision floating-point arithmetic is necessary in CUDA. CUMP supports addition, subtraction and multiplication. A recent study shows that the performance of CUMP is significantly better than that of GARPREC [16]. Joldes et al. [8] implemented CAMPARY, which supports extended (up to a few hundred bits) precision on both CPUs and GPUs. Like double-double and quad-double formats, the precision is extended by representing real numbers as the unevaluated sum of several standard machine precision floating-point numbers. The library contains "certified" algorithms with rigorous error bounds, as well as "quick-and-dirty" algorithms that perform well for the average case, but do not consider the corner cases.

3   Data Representation

We represent an arbitrary length floating-point number x as follows:

$$x = \langle s, X, e, I(X/M) \rangle, \tag{1}$$

where s is the *sign* of x, X is the multiple-precision *significand*, e is the *exponent*, and $I(X/M)$ is the *interval evaluation* of the significand.

The significand X is represented in the residue number system (RNS) [14] by the residues $(r_0, r_1, \ldots, r_{n-1})$ relative to the moduli set $\{m_0, m_1, \ldots, m_{n-1}\}$ and is considered as an integer in the range of 0 to $M - 1$, where $M = \prod_{i=0}^{n-1} m_i$. The residues $r_i = X \bmod m_i$ are machine integers.

The interval evaluation $I(X/M)$ is defined by two bounds, $\underline{X/M}$ and $\overline{X/M}$, that localize the value of X scaled by M. The bounds are represented in an extended-range floating-point format in order to avoid underflow when M is large ($\sim 2^{1000}$ or more). To compute $\underline{X/M}$ and $\overline{X/M}$, only modulo m_i operations and standard floating-point operations are required. The interval evaluation is used to quickly obtain a magnitude order of the residue significand (for sign identification, overflow detection, rounding, etc.). More details concerning the interval evaluation of fractional representations in RNS can be found in [5].

Let x has the form (1). To compute the binary representation of x, we can use the Chinese Remainder Theorem [14]: $x = (-1)^s \times \left| \sum_{i=0}^{n-1} M_i \left| r_i \alpha_i \right|_{m_i} \right|_M \times 2^e$, where $M_i = M/m_i$, and α_i is the modulo m_i multiplicative inverse of M_i.

Under the described number representation, the precision of arithmetic operations in bits is equal to $p = \lfloor \log_2 \sqrt{M} \rfloor - 1$. For example, if 1024-bit computations are required, then the moduli set must be such that $M \geq 2^{2050}$.

Now we consider the GPU implementation of the WAXPBY operation for floating-point vectors whose entries are numbers of the form (1).

4   Multiple-Precision GPU-Based WAXPBY

We assume that the input data (vectors x and y of length N, scalars α and β) are loaded into the GPU global memory. Our multiple-precision WAXPBY algorithm consists of a sequence of GPU kernel launches, as shown in Fig. 1.

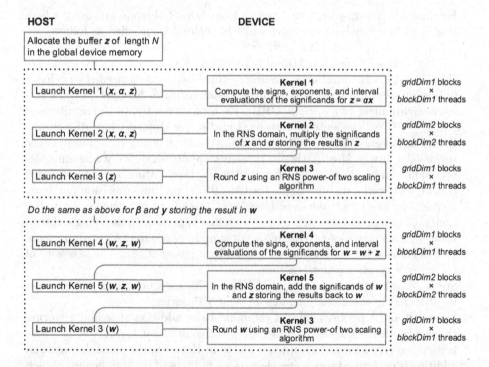

Fig. 1. Flowchart of the proposed multiple-precision WAXPBY algorithm. Kernels 1–5 are the `global` functions, which are called from the host side and execute on the GPU.

The following is a description of the algorithm. We denote the elements of the vectors x, y, z and w by x_i, y_i, z_i, and w_i, respectively. The symbol "." is used to access the fields of a multiple-precision number.

1. The host allocates the intermediate multiple-precision vector z of length N in the GPU global memory.
2. Kernel 1 in *gridDim1* blocks of *blockDim1* threads computes the signs, exponents, and interval evaluations of the significands for the elements of z:

$$z_i.s \leftarrow (\alpha.s + x_i.s) \bmod 2$$
$$z_i.e \leftarrow \alpha.e + x_i.e$$
$$z_i.I(X/M) \leftarrow \alpha.I(X/M) \times x_i.I(X/M)$$

Here *gridDim1* and *blockDim1* are tunable parameters. Interval arithmetic is used to compute $z_i.I(X/M)$.
3. Kernel 2 in a fully parallel way multiplies the significand of α by the significand of each element of the vector x and stores the results in z:

$$z_i.r_j \leftarrow (\alpha.r_j \cdot x_i.r_j) \bmod m_j \qquad 0 \le i \le N-1,\ 0 \le j \le n-1.$$

For multiplication, *gridDim2* blocks of *blockDim2* threads are used, where *gridDim2* is a tunable parameter while *blockDim2* is specified as follows:
   - if $n \ge MinBS$ then $blockDim2 = n$;
   - if $n < MinBS$ then $blockDim2 = \lfloor MinBS/n \rfloor \times n$.

Here *MinBS* is the minimum number of threads per block needed to achieve full occupancy of a streaming multiprocessor (*MinBS* depends on the hardware environment of a specific GPU). Each block multiplies the significands for $N/gridDim2$ elements of x using a grid-stride loop. Within a block, ith thread is assigned for modulo m_j computations, where $j = i \bmod n$. More specifically, if $n \ge MinBS$, then n threads compute all digits of one multiple-precision number in parallel; if $n < MinBS$, then $\lfloor MinBS/n \rfloor \times n$ threads concurrently compute all n digits for $\lfloor MinBS/n \rfloor$ multiple-precision numbers.
4. Kernel 3 in *gridDim1* blocks of *blockDim1* threads rounds the vector z. For each multiple-precision entry z_i, the rounding is performed as a single thread using an RNS power-of-two scaling algorithm. In particular, Algorithm 2 from [6] can be used for scaling the significand by a power of two. The necessity of rounding is checked using $I(X/M)$.
5. Steps 2–4 are repeated for β and y. The results are stored in w, which must be allocated in the global memory of the GPU device.
6. Kernels 4 and 5 perform parallel componentwise addition of w and z storing the results in the vector w: $w_i \leftarrow w_i + z_i$ for all $i = 0, 1, \ldots, N-1$. These kernels are very similar to Kernels 1 and 2. In Kernel 5, the multiple-precision addition is performed simultaneously on all residues of the significand: in each block, one thread calculates one residue of the significand.
7. Finally, Kernel 3 rounds the output vector w.

In principle, *gridDim1*, *blockDim1* and *gridDim2* can also be tuned automatically for a specific device, precision and/or problem size. However, for better performance, tuning these parameters through a series of experimental kernel launches seems to be more preferable.

If the GPU has enough resources, then αx and βy can be computed simultaneously. In order to exploit this type of concurrency, the kernels in steps 2–4 and in step 5 must be launched in different CUDA streams with synchronization immediately before running Kernel 4 in step 6. It is also possible to concurrently execute Kernels 1 and 2, since there are no data dependencies between them.

Data Layout. We use the "Structure of Arrays" layout to store multiple-precision floating-point vectors. For a vector of length N, all digits (residues) of the multiple-precision significands are stored as an integer array of length $N \times n$. All n residues used for the same multiple-precision number are located consecutively in the address space. Since the residues are computed in a parallel fashion, the access pattern for the threads in a warp are coalesced. Additional arrays are used for storing signs, exponents, and interval evaluations.

In addition to WAXPBY, we have also implemented the AXPBY routine on GPUs. The AXPBY algorithm is similar to the considered algorithm for WAXPBY with updating y instead of writing the results in w.

5   Experimental Evaluation

We compared the performance of our WAXPBY function, denoted as "Proposed", with counterparts based on the latest versions of the ARPREC [1] and MPFR [4] libraries for CPU, and the CUMP library [13] for GPU. For comparison purposes, we also provide a CPU-based multiple-precision WAXPBY implementation that uses sequential algorithms for multiplying and adding floating-point representations of the form (1). We denote this implementation on figures as "Baseline". To study the performance impact from computations with multiple precision, we also evaluated some of the currently available BLAS packages: OpenBLAS 0.3.5, cuBLAS 10.1, and XBLAS 1.0.248. OpenBLAS and cuBLAS support double precision, while XBLAS provides double-double precision.

Evaluation environment: Intel Core i5 4590 (3.3 GHz, Quad-Core)/16 GB DDR3 RAM/NVIDIA GeForce GTX 1060 (1.76 GHz, 1280 CUDA Cores, 6 GB GDDR5)/Ubuntu 19.04/GCC 7.4.0 (-O3)/CUDA Toolkit 10.1.105.

We note that all CPU-based implementations, excluding XBLAS, are developed using OpenMP and performed in parallel on 4 threads with 4 physical cores. XBLAS does not support parallel processing.

We measured the performance in Mflop/s (million floating-point operations per second). For our implementations, ARPREC, MPFR and CUMP, by one flop we mean one operation using multiple precision.

In all experiments, the input sets were composed of uniform random numbers. For each test case, we selected kernel execution configurations (*gridDim1*, *blockDim1*, and *gridDim2*) that provide better performance.

In the first experiment, all vectors have length $N = 1\,000\,000$ and the precision of computations p varies from 120 to 1200 bits (16 to 160 RNS moduli were used for our WAXPBY). The results are shown in Fig. 2.

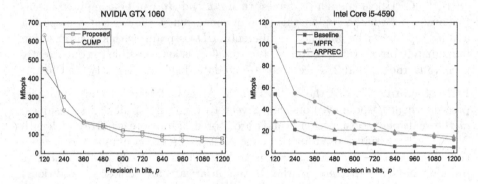

Fig. 2. Performance of multiple-precision WAXPBY functions at $N = 1\,000\,000$

At 120-bit precision, the performance of the proposed WAXPBY is 453 Mflop/s, which is 1.4x lower than that of CUMP. This is mainly due to the overhead incurred by the memory writes for the intermediate vector z. In addition, in the proposed WAXPBY, 9 kernel launches are performed, while for CUMP only one kernel is launched. However, our implementation is less dependent on the precision than CUMP and at 1200-bit precision it is faster than CUMP. Compared to the parallel CPU-based implementations using MPFR and ARPREC, our WAXPBY on the GPU gives a speedup of 4.6–6.6x and 5.4–15.6x, respectively. We also note that the proposed implementation clearly outperforms the quad-core baseline implementation with a speedup of up to 15.5x.

For the proposed WAXPBY, the measured memory bandwidth of the GPU is around 131.3 GB/s. This is 68% of the theoretical peak bandwidth (192 GB/s).

In the second experiment, the problem size N varied from 10 000 to 10 000 000, while the precision was fixed to $p = 240$ bits (this is slightly higher than octuple precision). To achieve this precision, 32 RNS moduli were used for our WAXPBY implementations. The `cblas_daxpby` and `BLAS_dwaxpby_x` routines were evaluated from OpenBLAS and XBLAS, respectively. In the case of cuBLAS, we used the `cublasDaxpy` function. Unlike WAXPBY, `cublasDaxpy` requires only $2N$ arithmetic operations, and this was taken into account in calculating the cuBLAS performance. Figure 3 presents the results of the experiment.

For multiple precision operations, the GPU-based implementations are significantly faster than their counterparts for CPUs. When $N = 10\,000\,000$, the proposed WAXPBY is 6.7x and 11.8x faster than MPFR and ARPREC, respectively. In turn, double-precision cuBLAS is 6.4x faster than OpenBLAS. We observe that the use of multiple-precision arithmetic results in a significant drop in performance compared to the standard double precision. When $N = 10\,000\,000$, the proposed WAXPBY is 40x slower than cuBLAS. In turn,

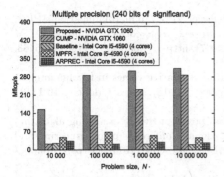
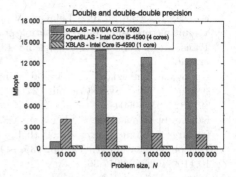

Fig. 3. Performance of WAXPBY and its counterparts as a function of problem size

MPFR is 42x slower than OpenBLAS. This is because the algorithms of multiple precision are much more complex than that of double precision. Compared to XBLAS, our WAXPBY is approximately 1.2x slower, but it provides 2x higher precision.

6   Conclusion

In this study, we implemented a multiple-precision WAXPBY operation on GPUs using CUDA. Our implementation is based on the representation of arbitrary length floating-point numbers using the residue number system, which eliminates carry propagation delays and introduces parallelism in arithmetic operations. We showed that the performance of the developed WAXPBY function is significantly higher than that of multi-threaded solutions based on the well-known multiple-precision arithmetic libraries for CPUs, ARPREC and MPFR.

The presented results were obtained with 16-bit RNS moduli. Using 32-bit moduli (with precautions to avoid overflow in intermediate calculations) will provide twice the precision without any performance penalties.

Acknowledgement. This work was supported by the Russian Science Foundation (grant number 18-71-00063).

References

1. Bailey, D.H., Hida, Y., Li, X.S., Thompson, B.: ARPREC: an arbitrary precision computation package. Technical report, Lawrence Berkeley National Laboratory (2002). https://www.osti.gov/servlets/purl/817634. Accessed 28 Jan 2019
2. Bailey, D., Borwein, J.: High-precision arithmetic in mathematical physics. Mathematics **3**(2), 337–367 (2015). https://doi.org/10.3390/math3020337
3. Blackford, L.S., et al.: An updated set of basic linear algebra subprograms (BLAS). ACM Trans. Math. Softw. **28**(2), 135–151 (2002). https://doi.org/10.1145/567806.567807

4. Fousse, L., Hanrot, G., Lefèvre, V., Pélissier, P., Zimmermann, P.: MPFR: a multiple-precision binary floating-point library with correct rounding. ACM Trans. Math. Softw. **33**(2), article no. 13 (2007). https://doi.org/10.1145/1236463. 1236468

5. Isupov, K., Knyazkov, V.: Interval estimation of relative values in residue number system. J. Circ. Syst. Comput. **27**(1), 1850004 (2018). https://doi.org/10.1142/S0218126618500044

6. Isupov, K., Knyazkov, V., Kuvaev, A.: Fast power-of-two RNS scaling algorithm for large dynamic ranges. In: IVth International Conference on Engineering and Telecommunication (EnT), pp. 135–139. IEEE, Moscow (2017). https://doi.org/10.1109/ICEnT.2017.36

7. Johnson-McDaniel, N.K., Shah, A.G., Whiting, B.F.: Experimental mathematics meets gravitational self-force. Phys. Rev. D **92**(4), 044007 (2015). https://doi.org/10.1103/PhysRevD.92.044007

8. Joldes, M., Muller, J.-M., Popescu, V., Tucker, W.: CAMPARY: cuda multiple precision arithmetic library and applications. In: Greuel, G.-M., Koch, T., Paule, P., Sommese, A. (eds.) ICMS 2016. LNCS, vol. 9725, pp. 232–240. Springer, Cham (2016). https://doi.org/10.1007/978-3-319-42432-3_29

9. Li, X.S., et al.: Design, implementation and testing of extended and mixed precision BLAS. ACM Trans. Math. Softw. **28**(2), 152–205 (2002). https://doi.org/10.1145/567806.567808

10. Lu, M., He, B., Luo, Q.: Supporting extended precision on graphics processors. In: Sixth International Workshop on Data Management on New Hardware (DaMoN 2010), pp. 19–26. ACM, Indianapolis (2010). https://doi.org/10.1145/1869389. 1869392

11. Mukunoki, D., Takahashi, D.: Implementation and evaluation of quadruple precision BLAS functions on GPUs. In: Jónasson, K. (ed.) PARA 2010. LNCS, vol. 7133, pp. 249–259. Springer, Heidelberg (2012). https://doi.org/10.1007/978-3-642-28151-8_25

12. Nakata, M.: Poster: Mpack 0.7.0: Multiple precision version of BLAS and LAPACK. In: 2012 SC Companion: High Performance Computing, Networking Storage and Analysis, pp. 1353–1353. IEEE, Salt Lake City (2012). https://doi.org/10.1109/SC.Companion.2012.183

13. Nakayama, T.: The CUDA multiple precision arithmetic library. https://github.com/skystar0227/CUMP. Accessed 30 Apr 2019

14. Omondi, A., Premkumar, B.: Residue Number Systems: Theory and Implementation. Imperial College Press, London (2007)

15. Simmons-Duffin, D.: A semidefinite program solver for the conformal bootstrap. J. High Energy Phys. **2015**(6), 174 (2015). https://doi.org/10.1007/JHEP06(2015)174

16. Sobyanin, P.: GPU multiple-precision arithmetic libraries (in Russian). Intellektual'nyye sistemy. Teoriya i prilozheniya **22**(3), 89–95 (2018). http://intsysjournal.org/pdfs/22-3/Sobyanin.pdf. Accessed 13 May 2019

HydroBox3D: Parallel & Distributed Hydrodynamical Code for Numerical Simulation of Supernova Ia

Igor Kulikov[✉], Igor Chernykh, Dmitry Karavaev, Evgeny Berendeev, and Viktor Protasov

Institute of Computational Mathematics and Mathematical Geophysics SB RAS, Lavrentjeva 6, 630090 Novosibirsk, Russia
kulikov@ssd.sscc.ru

Abstract. In the paper a new parallel & distributed hydrodynamical code HydroBox3D for numerical simulation of supernovae Ia type explosion was described. The HydroBox3D code is created on basis of combination the adaptive nested mesh for hydrodynamical simulation of supernovae explosion and the regular mesh is second level of nested mesh for hydrodynamical simulation of nuclear reaction. The adaptive nested mesh code for shared memory architecture with using Intel Optane technology was developed. The second level of nested mesh code for Intel Xeon Phi KNL supercomputer was developed. The HydroBox3D code analysis is described. The results of numerical simulation of supernova Ia explosions on massive parallel supercomputers by means HydroBox3D code are presented.

1 Introduction

Supernovas are major sources of "life" elements—from carbon to iron. Type Ia supernovas (SNIa) are very bright and, therefore, they are used as "standard candles" to determine distances to galaxies and the expansion rate of the Universe. A major scenario [1] of supernova explosion is based on the merging of two degenerate white dwarfs with subsequent collapse of a new star when it reaches the Chandrasekhar mass, ignition of the carbon burning process, and type Ia supernova explosion. The goal of this paper is to determine the role of the ignition point in nuclear fuel burning and in the dynamics of the remnants of a degenerate dwarf explosion.

Numerical simulations plays a key role in the modern astrophysics. Perhaps, it is the only universal approach to study the nonlinear evolutional processes in the Universe. One of the main problems of astrophysics simulation is the scale ratio. By example, a typical galaxy can have the mass of 10^{13} Solar masses and the size of 10^4 parsecs, resulting in 13 order gap for the mass and 14 order gap for the size in comparison to the Sun. Therefore it is necessary to use best available supercomputers in order to simulate complex astrophysical processes with high resolution.

© Springer Nature Switzerland AG 2019
V. Malyshkin (Ed.): PaCT 2019, LNCS 11657, pp. 187–198, 2019.
https://doi.org/10.1007/978-3-030-25636-4_15

Nine of the top ten supercomputers listed in the 2018 November version of the Top 500 list are equipped with graphic accelerators and Intel Xeon Phi/Sunway accelerators. Most likely, the first ExaScale performance supercomputer will be built based on the hybrid approach. The code development for the hybrid supercomputers is not a solely technical problem, but an individual complex scientific problem, requiring co-design of algorithms during all stages of problem solving – from physical statement to development tools.

The problem of Mind the Gap of reproducing the nuclear front of heavy elements burning thin relatively to the star size, remains even when using top-level supercomputers when solving problems SNIa. One possible solution to such problems is the use of multi-level nested grids. The approach is to use adaptive nested grids to simulate hydrodynamics of the SNIa explosion and the dynamics of residuals. The next level of nesting of grids allows to reproduce the burning front more correctly. Using the resources of SSCC, we were able to partially solve the Mind the Gap problem by reproducing seven orders of magnitude. We hope that regular access to more productive supercomputers will allow us to advance several orders of magnitude. Following is a short review of codes, that allow you to use a high resolution.

AREPO [2]. The code is based on the technology of moving mesh based on Voronoi and Delaunay triangulation with Lloyd's regularization [3]. This approach allows you to adapt the mesh for the solution. In this case, unlike the SPH methods, the method is based on the Eulerian approach. With all the advantages of such an approach, it is rather difficult in terms of computational costs. The question remains about the quality of the solution in the areas described by less detailed grid cells. Nevertheless, the AREPO code is one of the most used in the World at the moment.

BETHE-HYDRO [4]. This code is based on an ALE-approach combining advantages of the Euler and Lagrange approaches. The equations of hydrodynamics are solved on an unstructured grid in nonconservative Lagrangian form. The numerical method is based on an operator approach which makes it possible to construct (and this is done in the present paper) balanced schemes to approximate the gradient and divergence operators. To solve the Poisson equation in one-dimensional statement, the tridiagonal matrix algorithm (or the Thomas method) is used. In two-dimensional statement the Poisson equation is solved by a conjugate gradient method. Then the potential is corrected to conserve the total energy (the sum of the kinetic, internal, and potential energies) of the system. It should be noted that the total energy of the system is not exactly conserved, but the error in the collapse problem is insignificant, about 10^{-2} per cent. Unfortunately the approach has not been extended to the three-dimensional case.

CHOLLA [5]. The software package is designed for GPU computational experiments and is based on a CTU (Corner Transport Upwind) method. The method is used to extend the upwind scheme to the multidimensional case [6,7]. A cell structure containing all hydrodynamic parameters is used to store the calculation grid on the GPU. Such data locality allows more efficient use of the graphics card global memory. Calculations of a time step are performed on graphic accelerators with the use of CUDA extensions. All numerical methods being used are described in detail in [5].

ENZO [8]. The software package is based on the solution of the equations of magnetic gas dynamics with allowance for cosmological expansion. An N-body model is used to simulate the collisionless component. The code includes a large number of subgrid processes: primordium chemical kinetics, cooling/heating functions, radiation transport, as well as star formation processes and effects resulting from supernova explosions. Several solvers are used to solve the hydrodynamic equations: PPM (implemented only for the equations of gas dynamics), MUSCL, and a finite difference method. An algorithm based on the fast Fourier transform is used to solve the Poisson equation. A so-called structured adaptive grid is also used. Here the basic idea is that the calculation grid has a minimum difference between the neighboring cells. This structure allows using regular trees where a subdomain is divided not more than two times, which increases the efficiency of using such calculation grids.

GADGET2 [9]. The code uses an SPH method as a basic method of solution. At present this is the most widely used code based on the SPH approach. However, the number of codes based on the SPH method decreases, and a major tendency is to use Lagrange–Euler approaches in combination with grid methods. A passage along a Peano–Hilbert curve is used to distribute the particles between the processes. Now it is a standard approach for the parallel implementation of SPH methods.

GAMER [10]. The code contains a solution of the gas dynamics equation using an AMR approach on graphic accelerators. A TVD approach is used to solve the gas dynamics equations, and a combination of a method based on the fast Fourier transform and a method of successive upper relaxation is used to solve the Poisson equation. It seems that a major peculiarity of this complex is the implementation of the AMR approach on graphics cards. In this way a regular structure of the grid is naturally projected onto the GPU architecture, whereas a tree structure needs special approaches. This approach is in using "octets" to define the grid by projecting onto a specific graphics card flow. A major problem here is the formation of fictitious cells for the octet, which takes about 63% of the time. However, this procedure can be performed for each of the octets independently.

GIZMO [11]. For this software code, a new mesh-free approach to solving equations of gravitational gas dynamics has been developed and implemented. The approach is based on a combination of classical grid methods and an SPH method. This method is in using the gas dynamics equations in Euler coordinates which, according to the variational Galerkin principle, are multiplied by test functions. A peculiarity of these functions is that they are linked not to the calculation grid, as in paper [12], but to individual particles [13] which are similar to SPH particles. To determine the values at the domain boundaries, a solution of the Riemann problem using the MUSCL scheme is used.

RAMSES [14]. The code employs a numerical solution of the gravitational gas dynamics equations using an AMR approach based a division into octets. A combination of a method based on the fast Fourier transform and the Gauss–Seidel method is used to solve the Poisson equation. Simple 5-point finite difference approximation is used to solve the Poisson equation. It was replaced by a more efficient 19-point approximation implemented in the form of an extension of the RAMSES code for the case of nonclassical gravitation (MOND) [15].

In the Sect. 2, we describe the concept of co-design, within which the computational model SNIa was developed. We also briefly summarize the information about numerical methods that was used. The Sect. 3 will be devoted to the parallel implementation of the HydroBox3D code. In the Sect. 4 the results of mathematical modeling of the SNIa noncentral explosion will be presented. The conclusion is given in the Sect. 5.

2   The Co-design of Numerical Model

As mentioned in the introduction, the development of software for supercomputers is a complicated scientific problem and it requires the co-design at all stages of the numerical model creation. We outline six co-design stages of numerical modeling Fig. 1. The main difference between the co-design and the classic design of the computational model is the possibility of returning to the previous development stage with the constraints at the current stage. This makes it possible to build in a short time an effective computational model that takes into account all the developments.

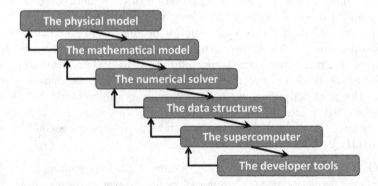

Fig. 1. The co-design conception of astrophysical problem solution method

The problem statement is studying the SNIa explosion during the perturbation of an individual white dwarf, which occurs before the merger of two white dwarfs. In this case, the SNIa explosion occurs at the periphery of the star.

The source of the perturbation is a companion, which is introduced into the physical model by a white dwarf perturbation displaced from the center. For the transition from deflagration to detonation, it is necessary to carefully take into account the combustion front at which nuclear combustion of carbon takes place (we will dwell on it in present study as the most energy efficient source of the explosion). The size of such a front is not resolvable for present day architectures, so we will focus on use of hydrodynamic modeling on multilevel nested grids. Next, we describe the organization of calculations, and then give a briefly description of mathematical model and numerical methods that are used.

2.1 The Parallel & Distributed Computing

The hydrodynamic numerical simulation of SNIa is performed on architecture with shared memory on adaptive nested meshes and is distributed using OpenMP tools within a single process. In our computational experiments we used an Intel Optane node which has 700 GB RAM for a single process. The nuclear reaction hydrodynamics of SNIa is performed on an architecture with distributed memory, with a software implementation based on a one-dimensional geometrical decomposition of a regular calculation domain by MPI tools and subsequent decomposition of the calculations into threads using OpenMP tools within a single process. A diagram of calculations organization is shown in Fig. 2. Regular grids at the second level of adaptive nested mesh are used to calculate hydrodynamic turbulence, which begins with a uniform density distribution corresponding to the cell. For a characteristic time step, one should not expect a

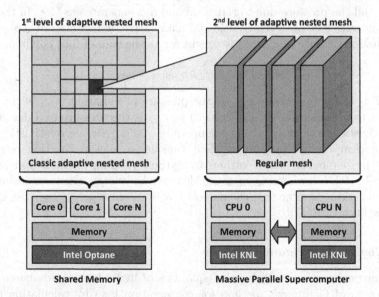

Fig. 2. The organization of parallel and distributed computing in HydroBox3d code

local increase in density by several orders of magnitude. Therefore, the use of regular grids on the second level is fully justified.

2.2 The Numerical Model

Consider the conservative form of the equations of gravitational gas dynamics of conservation of masses

$$\frac{\partial \rho}{\partial t} + \nabla \cdot (\rho \boldsymbol{u}) = 0, \tag{1}$$

conservation of momentum

$$\frac{\partial \rho \boldsymbol{u}}{\partial t} + \nabla \cdot (\rho \boldsymbol{u} \boldsymbol{u}) = -\nabla p - \rho \nabla \Phi, \tag{2}$$

and conservation of total mechanical energy

$$\frac{\partial}{\partial t} \left[E + \rho \frac{\boldsymbol{u}^2}{2} \right] + \nabla \cdot \left(\left[E + \rho \frac{\boldsymbol{u}^2}{2} \right] \boldsymbol{u} \right) = -\nabla \cdot (p \boldsymbol{u}) - (\rho \nabla \Phi, \boldsymbol{u}) + Q, \tag{3}$$

supplemented by the Poisson equation for the gravitational potential

$$\Delta \Phi = 4 \pi G \rho, \tag{4}$$

where ρ is the density, $\boldsymbol{u}$ is the velocity, p is the pressure, Φ is the gravitational potential, E is the internal energy of the gas, G is the gravitational constant, and Q is a source of energy due to nuclear reactions.

The equation of state for stars consists of the pressure of a nondegenerate hot gas and the pressure due to radiation and a degenerate gas [16]. In the case of a degenerate gas, both relativistic and nonrelativistic regimes are considered. The equation of state $p = (\rho, T)$ is sought for as the sum of four components:

$$p = p_{rad} + p_{ion} + p_{deg,nrel} + p_{deg,rel}, \tag{5}$$

where T is the temperature, p_{rad} is the pressure of radiation, p_{ion} is the pressure of a nondegenerate hot gas (ions), $p_{deg,nrel}$ is the pressure of a degenerate nonrelativistic gas, and $p_{deg,rel}$ is the pressure of a degenerate relativistic gas.

As nuclear carbon burning we first consider a nuclear reaction responsible for the bombardment of carbon by carbon yielding natrium and proton $12C (12C, p) 23Na$, where $Q = 2.24$ MeV is the energy released during the nuclear reaction. Assume that the nuclear reaction rate $k_{12C(12C,p)23Na}$ is known from the literature [17].

2.3 The Hydrodynamical Solver

The numerical method to solve the equations of hydrodynamics is based on a combination of Godunov's method for conservation laws by calculating fluxes through the boundaries [18], an operator splitting method to construct a scheme that is invariant with respect to rotation to approximate the advection terms

[19–21], and Rusanov's method to solve Riemann problems [22] for determining the fluxes with vectorization of the calculations [23]. A compact scheme for a piecewise-parabolic representation of the solution in each of the directions is used to solve the Riemann problems [24–26].

To solve the hydrodynamic equations, a modification of an original numerical method based on a combination of an operator splitting method, Godunov's method, and a Rusanov-type scheme is used. This method has all advantages of the above methods and a high degree of parallelization. The numerical scheme is considered in detail in paper [23]. The main idea of the method is in writing the equations of hydrodynamics in vector form:

$$\frac{\partial v}{\partial t} + \nabla \cdot f(v) = 0, \tag{6}$$

where v is the vector of conservative variables. For Eq. (6) we use the following numerical scheme in one of the directions:

$$\frac{v_i^{n+1} - v_i^n}{\tau} + \frac{F_{i+1/2} - F_{i-1/2}}{h} = 0, \tag{7}$$

where F is the solution to a Riemann problem. Omitting the details of derivation of the numerical scheme, which is based on adjoint equations and an operator splitting method, we have the final form of the solution to the Riemann problem:

$$F = \frac{f(v_L) + f(v_R)}{2} + \frac{c + \|u\|}{2} (v_L - v_R). \tag{8}$$

To determine the quantities $f(v_L)$, $f(v_R)$, v_L, and v_R, we use a piecewise-parabolic representation of the solution. The equations of hydrodynamics for the quantities will be calculated in the cells of the root and nested meshes. The Poisson equation for the root mesh will also be calculated in the cells. Then the solution will be projected onto the boundary nodes of the nested mesh. To solve the Poisson equation on the nested mesh the quantities of the potential (and density) will be arranged at the nodes of the nested mesh.

The equations of hydrodynamics (Riemann problems) are solved in two steps: (1) solving the Riemann problems on all boundaries of the nested mesh, and (2) solving Riemann problems at all internal interfaces of the nested mesh. Whereas the second part of solving the Riemann problems is rather trivial, in the first part the method of calculation depends on the sizes of cells of the two neighboring nested meshes. If the cell sizes are equal, the solution to the Riemann problem is the same as that of the Riemann problems at the internal interfaces of the nested mesh, and it is trivial. If a cell of the neighboring nested mesh is larger than the cell being considered the Riemann problem is solved at the interface between the reduced neighboring cell. If the cell being considered has a common boundary with several cells of the neighboring nested mesh the Riemann problems are solved at all interfaces, and then the fluxes are averaged [27]. To organize the satellite calculations, a regular mesh is used this is equivalent to using a root mesh.

2.4   The Poisson Solver

To solve the Poisson equation we use a combination of method based on the fast Fourier transform (for the root mesh) and method of successive over-relaxation (for nested meshes). The Poisson equation is solved in two steps:

1. Solve the Poisson equation on the root mesh by the fast Fourier transform.
2. Solve the Poisson equation on the nested mesh by the method of successive over-relaxation.

We will not consider the method at the first step of solving the Poisson equation (a detailed description of the method to solve it can be found in paper [26]), which is also used to solve the Poisson equation in the satellite calculations.

The method of successive over-relaxation (SOR) is an iterative process of finding the potential on a nested mesh with given initial and boundary conditions obtained by solving the Poisson equation on the root mesh. A similar approach to solve the Poisson equation has been proved to be efficient is some program codes, for instance, in the GAMER code [10].

3   The Performance Analysis

As noted above, the hydrodynamics numerical simulation of SNIa is made on architecture with shared memory. Therefore, we consider a parallel implementation of the second level of nested meshes based on domain decomposition [21]. The MPI tools are used to perform a one-dimensional geometrical decomposition of the calculation domain. In the case of Intel Xeon Phi processors the OpenMP tools are employed. When using Intel Xeon Phi (KNL) processors the calculations are vectorized with some low-level tools [23,28].

The speedup of the code on a mesh of size 512^3 has been studied. For this, the total numerical method time was measured in seconds at various numbers of threads. The speedup P was calculated as

$$P = \frac{Total_1}{Total_K},$$

where $Total_1$ is the calculation time using one thread, and $Total_K$ is the calculation time on K threads. The actual performance has also been estimated. The results of these investigations of the speedup and performance on the mesh of size 512^3 are shown in Fig. 3. A performance of 173 gigaflops and a 48x speedup are obtained on a single Intel Xeon Phi processor.

The scalability of the code on calculation grid size of a $512p \times 512 \times 512$ was studied using all threads for each of the processors, where p is the number of processors being used. Thus, a subdomain of size of 512^3 was used for each processor. To study the scalability, the total numerical method time was measured in seconds at various numbers of Intel Xeon Phi (KNL) processors. The scalability T was calculated as

$$T = \frac{Total_1}{Total_p},$$

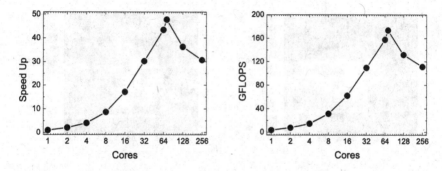

Fig. 3. Speedup and performance of the code on Intel Xeon Phi

where $Total_1$ is the calculation time with the use of one processor, and $Total_p$ is the calculation time with the use of p processors. The results of these investigations of the scalability are shown in Fig. 4. A 97% scalability is reached with 16 processors, which is a rather good result.

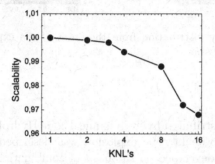

Fig. 4. Scalability of the code

4   The Numerical Simulation

Let's perform simulation of white dwarf with one solar mass and temperature $T = 10^9$ K and a normal distribution of the velocities with a variance of ten percent of the sound speed in the central part of the star. Fig. (5) shows the simulation results: density dynamics from the onset of the explosion to its passage through the bulk of the star. One can see from the simulation results (Fig. 5) that a periphery ignition of the white dwarf takes place when the critical densities for the onset of detonation carbon burning are achieved. As a limiting density for the onset of the process of carbon burning, we use the density of transition from deflagration to detonation from paper [29], which is $\rho_{DDT} = 10^{7.2}$ g cm^{-3}. From the distributed computing it is clear that carbon burning approx 80% complete. This statistics was used to simulate noncentral explosions. However, only the hydrodynamics can show the dynamics of real carbon burning.

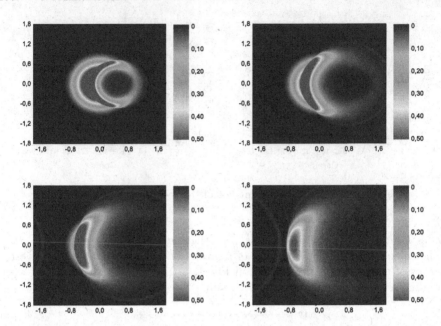

Fig. 5. Relative density distribution from the onset of the explosion to its passage through the bulk of the star

5   Conclusion

The new parallel & distributed hydrodynamical code HydroBox3D for numerical simulation of supernovae Ia type explosion was described in the paper. The HydroBox3D code is developed on the basis of combination of adaptive nested mesh for hydrodynamical simulation of supernovae explosion and regular mesh that is a second level of nested mesh for hydrodynamical simulation of nuclear reaction. A performance of 173 gigaflops and a 48x speedup are obtained on single Intel Xeon Phi processor. A 97% scalability is achieved on 16 processors. Results of numerical simulation of supernova Ia explosions on massive parallel supercomputers obtained with help of the HydroBox3D code are presented.

We developed the HydroBox3D code for a specific problem of supernova of Ia type. Requirements for describing the process of carbon nuclear burning are also was initiated by the features of the problem. However, as the result the technology for solving problems of different-scale gravitational hydrodynamics was developed. So staying within the framework of the implemented hydrodynamic model, we can perform simulation of the star formation process in the interstellar medium in the problems of galaxies collisions and evolution. Also we can perform simulation of the explosion hydrodynamics of supernovae of type II with explosion source – core-collapse, as well as model all the hierarchy of cosmological modeling "observed Universe – cosmic web – clusters of galaxies – and galaxies interaction". The code extension for that hyperbolic models, such as magnetic hydrodynamics, relativistic hydrodynamics and collisionless fluid dynamics allows one to use program code like a technology to solve a wide class of astrophysics problems. In the future, we plan to use the developed technology for actual problems of astrophysics.

Acknowledgements. The research work was supported by the Grant of the Russian Science Foundation (project 18-11-00044).

References

1. Iben, I., Tutukov, A.: On the evolution of close triple stars that produce type Ia supernovae. Astrophys. J. **511**(1), 324–334 (1999)
2. Springel, V.: E pur si muove: Galilean-invariant cosmological hydrodynamical simulations on a moving mesh. Mon. Not. Royal Astron. Soc. **401**, 791–851 (2010)
3. Lloyd, S.: Least squares quantization in PCM. IEEE Trans. Inf. Theory **28**(2), 129–137 (1982)
4. Murphy, J., Burrows, A.: BETHE-hydro: an arbitrary Lagrangian-Eulerian multidimensional hydrodynamics code for astrophysical simulations. Astrophys. J. Suppl. Ser. **179**, 209–241 (2008)
5. Schneider, E., Robertson, B.: Cholla: a new massively parallel hydrodynamics code for astrophysical simulation. Astrophys. J. Suppl. Ser. **217**(2), 24 (2015)
6. Collela, P.: Multidimensional upwind methods for hyperbolic conservation laws. J. Comput. Phys. **87**, 171–200 (1990)
7. Gardiner, T., Stone, J.: An unsplit Godunov method for ideal MHD via constrained transport in three dimensions. J. Comput. Phys. **227**, 4123–4141 (2008)
8. Bryan, G., et al.: ENZO: an adaptive mesh refinement code for astrophysics. Astrophys. J. Suppl. Ser. **211**(2), 19 (2014)
9. Springel, V.: The cosmological simulation code GADGET-2. Mon. Not. Royal Astron. Soc. **364**, 1105–1134 (2005)
10. Schive, H., Tsai, Y., Chiueh, T.: GAMER: a GPU-accelerated adaptive-mesh-refinement code for astrophysics. Astrophys. J. **186**, 457–484 (2010)
11. Hopkins, P.: A new class of accurate, mesh-free hydrodynamic simulation methods. Mon. Not. Royal Astron. Soc. **450**(1), 53–110 (2015)
12. Mocz, P., Vogelsberger, M., Sijacki, D., Pakmor, R., Hernquist, L.: A discontinuous Galerkin method for solving the fluid and magnetohydrodynamic equations in astrophysical simulations. Mon. Not. Royal Astron. Soc. **437**(1), 397–414 (2014)
13. Gaburov, E., Nitadori, K.: Astrophysical weighted particle magnetohydrodynamics. Mon. Not. Royal Astron. Soc. **414**(1), 129–154 (2011)
14. Teyssier, R.: Cosmological hydrodynamics with adaptive mesh refinement. A new high resolution code called RAMSES. Astron. Astrophys. **385**, 337–364 (2002)
15. Candlish, G., Smith, R., Fellhauer, M.: RAyMOND: an N-body and hydrodynamics code for MOND. Mon. Not. Royal Astron. Soc. **446**(1), 1060–1070 (2015)
16. Timmes, F.X., Arnett, D.: The accuracy, consistency, and speed of five equations of state for stellar hydrodynamics. Astrophys. J. Suppl. Ser. **125**, 277–294 (1999)
17. Spillane, T., et al.: $^{12}C + {}^{12}C$ fusion reactions near the Gamow energy. Phys. Rev. Lett. **98**, 122501 (2007)
18. Godunov, S., Kulikov, I.: Computation of discontinuous solutions of fluid dynamics equations with entropy nondecrease guarantee. Comput. Math. Math. Phys. **54**, 1012–1024 (2014)

19. Kulikov, I., Chernykh, I., Snytnikov, A., Protasov, V., Tutukov, A., Glinsky, B.: Numerical modelling of astrophysical flow on hybrid architecture supercomputers. In: Tarkov, M. (ed.) Parallel Programming: Practical Aspects, Models and Current Limitations, pp. 71–116 (2014)

20. Vshivkov, V., Lazareva, G., Snytnikov, A., Kulikov, I., Tutukov, A.: Computational methods for ill-posed problems of gravitational gasodynamics. J. Inverse Ill Posed Probl. **19**(1), 151–166 (2011)

21. Kulikov, I., Lazareva, G., Snytnikov, A., Vshivkov, V.: Supercomputer simulation of an astrophysical object collapse by the fluids-in-cell method. In: Malyshkin, V. (ed.) PaCT 2009. LNCS, vol. 5698, pp. 414–422. Springer, Heidelberg (2009). https://doi.org/10.1007/978-3-642-03275-2_41

22. Rusanov, V.V.: The calculation of the interaction of non-stationary shock waves with barriers. Comput. Math. Math. Phys. **1**, 267–279 (1961)

23. Kulikov, I.M., Chernykh, I.G., Tutukov, A.V.: A new parallel intel xeon phi hydrodynamics code for massively parallel supercomputers. Lobachevskii J. Math. **39**(9), 1207–1216 (2018)

24. Popov, M., Ustyugov, S.: Piecewise parabolic method on local stencil for gasdynamic simulations. Comput. Math. Math. Phys. **47**(12), 1970–1989 (2007)

25. Popov, M., Ustyugov, S.: Piecewise parabolic method on a local stencil for ideal magnetohydrodynamics. Comput. Math. Math. Phys. **48**(3), 477–499 (2008)

26. Kulikov, I., Vorobyov, E.: Using the PPML approach for constructing a low-dissipation, operator-splitting scheme for numerical simulations of hydrodynamic flows. J. Comput. Phys. **317**, 318–346 (2016)

27. Kulikov, I.: The numerical modeling of the collapse of molecular cloud on adaptive nested mesh. J. Phys. Conf. Ser. **1103**, 012011 (2018)

28. Kulikov, I.M., Chernykh, I.G., Glinskiy, B.M., Protasov, V.A.: An efficient optimization of Hll method for the second generation of Intel Xeon Phi processor. Lobachevskii J. Math. **39**(4), 543–551 (2018)

29. Willcox, D., Townsley, D., Calder, A., Denissenkov, P., Herwig, F.: Type Ia supernova explosions from hybrid carbon - oxygen - neon white dwarf progenitors. Astrophys. J. **832**(1), 13 (2016)

GPU Implementation of ConeTorre Algorithm for Fluid Dynamics Simulation

Vadim Levchenko[1], Andrey Zakirov[2], and Anastasia Perepelkina[1,2]

[1] Keldysh Institute of Applied Mathematics, 4, Miusskaya Sq., Moscow, Russia
lev@keldysh.ru, mogmi@narod.ru
[2] Kintech Lab Ltd., Moscow, Russia
zakirov@kintechlab.ru

Abstract. LRnLA algorithms allow simulation of large problems with performance that exceeds the memory-bound limit of the traditional stepwise algorithms, that is, algorithms without any kind of temporal blocking. We show how the ConeTorre LRnLA algorithm that was successfully implemented for various CPU codes may be ported to work with CUDA framework and implemented the Lattice-Boltzmann Method (LBM) for fluid dynamics. As the standard tools and guidelines do not comply with the LRnLA paradigm, we have performed manual optimization of the communication between main memory levels of GPU and reduce overhead for data access patterns. We have made the performance estimate of the LRnLA implementation with the use of the Roofline model. The computation remains memory-bound, but with the ConeTorre algorithm the operational intensity is increased several times, and the maximum achievable performance for the chosen algorithm parameters is 9 billion cell updates per second on Tesla V100. We have achieved more than 66% of the estimate. As a result, we have developed a fluid simulation code based on the Lattice-Boltzmann method with a performance that surpasses state-of-the-art solutions.

Keywords: LRnLA · Lattice-Boltzmann · Temporal blocking · Wavefront blocking

1 Introduction

One way to increase the performance of the stencil computations is to increase the operational intensity since this kind of problems is essentially memory-bound. In most codes, the computational domain is synchronized after each time step. We denote this method as 'stepwise'. In temporal blocking algorithms [5,10,13], with the given amount of data, as much as possible operations are performed. However, the question now is not just about the spatial, but about the space and time decomposition of computations. Locally Recursive non-Locally Asynchronous (LRnLA) algorithms [8] provide a theory for some optimal decomposition, the effect on the operational intensity, and the estimation of the efficiency of the chosen algorithm on the given hardware.

© Springer Nature Switzerland AG 2019
V. Malyshkin (Ed.): PaCT 2019, LNCS 11657, pp. 199–213, 2019.
https://doi.org/10.1007/978-3-030-25636-4_16

Computational Fluid Dynamics (CFD) is one of the fields where the demand for high performance codes exist. CFD simulation aids in modern science and technology. Among the CFD methods, the Lattice-Boltzmann Method (LBM) has a good balance of parallel efficiency and physical correctness, which is why it is heavily used in modern applications [4,11,14,20]. It was also a subject of several temporal blocking developments [6,13,20].

With the popularization of general purpose GPU computing with the appearance of the CUDA programming model, many LBM code developers have used it as a primary computational platform [3,14,22]. The GPU is preferable since it has both higher memory bandwidth and compute rate than CPU, and LBM is a readily parallelizable CFD scheme. Thus, GPU implementations are usually faster than CPU implementations by an order of magnitude [19,25]. The highest documented one GPU performance we have currently found is $2.96 \cdot 10^9$ cell updates per second [19].

Previously the LBM was implemented with LRnLA algorithms on CPU [16]. On GPU, some successful LRnLA implementations are based on the fact that the used numerical scheme has a cross-shaped stencil [7,26], which is not the case for relevant LBM schemes. LBM stencil is better fitted to a cube, so ConeFold-based algorithms are preferred.

Here we present the recent developments in our first attempt to implement the ConeFold-based algorithm on NVidia GPU. The CUDA programming model provides a useful tool that made the first GPU LRnLA codes possible [26]. However, the model is more suited to stepwise domain decomposition than to any kind of time skewing. For example, the cell updates may be distributed between CUDA threads, and the distribution remains static. In LRnLA, the position of the domain, allocated to an SM, dynamically shifts inside the computational region. The thread allocation and data access pattern are to be manually programmed, which results in a cumbersome code even for the simplest schemes, and the overhead may surpass the computation cost.

The challenge to make an efficient temporal blocking on GPU can be simplified, if the decomposition of the space-time domain occurs only in time and one spatial direction [9,10,23]. However, full 3D1T decomposition is preferable, since it provides more locality, and the data in the main computation loop may be localized in the on-chip memory [17,18].

As a result, to make an efficient code, the understanding of the LRnLA theory should be complemented by the proficiency in CUDA tools and optimization techniques. In this paper, we provide a thorough explanation of the ConeTorre algorithm, implementation ideas, and also put an emphasis on the programming details.

2   Numerical Method

Among its numerous variations, the specific Lattice Boltzmann Method [21] is defined by:

A set of discrete speeds c_i, $i = 1, 2, .., Q$ the links between cells of the numerical grid. Each cell has the number of Discrete Distribution Functions (DDF) f_i equal to the number of discrete speeds ($i = 1, 2, ..Q$). In the *streaming* step, each f_i is copied to the cell in the c_i direction.

The collision operator Ω, which locally transforms f_i.

The equilibrium function that is used in most types of the collision operators.

The cell update is

$$f_i(\boldsymbol{x} + \boldsymbol{c}_i, t + 1) = f_i^*(\boldsymbol{x}, t); \quad i = 1, ..19; \tag{1}$$

$$f_i^*(\boldsymbol{x}, t) = \Omega(f_1(\boldsymbol{x}, t), f_2(\boldsymbol{x}, t), ... f_{19}(\boldsymbol{x}, t)). \tag{2}$$

For the purpose of the performance benchmark we take the most common variation of LBM, with $Q = 19$ speeds (D3Q19), collision term in BGK form, and an equilibrium function as a polynomial of order 2. This is one of the most computationally cheap collision operators, so the memory bound property of the method is heavily pronounced.

For many stepwise codes, the collision and streaming operations are merged. Among these, the algorithm where the data is read and stored in place, are preferable for parallel implementation, especially on GPU, so as to avoid write conflicts. We choose the AA-pattern in the current code [2] (Fig. 1). The α step of the AA-pattern involves only one collision. It is local in a way that it reads and stores the f_i inside only one cell. In the β step of AA-pattern, for each collision, the data, necessary for the collision, is pulled from the neighboring cells, collision is performed, and the updated f_i propagate further and are stored into the same cells from which the data was taken beforehand. Thus, the β step combines 2 streaming steps and a collision step.

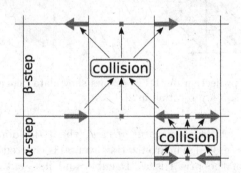

Fig. 1. The AA pattern of streaming propagation. The 1D slice is similar in all directions.

In α and β steps the f_i data is swapped back and forth between the storage points of f_i and $f_{i'}$ which corresponds to the opposite direction of c_i.

3   LRnLA Algorithm ConeTorre

3.1   LRnLA Algorithm Construction

LRnLA algorithms are built as a hierarchical recursive decomposition of the dependency graph (DG). The DG can be aligned with coordinates in $(d + 1)$-dimensions, where d is the dimensionality of simulation space. This dimensionality is denoted dD1T, to emphasize the cases where the time axis is included. In the implementation discussion we take $d = 3$, smaller d is used only for illustration of the concepts. The recursive definition of the LRnLA algorithm states that it is a shape in dD1T space with rule of its decomposition into smaller shapes. The shapes after the decomposition should have only unilateral dependencies. A shape represents the computation of all DG points inside it, in the order, that is determined by the dependencies between its parts. The ability to parallelize portions of operations is determined by tracing these dependencies.

3.2   ConeTorre

For GPU implementation the prism-based shapes were successfully used before [7, 26] and seem to be a more viable option this time as well. Let us take the simulation domain as a cube with side N. Other shapes of the domains may be tiled by cubes, so there is no loss in generality. The ConeTorre (CT) algorithm is initially built by stacking ConeFold [15] shapes on top of each other and defining the new decomposition rule for the resulting shape. Hence, the ConeFold illustration (Fig. 2) helps to visualize the ConeTorre construction.

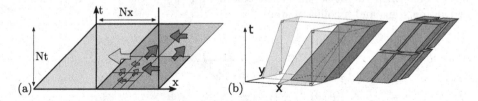

Fig. 2. (a) ConeFold projection in 1D1T. Arrows show data dependencies. (b) Cone-Fold and its decomposition projection in 2D1T.

From the algorithm construction point of view, the 3D1T domain between two chosen time instants (global synchronization events) is covered by slanted prisms (Fig. 3). The base of the prism is a 3D cube, and its slant is parallel to the $(\Delta x, \Delta y, \Delta z, c\Delta t)$ direction. Here, c is the half-width of the stencil, which is equal to one in the chosen scheme. The prism is the ConeTorre. The size of the cube in the base and the height of the prism are the parameters of the algorithm, denoted by TS and NT respectively. To cover the 3D1T simulation domain by prisms, the prisms which have the lower or upper bases outside of the domain should be included, so $((N + N_T)/TS)^d$ CTs are required. The portions that lie outside of the domain, are empty and do not include cell updates.

The CTs are divided into prisms with a lower height, `CFsteps` time steps each. The smaller prisms are divided into flat (in time) layers with the height equal to two time steps. In the current scheme, these two steps are the α and β steps of the AA algorithm. The computation in one layer is divided into portions of one cell update. At each subdivision, the dependencies between shapes are traced. The shapes which have no dependencies may be executed in parallel. If the dependency exists, the synchronization must occur.

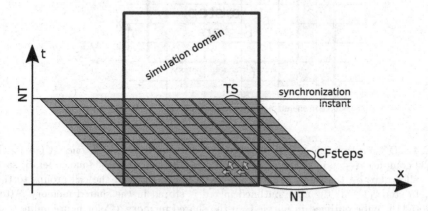

Fig. 3. ConeTorre illustration in 1D1T case.

From the GPU implementation point of view, one CT is assigned to one CUDA-block on a Streaming Multiprocessor (SM). The cell updates are distributed between CUDA-threads in the block, one cell per thread. The data for the updates is in the shared memory, and forms a cube with $(TS + 2)$ edge length (Fig. 4). After each 2 time steps the data in the $(-1, -1, -1)$ direction is loaded from the device memory into the shared memory, and the data in the $(1, 1, 1)$ direction is saved to the device memory and deleted from the shared memory. Thus,

- the shared memory stores a cube of data that travels inside the computation domain in one CT computation.
- the communication of SM with device memory per $2 \cdot TS^3$ cell updates amounts to save and load of $(TS + 2)^3 - TS^3$ cell data.

The latter illustrates the advantage of the temporal blocking: in the stepwise approach, for each cell update its data should be loaded and stored once.

The computation starts with one CT at the corner of the domain. Its slant is directed towards the domain boundary, so it has no dependencies with any other CT. The CTs that are adjacent to it may start after it has progressed several steps. The CUDA-block synchronization event occurs each `CFsteps`. It is implemented with an array of semaphores. The CT base has 3 semaphores assigned to it, since each CT is influenced by 3 CTs to its left and influences 3 CTs to its right. The semaphore is implemented as an integer. It is considered

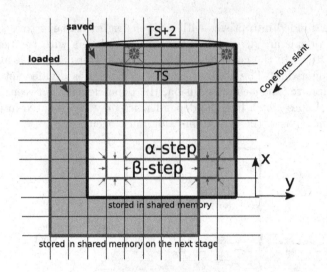

Fig. 4. 2D ConeTorre slice. The algorithm progression in à loop in one CUDA-block is: (1) compute α-step in the green cube; (2) compute β-step in the blue cube; (3) save the cell groups on the right; (4) wait for semaphores; (5) load the cell groups to the left. In steps (1)–(3) the cube outlined in red is stored in the shared memory. After step (5) the cube outlined in purple is in the shared memory. (Color figure online)

locked if it is zero, and unlocked if it is positive. After CFsteps, the semaphores to the right of the current CT are incremented by 1, and the semaphores to the left are decremented by 1.

4   Performance Analysis

The Roofline model [24] presents the performance limitation based on the operational intensity of the algorithm. Operational intensity represents how many operations may be performed per byte of data throughput. If it is high, the performance is limited by the horizontal roof of the peak performance (compute-bound). Memory bound problems are limited by the inclined slope. Thus the central task for the implementation of memory-bound problems is the increase of the operational intensity, since at some point it affects the performance more than reducing the computational overhead and than hardware optimization.

In [8], it is shown that the construction of the LRnLA algorithms simplifies the estimations of the location of the algorithm under the Roofline. This may be summarized as follows. For the whole problem, the operational intensity is the ratio of the total number of operations required for the simulation to the total data of the simulation. This ratio may also be arbitrarily large. On the other hand, under some assumptions (for example, if the data load/store operations and floating point calculations are assumed to be parallel), the time of the algorithm execution is the accumulated time of execution of its parts. Thus the

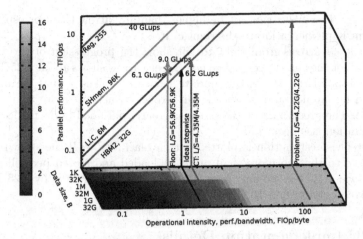

Fig. 5. The Roofline for Tesla V100 GPU. The colored arrows correspond to the sub-tasks of the implemented algorithm for D3Q19 LBM. The color of the arrow corresponds to the color of the Roofline it is limited by, and this is determined by the data size of the task. The required Load/Store data is printed on the arrow and pointed in the third, 'Data size', axis. The floor is a color map of the Roofline for several localization sites. (Color figure online)

performance is limited by the minimal operational intensity which is encountered at some level of subdivision. In a stepwise algorithm, a loop over the domain on one time step is repeated several times, so the operational intensity of one time iteration limits the performance in case of an ideal implementation (one load and store per f_i per time step).

For LRnLA algorithms there are several steps where a task is subdivided into sub-tasks. If at some decomposition level the task data is localized in the higher level of memory hierarchy, the performance of its sub-tasks is limited by the bandwidth of this level.

$$\Pi \leq min\left(\Pi_{GPU}, \Theta_{HBM2}\frac{O(\text{Problem})}{S(\text{Problem})}, \Theta_{HBM2}\frac{O(\text{CT})}{S(\text{CT})}, \Theta_{SHmem}\frac{O(\text{floor})}{S(\text{floor})}, \Theta_{Reg}\frac{o}{s}\right). \quad (3)$$

Here, $O(x)$ is the number of arithmetic operations in a task x, where x is `Problem` for the whole simulation, `CT` for one ConeTorre, `Floor` for `CFsteps` updates in a CT between block synchronizations. $S(x)$ is the amount of data in the load and store operations in a task x. Π_{GPU} is the peak performance of the device; Θ_M is memory throughput of the memory storage M. Since data of CT are localized in the shared memory, the next argument (`floor`) is limited by Θ_{SHmem}. o and s are the operations and data of one cell update. The formula is devised in [8], with the procedure of plotting arrows under the roofline from right to left. Here, we plot it in 3D, adding the data localization axis (Fig. 5). The heat map at the bottom of the plot represents the compute-bound domain in yellow, and memory bound in darker shades. The color of the arrow shows what Roofline it is limited by. The arrow can not be higher than any arrow that corresponds to the current implementation to the right of it.

The colored arrows from right to left show the progression of the LRnLA subdivision for the current code. The last (green) arrow shows about 3.25 TFlops, which corresponds $9 \cdot 10^9$ lattice update per second. The actual performance is reduced by latency and overhead, but expect to achieve more than 50% of that value. The green marker shows the currently obtained performance in our implementation, see Sect. 5.6.

The stepwise estimation is plotted with a black arrow. If the stepwise code is optimized so that no more than one f_i is loaded and stored per cell update (i.e. every value is accessed only once), its performance is limited by 2.25 TFlops which corresponds to $6.2 \cdot 10^9$ lattice update per second.

5   GPU Implementation Details

5.1   Data Structure

In the data structure in the global memory, 8 cells in a $2 \times 2 \times 2$ cube are combined in groups.

```
struct Group {float f[Q][8]; float rho[8];};
```

Here $Q = 19$ is the number of f_i in the cell for D3Q19. The density is also stored so that the data block is aligned to 128B. It may be used for results visualization. This guarantees coalesced thread access to the device memory if one CUDA-block is assigned to read one group data.

The groups in device memory are stored in a Z-curve array [12]. Thus the size of the cube-shaped domain is parametrized by maximal rank MR, where $N = 2^{MR}$ is the linear size of the domain.

The $(TS + 2)^3$ cells are loaded into the shared memory of an SM. Here, the data is not stored in groups. The data organized as a simple 3D array of cells. If $TS = 6$, 40KB of data is required in shared memory of the SM. This fits most modern NVidia GPU architectures shared memory capacity. Thus, hereafter, the TS parameter is fixed to 6.

5.2   Cell Updates

The computation of TS^3 cells is distributed among the CUDA-threads. The α step is performed in the TS^3 cube, then the β step is performed in the cube shifted by $(-1, -1, -1)$ cells. $TS^3 = 216$ CUDA-threads are required. We decide that the optimum number of threads to start in the kernel is 256. Only 216 of these are performing calculations. All started threads are involved in the load and store operations. This way, each thread has access to 255 registers, which allows to store of all necessary data.

5.3   Data Communication

After the two steps, the two cell halo layer of the TS^3 cube in the $(1, 0, 0)$, $(0, 1, 0)$ and $(0, 0, 1)$ directions is to be saved to the device memory. These cells will not be updated anymore in the current CT, and will not be used in further updates. Thus, the place they take in shared memory may be freed for future use.

In their place, the 2 cell layer in the $(-1, 0, 0)$, $(0, -1, 0)$ and $(0, 0, -1)$ direction is loaded from the device memory into the shared memory. Thus, the shared memory stores the copy of cell block from the whole region, that travels in time in the $(-1, -1, -1)$ direction.

Since the shared memory is limited, the data that is saved to device is exchanged in place by the newly loaded data. Thus, the shared memory array is in a cyclic order. The indexes of cells may be retrieved by the modulo operation.

5.4   Main Calculation Kernel

Roughly, the CT kernel three parts are: (1) preliminary loading of the TS^3 cell data cube into the shared memory; (2) loop over time iterations $N_T/2$ times (or less if a boundary intersection occurs); here only the 2 cell halo layer is stored and loaded at each iteration and α and β steps of the AA-pattern are performed; (3) save the cell data that was not saved during the loop to the device memory.

5.5   Data Access

At this step the main difficulty of the implementation of LRnLA algorithms is encountered. The offsets of the required data in the shared memory (during the cell update, and during the rewrite of the shared memory array) and in the global memory (during the load and store operations) need to be computed.

At the naive approach, many integer operations and conditional statements are put in a code. Most GPU devices are not optimized for this kind of operations, so the overhead may become overwhelming. While this remains the main challenge for 3D LRnLA algorithm implementation, we list the considerations that help to optimize the data offset calculation.

Base Point. The CT is defined by the coordinate of its lower base. Each CUDA-thread executes calculation for a cell in this cube, and then for the (NT-1) cells shifted by $(-1, -1, -1)$ each. The behavior of the thread, that is, which cell updates it performs, and what data it fetches, is defined by the base point of the CT and the thread index.

Cyclic Shared Memory Access. Each CUDA-thread at each time step requires an integer offset of the cell it currently updates in the shared memory. The array in the shared memory is cyclic, so the 3D coordinates are wrapped around. The offset is one integer value that is stored for the thread. It is updated at each step of the calculation to point to the cell shifted by $(-1, -1, -1)$ from the current position. Storing less integers in the registers is advantageous to the computation. If the number of registers in use is less than 128, two CUDA-blocks may fit an SM on some architectures, such as Pascal or Volta. The wrap-around shift for 3 coordinates requires at least 16 integer operations, so the shift is implemented directly on the offset.

The offset is encoded with the two's complement approach. First of all, we fix the array size $TS + 2 = 8$. Thus, 3 bits are required for each of the x, y and z position of the current cell in the array. We take 1 integer and put these bits to the 0–2, 6–8, and 12–14 bits of the integer. The spaces between are filled by ..100.., such as

```
int id = ...100yyy100zzz100xxx
```

This integer is decremented by 001000001000001. Then, the offset is collected from the id as zzzyyyxxx. This way, the offset update takes 5 integer operations. Note that the id should be renewed at least once in 8 time steps.

Offset of the Stencil Points. For a cell update, the location of the current cell in the shared memory, as well as the locations of its 18 neighbors should be known. The 18 shifts to the offset described above are required. They are implemented as constants, so that they are known at the compilation stage. In this case, the loop over neighboring cells is unrolled. The compiler optimization ensures that there are no redundant shifts, loads or calculations. The accumulated cost cut of two's complement approach above becomes considerable.

Domain Boundary. The thread should exit the computation loop if the cell assigned to it on the current step falls outside of the domain. On the other hand, the CT that had started outside the domain boundary may cross it at later steps. In this case it is required to begin the computations at some point. We find that a better way to implement this is to calculate the time step, at which the thread enters the domain and the time step, at which the thread exits the domain, using the location of the base point of the currents CT.

Similarly, when the cell is close to the boundary, the boundary condition should apply. At these steps, the shared memory array is partially empty. All f_i in the stencil are checked if they are required from a cell outside of the domain. In the naive approach, the conditional statement should be applied on each f_i. To avoid this, first of all, the special treatment is only applied on at most two time steps (CT entering and leaving the domain). These steps are known beforehand from the base point. Secondly, the type of boundary case (x-boundary, y-boundary, xy-corner, etc) is also computed beforehand and stored as a mask.

Thus, the conditional statements are minimized in the code. This is a general rule for a CUDA-kernel and helps to avoid significant decrease in performance.

Balancing of the Load and Store Operations. After the two computation step the store, block synchronize and load operations are performed in a halo around the TS^3 cube. The communication with device memory is executed by whole groups (see Sect. 5.1). There are $4^3 - 3^3 = 37$ groups to be loaded.

One group contains 160 float values. For the coalesced access, the data is loaded in portions of whole groups. Each one group is loaded by a warp, 5 values per thread. The load and store operations are distributed as evenly as possible among the 8 warps in the thread block. This amounts to approximately five groups for save and load in total per warp.

It is important to remark that near the outer corners of a CT the cell groups that are written may be updated by other CUDA-blocks. There are no data conflicts, thanks to the properties of the AA-pattern. In these cells, only a portion of f_i should be written. To implement this, the data save operation is performed with a mask.

Global and Shared Memory Offset Compression. A thread needs several offsets: the shared memory offset of a cell that it updates, and the shared memory and the global memory offsets of the cells the data from which it loads and stores. About 5 groups are loaded and stored by a warp, so this takes 10 shared memory offsets and 10 global memory offsets. This many values is too much to compute each time or to store. However, since these values concern a whole warp, there is a way to use only one register per thread to store them. The 10 offsets are computed by ten different threads in the warp, and are accessed by other threads by shuffle operations. After the initialization, the update of the offsets due to the $(-1, -1, -1)$ shift is identical for all these values. Since they are stored in one register, one instruction is enough to update them all.

The data in the global memory is stored in groups, and a whole warp loads a group. Each thread in a warp should be assigned an offset to a value in a group that it loads. This offset is constant for each load operation: for each of the 5 groups the warp loads, and for each time step. This is due to the fact that in the main kernel the shift is $(-2, -2, -2)$ after the two α and β steps.

5.6 Semaphore Implementation

Semaphores are stored in a separate Z-curve array. The cell of the array corresponds to the TS^3 cell data block, which is the base of some CT. Note that many of these are outside of the domain Fig. 3. The semaphore array cell contains three integers: the semaphores to the next CT base in x, y, and z axes. Each CT on the block synchronization step requires six semaphores which are stored in the four cells of the semaphore array. Namely, the three semaphores that are unlocked in the progression of the corresponding CT are stored in one cell of the semaphore array; the semaphores that are to be locked are taken from the three neighboring cells. The semaphore read is one L2 access in the CUDA-block synchronization event. The semaphores are accessed in parallel.

The Z-curve traversal rule ensures that at the start of the CT it is already unlocked most of the times, so the wait does not take additional time.

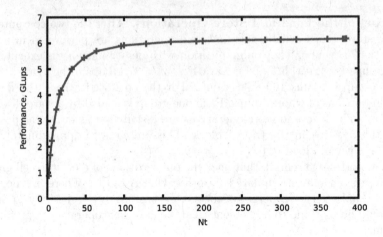

Fig. 6. Performance dependency on the ConeTorre height NT.

6   Performance Results

For the performance test we used GPU devices on the K60gpu [1] cluster: nVidia Volta GV100GL. The code was compiled with `nvcc` v.10.0 with optimization level `-O3`. The performance was tested on a problem that uses as much as possible of the device memory. That is, $MR = 6$, and $(TS \cdot 2^{MR})^3$ cells in the cubic domain.

In (Fig. 6) the performance dependency on N_T parameter is shown, with $CFsteps = 2$. As is expected for LRnLA algorithms, the performance at small N_T is low. At the start (and at the end) of the CT the load (and store) of the whole TS^3 cell data is performed, so there is no gain from temporal blocking. Moreover, the preparation of the offsets is performed, so the algorithm is more computationally intensive than the traditional stepwise implementation. With higher N_T the performance increases. We see, that $N_T \sim 100$ is enough for good efficiency.

In (Fig. 7) we show the dependency of the performance on the `CFsteps` parameter. `CFsteps` is the number of time steps between block synchronizations, which are operated by semaphores. N_T is a multiple of `CFsteps`. If the `CFsteps` parameter is low, more asynchronous computation portions exist in the algorithms, and higher parallelism is possible. If the parameter is high, less synchronizations occur. The optimal value should be a compromise between these two considerations. In (Fig. 7) we see that this value has little effect on the performance, and lower values lead to more efficiency. This shows that the asynchrony of the algorithm is slightly less than enough in the current environment.

Since one CT occupies one CUDA-block, its kernel may be started for all $N_{blk} = (2 \cdot 2^{MR})^3$ CT simultaneously. On the other hand, the number of the asynchronous block N_A may be controlled. The same CTs may be started in a loop of N_{blk}/N_A iterations, where each kernel is started with N_A CUDA-blocks. In (Fig. 8) we show the dependency of the performance on N_A. The V100 GPU has 80 SM. Thus, up to $N_A = 80$ the low occupancy is expected. The full performance is reached at $N_A = 160$.

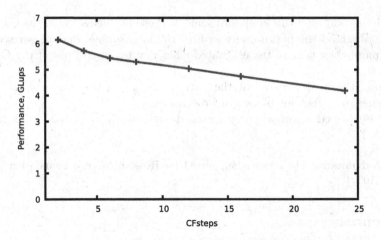

Fig. 7. Performance dependency on the CFsteps parameter

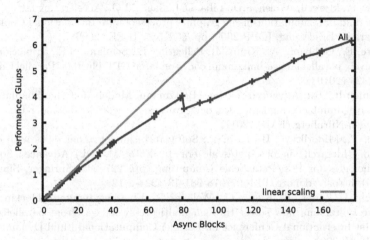

Fig. 8. Performance dependency on the asynchronous blocks number

7   Conclusion

We have implemented the ConeTorre LRnLA algorithm on CUDA GPU. While previous LRnLA algorithm implementations on GPU relied on the diamond blocking in space [7,26], the necessity of the use of the CT shape was imposed by the consideration of LBM scheme which has the cube-shaped stencil.

The algorithm was constructed for CPU in the previous works [15]. To obtain the desired efficiency in a GPU implementation, the difficulties of the implementation overhead had to be overcome. One of this difficulties in the large cost of the data offset storage or calculation, coalesced data access and balancing the load/store operation between CUDA-threads. We note that full 3D1T decomposition is sometimes avoided in favor of 1D1T or 2D1T [7] decomposition of a

3D1T simulation, and this simplifies some of these problems. Nevertheless, here we have obtained the performance of $6.1 \cdot 10^9$ lattice node update per second, that is more than 66% of the estimated efficiency by careful use of the CUDA tools.

Among the achievements of the current work are the implementation of semaphores in CUDA for block synchronization.

The developed approach may be used for other codes with cube-shaped stencils.

Aknowledgement. The work is supported by Russian Science Foundation, grant # 18-71-10004.

References

1. Computer system K-60 (2018). http://kiam.ru/MVS/resourses/k60.html
2. Bailey, P., Myre, J., Walsh, S.D., Lilja, D.J., Saar, M.O.: Accelerating lattice Boltzmann fluid flow simulations using graphics processors. In: International Conference on Parallel Processing, ICPP 2009, pp. 550–557. IEEE (2009)
3. Calore, E., Gabbana, A., Kraus, J., Pellegrini, E., Schifano, S.F., Tripiccione, R.: Massively parallel lattice-boltzmann codes on large GPU clusters. Parallel Comput. **58**, 1–24 (2016)
4. Degenhardt, R.: Advanced Lattice Boltzmann Models for the Simulation of Additive Manufacturing Processes. doctoralthesis, Friedrich-Alexander-Universität Erlangen-Nürnberg (FAU) (2017)
5. Endo, T., Midorikawa, H., Sato, Y.: Software technology that deals with deeper memory hierarchy in post-petascale era. In: Sato, M. (ed.) Advanced Software Technologies for Post-Peta Scale Computing, pp. 227–248. Springer, Singapore (2019). https://doi.org/10.1007/978-981-13-1924-2_12
6. Habich, J., Zeiser, T., Hager, G., Wellein, G.: Enabling temporal blocking for a lattice Boltzmann flow solver through multicore-aware wavefront parallelization. In: 21st International Conference on Parallel Computational Fluid Dynamics, pp. 178–182 (2009)
7. Levchenko, V., Perepelkina, A., Zakirov, A.: Diamondtorre algorithm for high-performance wave modeling. Computation **4**(3), 29 (2016)
8. Levchenko, V., Perepelkina, A.: Locally recursive non-locally asynchronous algorithms for stencil computation. Lobachevskii J. Math. **39**(4), 552–561 (2018)
9. Malas, T., Hager, G., Ltaief, H., Stengel, H., Wellein, G., Keyes, D.: Multicore-optimized wavefront diamond blocking for optimizing stencil updates. SIAM J. Sci. Comput. **37**(4), C439–C464 (2015)
10. Maruyama, N., Aoki, T.: Optimizing stencil computations for NVIDIA kepler GPUs. In: Proceedings of the 1st International Workshop on High-Performance Stencil Computations, Vienna, pp. 89–95 (2014)
11. Montessori, A., et al.: Chapter 20 - multicomponent lattice Boltzmann models for biological applications. In: Cerrolaza, M., Shefelbine, S.J., Garz-Alvarado, D. (eds.) Numerical Methods and Advanced Simulation in Biomechanics and Biological Processes, pp. 357–370. Academic Press (2018). https://doi.org/10.1016/B978-0-12-811718-7.00020-4, http://www.sciencedirect.com/science/article/pii/B9780128117187000204

12. Morton, G.M.: A computer oriented geodetic data base and a new technique in file sequencing (1966)
13. Nguyen, A., Satish, N., Chhugani, J., Kim, C., Dubey, P.: 3.5-D blocking optimization for stencil computations on modern CPUs and GPUs. In: High Performance Computing, Networking, Storage and Analysis (SC), pp. 1–13. IEEE (2010)
14. Niedermeier, C.A., Janßen, C.F., Indinger, T.: Massively-parallel multi-GPU simulations for fast and accurate automotive aerodynamics. In: 7th European Conference on Computational Fluid Dynamics (2018)
15. Perepelkina, A.Y., Levchenko, V.D., Goryachev, I.A.: Implementation of the kinetic plasma code with locally recursive non-locally asynchronous algorithms. J. Phys. Conf. Ser. **510**, 012042 (2014)
16. Perepelkina, A., Levchenko, V.: LRnLA algorithm ConeFold with non-local vectorization for LBM implementation. In: Voevodin, V., Sobolev, S. (eds.) RuSCDays 2018. CCIS, vol. 965, pp. 101–113. Springer, Cham (2019). https://doi.org/10.1007/978-3-030-05807-4_9
17. Perepelkina, A., Levchenko, V., Khilkov, S.: The DiamondCandy LRnLA algorithm: raising efficiency of the 3D cross-stencil schemes. J. Supercomputing (2018). https://doi.org/10.1007/s11227-018-2461-z
18. Perepelkina, A., Levchenko, V.: The DiamondCandy algorithm for maximum performance vectorized cross-stencil computation. Keldysh Institute Preprints (225) (2018)
19. Riesinger, C., Bakhtiari, A., Schreiber, M., Neumann, P., Bungartz, H.J.: A holistic scalable implementation approach of the lattice Boltzmann method for CPU/GPU heterogeneous clusters. Computation **5**(4), 48 (2017)
20. Shimokawabe, T., Endo, T., Onodera, N., Aoki, T.: A stencil framework to realize large-scale computations beyond device memory capacity on GPU supercomputers. In: 2017 IEEE International Conference on Cluster Computing (CLUSTER), pp. 525–529. IEEE (2017)
21. Succi, S.: The Lattice Boltzmann Equation: For Fluid Dynamics And Beyond. Oxford University Press, Oxford (2001)
22. Tomczak, T., Szafran, R.G.: A new GPU implementation for lattice-Boltzmann simulations on sparse geometries. arXiv preprint arXiv:1611.02445 (2016)
23. Vizitiu, A., Itu, L., Niṭă, C., Suciu, C.: Optimized three-dimensional stencil computation on Fermi and Kepler GPUs. In: 2014 IEEE High Performance Extreme Computing Conference (HPEC), pp. 1–6. IEEE (2014)
24. Williams, S., Waterman, A., Patterson, D.: Roofline: an insightful visual performance model for multicore architectures. Commun. ACM **52**(4), 65–76 (2009)
25. Wittmann, M.: Hardware-effiziente, hochparallele implementierungen von lattice-boltzmann-verfahren für komplexe geometrien (2016)
26. Zakirov, A., Levchenko, V., Perepelkina, A., Zempo, Y.: High performance FDTD algorithm for GPGPU supercomputers. J. Phys. Conf. Ser. **759**, 012100 (2016)

GPU-Aware AMR on Octree-Based Grids

Pavel Pavlukhin[1,2(✉)] and Igor Menshov[1]

[1] Keldysh Institute of Applied Mathematics, Moscow 125047, Russia
{pavelpavlukhin,menshov}@kiam.ru
[2] Research and Development Institute "Kvant", Moscow 125438, Russia

Abstract. Algorithms for refinement/coarsening of octree-based grids *entirely* on GPU are proposed. Corresponding CUDA/OpenMP implementations demonstrate good performance results which are comparable with p4est library execution times. Proposed algorithms permit to perform *all* dynamic AMR procedures on octree-based grids *entirely* in GPU as well as solver kernels without exploiting CPU resourses and pci-e bus for grid data transfers.

Keywords: AMR · CUDA · OpenMP · Octree

1 Introduction

Exploiting GPU for calculations of problems with dynamic adaptive mesh refinement (AMR) when the computational grid is locally refined or coarsened depending on the solution is quite limited. This is basically performed in the following manner: grid data is first copied from GPU to CPU memory then modified on CPU generally in sequential mode and after that transferred back to GPU (for example [1,2]). In this conventional scheme GPU stalls are unavoidably happened because of pci-e transfers with quite low bandwidth and CPU grid modification.

In the present paper we consider a multi-GPU gas dynamics solver [3,4] based on Cartesian regular grids and aim to extend this solver to locally adaptive octree-based grids coming from the AMR procedure. To authors knowledge there is only one ongoing work in which all operations related with the grid adaptation are performed *entirely* on GPU [5]. However details of implementation and performance results are not published. In other works [6,7] CPU is utilized along with GPU for grid refinement and coarsening that also leads to decrease in overall performance. A common drawback in the mentioned papers [5–7] is a coarse-graded AMR procedure which is implemented on entire structured grid blocks.

In the approach considered in the present paper the AMR function is fine-graded, i.e., grid modification procedures are carried out at each grid cell separately. To improve the overall GPU performance, all grid modifications are carried out *entirely* in GPU, so that CPU is only exploited to exchange data between processes. In what follows we discuss AMR-related algorithms of octree-based grids that run on only GPU.

© Springer Nature Switzerland AG 2019
V. Malyshkin (Ed.): PaCT 2019, LNCS 11657, pp. 214–220, 2019.
https://doi.org/10.1007/978-3-030-25636-4_17

2   Algorithms for Octree-Based Grids on GPU

At initial, a Cartesian regular grid consisting from *base* cells is generated in the whole computational domain. Then AMR procedure is performed, and some base grid cells become root nodes for corresponding octrees (Fig. 1. Here, for simplicity the sketch is given for quadtrees which are 2-dimensional analogs of octrees). It is assumed the 2-to-1 balance property in the AMR that means that each computational cell may have no more than 4 neighbors over any its face in the 3D case and no more than 2 neighbors in the 2D case.

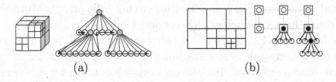

(a)   (b)

Fig. 1. (a) Octree-based grid (left) and corresponding graph representation with space-filling curve (right); (b) Grid after AMR procedure (left) and corresponding graph representation (right).

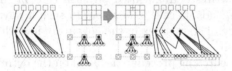

Fig. 2. Grid, corresponding graph and arrays before (left) and after (right) coarsening/refining.

Basic AMR-related grid modification procedures are refining when a cell is divided into 8 subcells and coarsening when 8 cells with a common parent in the octree are united into one cell). For these procedures, all grid octrees are stored in three arrays allocated in GPU (Fig. 2): base/root cell array (denoted as "□"), anchored/*virtual* nodes (denoted as "●") and (denoted as "○") dangling/*physical* nodes (which are real grid cells) attributed with gas dynamical state vectors. CFD solvers actually exploit only the last array. All octree node pointers (denoted as arrows) are initialized. Base cells are stored in accordance with the Z-SFC (space filling curve) order with pointers to its neighbors over 6 faces.

Each thread in parallel performs solution analysis for coarsening/refinement on physical cells. If cells are pointed for performing coarsening/refinement, their locations in the array of "○" are marked as empty and new corresponding cells are written to the reserved free space at the end of the array by each thread. Since some writings may occur simultaneously, one should use atomic addition memory operations on the variable storing current last used index in array of "○". The same modification procedure is performed with the array of "●". In the final coarsening/refinement stage all necessary octree pointers are updated fully independently in parallel by each thread.

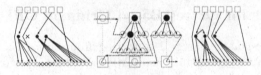

Fig. 3. Grid arrays before defragmentation (left), cell reordering based on Z-space SFCs (center) and grid arrays after defragmentation (right).

One can see that coarsening/refinement procedures lead to appearance of holes (denoted as "×" on Fig. 2) in physical/virtual cells arrays. Multiple callings these holes result in poor efficiency (less opportunities for coalescing load/store transactions from/to GPU RAM) and exhaustion of available memory. Therefore, a kind of the array defragmentation must be performed. In our approach this operation is also executed in GPU entirely. Each thread treats one base cell and traversing via its octree is performed over all physical cell based on the local Z-SFC (Fig. 3). Physical cells are copied to a new array in traversing order by each thread. To improve memory locality, all octrees with physical cells must be written in order given by the Z-SFC over base cells. According to this order first thread initially writes defragmented physical cells in the new array, then the second thread appends its cells, and so on. In such a way, thread serializing is performed - defining of start index in the new array in each thread for further writing of physical cells.

Finally, neighbor searching procedure is based on the octree coordinates storing in each physical cell. At each octree layer each its node has local number from 1 to 8 meaning corresponding geometrical position in the $2 \times 2 \times 2$ cube. Thus, the physical cell in a layer n is uniquely defined by n such local numbers named octree coordinates. By using only these coordinates, corresponding neighbor octree coordinates for each physical cell are defined quite straightforward; neighbors are searched by simple traversing over octree according to these coordinates. All these operations are performed in each thread for each physical cell independently in parallel.

3   Implementation Details

Two versions of AMR procedures are developed: for GPU using CUDA Toolkit and for CPU using OpenMP pragmas. They are based on the same code excepting some parts that are described below.

Procedures performed over all physical cells in CUDA version exploits grid/block indexes for computing unique cell indexes whereas the iteration number is used as the cell index in OpenMP loops.

Atomic addition memory operations used in refining/coarsening procedures represent as *atomicAdd()*/*__atomic_fetch_add()* built in function in CUDA/icpc compiler.

In the defragmentation procedure the variable serving for sequential assignment of indexes in new physical cell array between threads is declared as volatile and threads after its updating perform memory fence operation via _threadfence()/_sync_synchronize() builtin function in CUDA/icpc compiler.

To avoid deadlocks in this procedure pragma for the loop over base cells is used with static scheduling and *chunk_size = 1* in OpenMP version. For the same reason in each warp only one thread is used in CUDA version.

4 Results

For evaluating performance of the implemented algorithms the following test is used. A half part of a $8 \times 8 \times 8$ cube consisting of base cells is divided in $(8 \times 8 \times 4) \times 8 = 2048$ physical cells whereas another part corresponds to $8 \times 8 \times 4$ $= 256$ physical cells. With such a grid, the following procedure is iteratively (for some iteration number) performed in given order: (1) refining each physical cell into 8 subcells; (2) defragmentation of physical cells; (3) neighbors searching for each physical cell. After that the resulting grid is iteratively (for some iteration number) transformed by following procedures in given order: (1) coarsening each 8 physical cells with common parent into 1 cell; (2) defragmentation of physical cells; (3) neighbors searching for each physical cell. Each physical grid has only 1-dimensional state vector (float) which is simply copied into 8 child subcells or arithmetic mean of 8 cells is assigned to new cell in the case of coarsening.

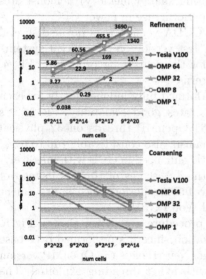
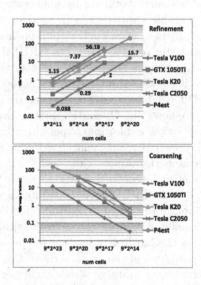

Fig. 4. Results for refining and coarsening procedures.

Measurements are performed on Nvidia Tesla C2050 (CUDA Toolkit 7.5), Tesla K20 (CUDA Toolkit 7.5), GTX 1050 Ti (CUDA Toolkit 9.0), Tesla V100 (32 GB) (CUDA Toolkit 10.0) and on a dual-socket system with 2 Intel Xeon Gold 6142 (OpenMP implementation with 1, 8, 32, 64 threads, Intel icpc 18.0

compiler). For comparison this test is also implemented with p4est [8] and run on dual-socket system with 2 Intel Xeon E5-2690 v4. Although test servers are based on different CPUs their technical specifications are matching so comparision is correct. As mentioned above P4est lacks multithreaded AMR capabilities. Therefore, to use multiple CPU cores, the test is run with 28 MPI ranks in the system. Since there is no defragmentation, neighbor searching procedures explicitly available in P4est refining, coarsening and dynamic load balancing procedures are only measured.

Results for refining procedure are shown in Fig. 4. Hereafter, numbers on the horizontal axis mean the order number of physical cells before corresponding procedure. As described in the previous section, OpenMP implementation for this procedure is very naive and straightforward: atomic addition memory operation is called for *every* physical cell although chunk of them is performed sequentially in thread. So OpenMP version demonstrates primarily slowdown performance of atomic addition memory operations implemented in CPU especially in multithreaded mode. P4est performs refining independently in different processes, and thus delivering performance order of magnitude is larger in comparison with the OpenMP version. The subsequent dynamic load balancing procedure takes small fraction (less than 10%) of refining time.

Quite naive CUDA implementation with the atomic addition memory operation performed in *every* physical cell/thread (although there is a big room for great reducing total its number via exploiting shared memory) demonstrates considerable improvement in atomicAdd() performance in newer GPU architectures. In fact, even current non-optimized CUDA version is performed several times faster (on Pascal, Volta GPU) then P4est version on dual-socket system.

All previous statements are also valid for the coarsening procedure, Fig. 4, since it exploits similar memory access patterns.

The neighbor searching procedure is performed fully independently for each cell so now the OpenMP version demonstrates good scalability, Fig. 5. It's notable that exploiting hyper-threading (64 OpenMP threads on 32 physical CPU cores) as memory latency hiding mechanism also considerably (up to 64%) increase performance. This mechanism greatly enhanced in GPU with massively-parallel architecture (ability to serve "on-the-fly" scores of threads instead of only two in CPU hyper-threading) leads to further performance improvement by about an order of magnitude (for Pascal, Volta GPU).

Finally, results of the defragmentation procedure are shown also on Fig. 5. This is the hardest procedure for parallelization since it's performed over base but not physical cells with serialization stage. Nevertheless the OpenMP version demonstrates quite good scalability even with hyper-threading exploited. The CUDA version has very low SM utilization since in each warp only one thread is used (again, there is a big room for further optimization of this procedure) which leads to performance (on all GPUs) comparable with only 1-thread OpenMP version.

By summing times of refining/coarsening, neighbor searching, defragmentation for GPU and comparing with times of refining/coarsening + dynamic load

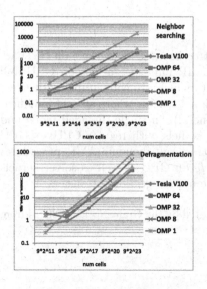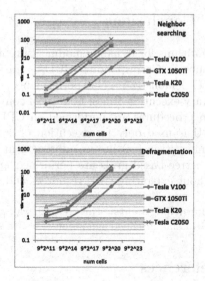

Fig. 5. Results for neighbor searching and defragmentation procedures.

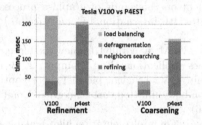

Fig. 6. Refinement procedure on 9×2^{20} cells grid (left) and coarsening procedure on 9×2^{23} cells grid (right).

balancing for p4est, one can see that GPU (namely Tesla V100) performance only slightly lower than CPU one (Fig. 6) in refinement procedure and considerably higher in coarsening process. The most time consuming part for GPU is defragmentation which will be optimized in ongoing work.

Therefore, the main conclusion can be inferred as follows: it's possible to efficiently perform *all* dynamic AMR procedures on octree-based grids *entirely* on GPU without pci-e transfers and CPU processing on AMR grid. In other words, eliminating CPU - GPU pci-e transfers in our GPU-only AMR processing significally decreases overall AMR processing time compared to processing on CPU with p4est even in "the worst" case when *all* grid cells need to be refined/coarsen. In real CFD applications only small fraction of cells is modified during one refining/coarsening procedure call so it would take insignificant time compared to GPU - CPU *entire* grid moving for AMR processing on CPU.

5   Conclusions

Algorithms for dynamic modification of octree-based grids *entirely* on GPU are proposed. Corresponding CUDA and OpenMP implementations on GPU and CPU demonstrate good performance results which are comparable with p4est library execution times. Instead of conventional scheme where all grid modification procedures are performed on CPU generally in single-threaded mode and GPU is used only for execution of solver kernels with regular CPU $\leftrightarrow$ GPU memory transfers in our approach, in contrast, *all* dynamic AMR procedures on octree-based grids are performed *entirely* on GPU as well as solver kernels without exploiting CPU resources and pci-e bus for grid data transfers.

Acknowledgments. This research was supported by the Grant No 17-71-30014 from the Russian Science Foundation.

References

1. Beckingsale, D., Gaudin, W., Herdman, A., Jarvis, S.: Resident block-structured adaptive mesh refinement on thousands of graphics processing units. In: Parallel Processing (ICPP), 2015 44th International Conference on Parallel Processing, pp. 61–70. IEEE (2015)
2. Lawlor, O.S., et al.: ParFUM: a parallel framework for unstructured meshes for scalable dynamic physics applications. Eng. Comput. **22**(3–4), 215–235 (2006)
3. Menshov, I.S., Pavlukhin, P.V.: Efficient parallel shock-capturing method for aerodynamics simulations on body-unfitted cartesian grids. Comput. Math. Math. Phys. **56**(9), 1651–1664 (2016)
4. Pavlukhin, P., Menshov, I.: On implementation high-scalable CFD solvers for hybrid clusters with massively-parallel architectures. In: Malyshkin, V. (ed.) PaCT 2015. LNCS, vol. 9251, pp. 436–444. Springer, Cham (2015). https://doi.org/10.1007/978-3-319-21909-7_42
5. Brown, A.: Towards achieving GPU-native adaptive mesh refinement. Oxford e-Research Centre (2017). https://www.oerc.ox.ac.uk/sites/default/files/uploads/ProjectFiles/CUDA//Presentations/2017/A_Brown_8th_March.pdf
6. Sætra, M.L., Brodtkorb, A.R., Lie, K.A.: Efficient GPU-implementation of adaptive mesh refinement for the shallow-water equations. J. Sci. Comput. **63**, 23 (2015). https://doi.org/10.1007/s10915-014-9883-4
7. Xinsheng, Q., Randall, L., Michael, R.M.: Accelerating wave-propagation algorithms with adaptive mesh refinement using the Graphics Processing Unit (GPU) (2018). https://arxiv.org/abs/1808.02638
8. Burstedde, C., et al.: Extreme-scale AMR. In: Proceedings of the 2010 ACM/IEEE International Conference for High Performance Computing, Networking, Storage and Analysis, pp. 1–12. IEEE Computer Society (2010)

Performance and Energy Efficiency of Algorithms Used to Analyze Growing Synchrophasor Measurements

Aleksandr Popov[1,2] ⓘ, Kirill Butin[1,2] ⓘ, Andrey Rodionov[2] ⓘ,
and Vladimir Berezovsky[1(✉)] ⓘ

[1] Northern (Arctic) Federal University,
Severnaya Dvina Emb.17, 163002 Arkhangelsk, Russia
v.berezovsky@narfu.ru
[2] Engineering Center Energoservice,
Kotlasskaya St., 26, 163046 Arkhangelsk, Russia

Abstract. The development of synchrophasor measurement technology opens new possibilities in solving the problems of ensuring the proper functioning of energy systems. The timely processing of large volumes of measurement data is required. One of the current applications of synchrophasor measurement technology is the analysis of the oscillatory stability of the power systems. Many signal processing procedures can be represented as a set of related typical subtasks. In this paper an approach to high-level description of signal processing schemes in the form of generalized graph structures with the possibility of varying the applied methods for solving subtasks is presented. The program implementation of this approach is presented. The ways to paralleling such schemes on the general level are reviewed and one of them is implemented and analysed in details from performance and energy efficiency points of view. The implementation shows satisfactory performance and parallel scaling. The energy-efficient regimes of its parallel execution were found. Ways of further optimization are identified. The results of numerical experiments are presented.

Keywords: High-performance computing · Digital signal processing · Synchrophasor measurements · Energy efficient computing · Green computing

1 Introduction

The synchrophasor measurement technology [1] at the present level of development allows to analyze the dynamic processes in the electrical network and creates the opportunities for the implementation of new methods of control and management. Deploying in electrical grid wide area measurement system (WAMS), based on this technology, provide voltage and current phasors, frequency and other electrical parameters with high data rate. Nowadays in the production systems the data rate equals to 50 frames per second. The high-precision sources of time synchronization (GLONASS/GPS or like other) are used for the phasor measurement units (PMUs). Electrical parameters are measured discretely and relative to the start of the second (1PPS) [2]. This provides time synchronism of measures between all PMUs at any point of electrical grid.

© Springer Nature Switzerland AG 2019
V. Malyshkin (Ed.): PaCT 2019, LNCS 11657, pp. 221–231, 2019.
https://doi.org/10.1007/978-3-030-25636-4_18

As of the end of 2018 year, 740 PMU devices have been installed in the power system of Russia, and their number is increasing annually [3]. Computational complexity of the algorithms used to analyze growing synchrophasor measurements data is causes need to apply new approaches to implementation of computational operations.

One of the current applications of synchrophasor measurement technology is the analysis of the oscillatory stability of the power systems [4]. Poor damped low frequency oscillations degrading the stability of system are especially interesting [5–7]. One of the main reasons for their occurrence is a significant power imbalance, and one of the factors of their continuation and development is the incorrect work of the regulators of the excitation system of electric generators [6]. The poor damping of these oscillations leads to negative consequences, for example, the degradation of power quality, undue stress on equipment, system separation, blackouts and others.

The mentioned oscillations are observed in the low frequency part of spectrum (to 4 Hz) of the measured quantities (power, frequency). The corresponding components of the measurement signals are called low frequency modes. The problems of analysis of low frequency modes include: selecting modes from input signal, calculation of their dynamic parameters (phase, frequency, amplitude, number of damping characteristics), clustering of power system objects involved to the oscillations, identifying the source of oscillations, analysis of further development of process and others. The calculations for selecting modes and getting their parameters are viewed because the solving of these tasks is the basis of all following analysis.

The example of measured signal and two low-frequency modes are shown on Fig. 1.

This article describes applying of high-performance computing (HPC) in processing synchrophasor measurement data using a low-frequency oscillation problem as an example.

HPC clusters are indispensable in a variety of applications, energy-intensive infrastructures that run large-scale programs. Their energy models play a key role in the design and optimization of energy-saving operations to reduce over-power in HPC clusters. This can be achieved through the efficiency of algorithms, proper allocation of resources and virtualization. The modern methods of energy saving include algorithmic efficiency, virtualization, power management and equipment optimization. Energy efficiency in computing has recently been mentioned in the light of the so-called "green" computing and "green" technologies that relate to the environmentally responsible use of computers and any other resources related to technology [8]. Green computing includes the introduction of advanced technologies such as energy efficient central processing units (CPUs), peripherals, and servers. In addition, green technologies are aimed at reducing resource consumption and improving the disposal of electronic waste (e-waste).

Green computing is no longer just a white expression, like government schemes such as the Carbon Reduction Commitment (CRC) scheme (e.g. Russia adopted a domestic greenhouse gas emissions target that limits emissions to 75% of the 1990 level by 2020 [8]), the Climate Change Agreement (CCA). National commitments, national regulations or e.g. The European Union Emissions Trading Scheme (EU ETS) encourages companies to reassess their use of IT resources. Searching for new ways to improve energy efficiency is no longer a discussion on the board of directors, but a

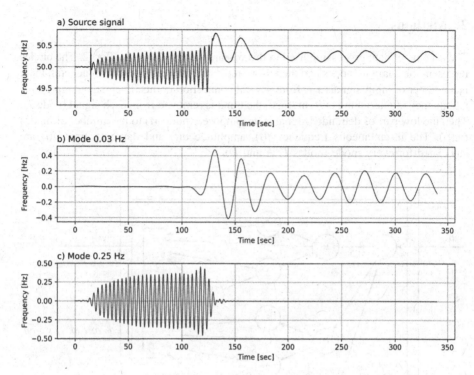

Fig. 1. Example of low-frequency mode: (a) source signal, (b) mode.

reality for many enterprises. The Carbon Reduction Commitment (CRC) scheme is designed to reduce carbon emissions. The CRC covers all forms of energy—electricity, gas, fuel, and oil—with the exception of fuel for transportation. For example, in 2005, the total energy consumption of the data center was 1% of the total energy consumption in the United States, and created as many emissions as a medium-sized nation, such as Argentina [9].

We survey the algorithmic efficiency that directly affects the amount of resources required for running computer functions. Because of this, changes such as moving from linear search to hashing or indexing can speed up processes, thereby reducing resource utilization. Closely linked to this is the aspect of resource allocation. If proper allocation of resources can be made in computing, one can reap the benefits, as it means their efficient utilization. It can also lead to reduction in costs for businesses.

In this paper an approach to high-level description of signal processing schemes in the form of generalized graph structures with the possibility of varying the applied methods for solving subtasks is presented. The ways to paralleling such schemes are reviewed and one of them is implemented and analysed in details from performance and energy efficiency points of view.

2   Methods

The scheme of data processing used in this research is shown on the Fig. 2. The circles represent the compute nodes and the arrows represent the data streams. In general, there is a number of input signals $x(t)$ from several connection points. Each signal represents a frequency of alternating current. First, the trend is removed from each signal (node T). Then the low part of detrended signal is decomposed (node M) to the number of modes ($m(t)$). The instantaneous frequency $f(t)$, amplitude $a(t)$ and damping time $d(t)$ are calculated for every mode (nodes F, A and D).

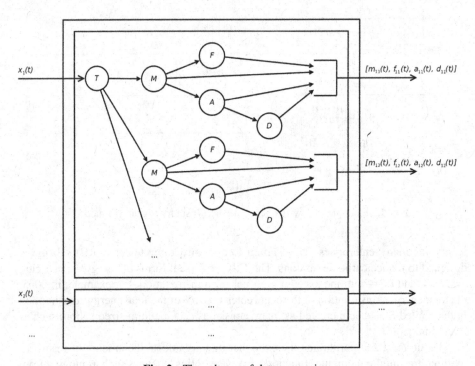

Fig. 2. The scheme of data processing

To make this scheme executable, it is necessary to assign an algorithm with each node. The **dsplab** [10] package developed by authors provides such a two-step design procedure: the system of linked works is built in the first step and the workers from library, specific to the subject area, are «put» to the works in the second one. Works are not connected directly but should be placed to the nodes. The system of linked nodes is called a plan in **dsplab**. The structure of node is shown on the Fig. 3.

In this work the following methods are used for signal processing. The trend line is calculated by smoothing filter with Hamming window and then subtracted from input signal. The three modes with frequency in bands 0.01–0.1 Hz, 0.1–0.2 Hz, 0.2–0.4 Hz are selected with FIR-filter with Hamming window. The amplitude of mode is calculated with digital Hilbert filter. The instantaneous frequency of mode is calculated using

IQ demodulation. The decay time is detected by fitting of fading exponents to the amplitude. The length of all used filters equals to 3001 samples (1 min). All calculating procedures are implemented in the package **es_analytics** [11].

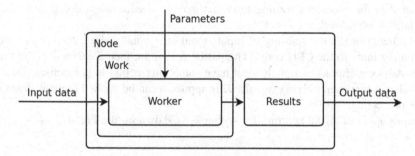

Fig. 3. The structure of node.

Due to the computational complexity and the large amount of data being processed, it becomes necessary to use HPC. In this work computing resources of Northern (Arctic) Federal University (NArFU) has been used. Computing cluster in NArFU with a peak performance of 17.6 TFLOPS is still one of the most high-performance system in the Barents region today. The cluster has a hybrid architecture consisting of twenty 10-core dual-processor nodes with an Intel Xeon E5-2680 processor having 64 GB RAM, eight of which have Intel Xeon Phi 5110P co-processors. The nodes are connected by a high-performance interconnect InfiniBand 56. The platform management interface is integrated into the cluster motherboard system.

For monitoring and collecting statistical information from cluster nodes Intelligent Platform Management Interface (IPMI) was used. IPMI is a platform management interface designed for offline monitoring and control of functions built directly into the hardware and firmware of server platforms. It provides the number of parameters which are necessary for monitoring the system.

IPMI works independently of the operating system and allows to manage a platform that does not have an operating system, even in cases when the server is turned off. The greatest interest in the process of technical control over the operation of the system is represented by such IPMI capability as monitoring the following parameters: temperature, voltage, fan speed, power supply status, bus errors, physical system safety.

3 Parallelizing

The three hours record of current frequency from single connection point was used in experiments with paralleling of calculations. The sample rate of processed signal is 50 Hz. So, the length of signal equals to 540,000 samples.

We can use fast parallel computation at several levels: Vector or array operations, which allows to distribute data chunks on several CPU cores and process them in parallel. In order to gain an advantage from this, some additional effort is usually required during implementation. With packages like NumPy [12] and Python's multiprocessing module [13] the additional work is manageable and usually pays off when compared to the enormous waiting time that may be needed when doing large-scale calculations inefficiently.

The main idea is the splitting of input signal into equally sized chunks and then distributing them to the CPU cores. The partial results are then combined to the final result. Adjacent chunks of signals must have some intersection to compensate for the losses that occur when filtering signals. This approach can be applied in many cases of signal processing.

An example of signal splitting for 4 chunks is shown on the Fig. 4.

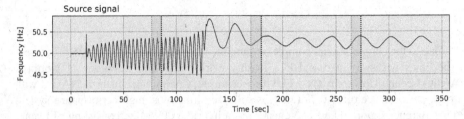

Fig. 4. Example of splitting signal for 4 chunks.

This approach should not necessarily speed up calculations. For example, with 10 min length signal 4-core machine took about 17% longer for calculating the results in parallel than with single processor execution. This is explained as follows. The optimized code already runs quite fast, even for large datasets. Splitting the data into chunks, starting the worker processes, distributing the data and then collecting and combining the results again introduces a lot of extra work ("overhead"). This extra work unfortunately does not pay off in this scenario, because the actual processing time for each chunk is quite low, compared to the time spend for the parallelization overhead (Table 1).

Table 1. Execution time for serial and parallel processing of various length signals.

Signal length (min)	10	100	200	300	400	500
Single process (time, sec)	1.37	4.43	7.79	11.22	14.76	17.72
4 processes (time, sec)	1.59	11.64	23.47	35.27	46.98	59.23

Still, there are many scenarios where parallelization *does* pay off despite the overhead. This is usually the case when the processing time for the data is high as

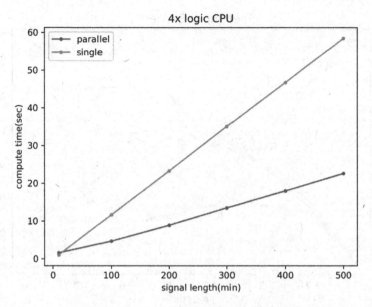

Fig. 5. Execution time versus signal length

compare to the parallelization overhead. If one has a relatively big data frame (100 min or larger, see Fig. 5), running this in parallel gives a speed up factor of ~3 on 4-core machine (again, the theoretical speed up of 4 is not reached because of overhead).

4 Results

A series of experiments was conducted to determine the acceleration and efficiency of the parallel implementation for the scheme of data processing, as well as the energy efficiency of task execution for solving the problem using HPC cluster. Scaling results for various signal length are shown on Fig. 6. In those tests the signal was splitted to equal chunks of the same length and then they distributed between cores for processing. Experiments show degradation of efficiency to 0.3–0.5 depending on the signal length, giving the worst results for larger signal, that corresponding larger chunk size. This situation indicates the emergence of competition between processes for the shared memory and i/o channels.

We also investigated runtime behaviors of the implementation. The utilization, energy consumption, temperature of the devices and other information obtained from the sensors onboard were measured. Figure 7 shows overall picture of the computing. The top graph represents the CPU utilization value in percent over time. The graph of power consumption in Watts at the same times is shown in the middle. There is also the bar diagram showing the specific energy consumption in Joules – energy consumed by hardware to processing signal with 1-minute length (3001 samples). The width of the bar is expended computation time. Speedup and efficiency of task execution at the same time are shown at the bottom. Note, that the x-axes positions of the markers on this

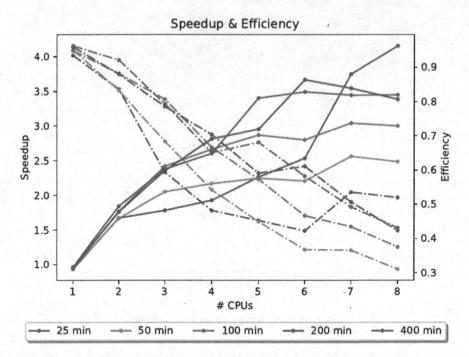

Fig. 6. Speedup and efficiency of parallel realization

graph correspond to the completion time. Specific energy consumption graph shows that best energy efficient execution regime corresponds to a rather mediocre efficiency of execution parallelism around values of 0.5.

Figure 8 presents the diagram of the temperature change for the processor and RAM modules when the task is running. Visible characteristic hysteresis, indicating the presence of an unbalanced number of calculations and memory accesses.

5   Discussion

Consideration of power consumption by hardware during the execution of calculations allows us to suppose the linear model as satisfactory for assessing the dependence of energy costs on computational complexity. Power breakdown across the components of server [14] are shown on the Fig. 9a. In Xeon based server, the CPUs are the main power consumers. Experiments on different workload shows the correlation in character of power consumption, CPU utilization and cache misses which in turn are proportional to the memory consumption of energy [15] (Fig. 9b). This leads us to a linear consumption model (1) (see Fig. 9c).

$$P = aU + b \qquad (1)$$

where P is power consumption, U – computation complexity in e.g. FLOP, a and b – arbitrary constants.

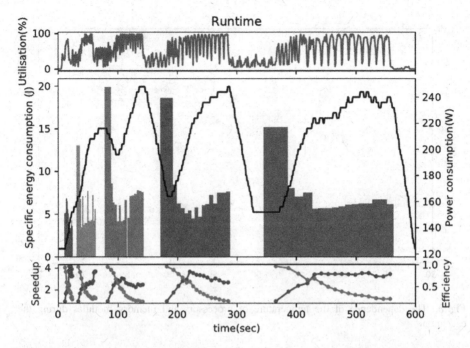

Fig. 7. Runtime behaviors of the realization. Overall CPU utilization during tests runs is shown on the top of the figure. There is speedup (green line) and efficiency (brown line) of currently ran case at the bottom. Specific energy consumption in Joules as bars with height equal to energy consumed to compute signal with 1-minute length and width equal to expended computation time for processing 25 (blue), 50 (orange), 100 (green), 200 (red) and 400 (orchid) minutes signals are shown in the middle. The current power consumption in Watts (black line) is also shown there. (Color figure online)

Performed energy consumption tests did not include the power consumption for network interaction (MPI) and input/output operations (work with the file system). The linear model of energy consumption assumes the idea that the best strategy for minimizing energy consumption will be the most complete use of the processor, and, therefore, the best algorithm in terms of optimizing energy consumption and its implementation will be those that solve given problem as soon as possible and at the same time most fully use computing hardware.

Ways of further optimization may be as follows. We have exploited data parallelism, but it is possible to get a gain with task parallelism. If there are many signals at the input and one need to do the same with them, then one can divide the initial data set into groups of signals and process them in parallel. In that case even the shared memory and exchanges are not needed during processing. This approach will increase the efficiency of both parallelism and power consumption.

Another approach consists in parallel execution of those nodes of the plan (scenario) for which the following conditions are fulfilled: the input data is ready and the node has not yet been executed (look at Figs. 2 and 3 for the guidance). There should

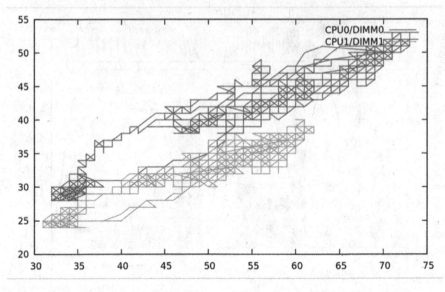

Fig. 8. The dependence of the temperature of processors and memory modules during the execution.

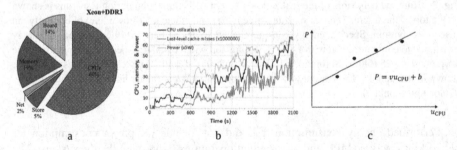

Fig. 9. (a) Power breakdown across the components of server [14]. (b) Characteristics of the real-world workload. (c) The linear power model

be a less gain. This may not be the case if the scenario is linear. Nothing is needed to know about the task for its correct execution, but in order to have a positive effect, you must still consider the specifics of the task.

6 Conclusion

A Python based framework for digital signal processing with its parallel implementation was developed to use it in analysis of growing synchrophasor measurements. The results of numerical experiments with implementation are presented. An implementation shows satisfactory performance and parallel scaling. This will allow the processing of data collected from power grids in real-time and carrying out post analysis of incidents in the shortest time. The energy-efficient regimes of the parallel execution were found. An interesting result was obtained that the optimal energy efficiency of computing is achieved with a mediocre parallel efficiency near value of 0.5. Ways of further optimization are identified.

Acknowledgements. All computing experiments were performed using the HPC environment at NArFU [16].

References

1. Mokeev, A.V.: Methods for analysis of PMU functioning during electromagnetic and electromechanical transient processes. In: 5th International Scientific and Technical Conference Actual Trends in Development of Power System Relay Protection and Automation, Sochi (2015)
2. IEEE Std C37.118.1-2011 IEEE Standard for Synchrophasor Measurements for Power Systems. https://doi.org/10.1109/61.660853
3. smpr.technology|WAMS. https://www.smpr.technology/. Accessed 16 Feb 2019
4. Antonello, M., Muscas, C., Ponci, F.: Phasor Measurement Units and Wide Area Monitoring Systems. Academic Press, Cambridge (2016)
5. Al-Ashwal, N., Wilson D., Parashar M.: Identifying sources of oscillations using wide area measurements. In: Proceedings of the CIGRE US National Committee 2014 Grid of the Future symposium, Houston, vol. 19 (2014)
6. Kovalenko, P.Yu.: Methods of analysis of low-frequency oscillations and the synchronizing action of a generator based on vector measurements: dissertation for the degree of candidate of technical sciences: 05.14. 02 (Diss.) (2016). (in Russian)
7. Rodionov, A.V., Popov, D.N., Sosnin, A.S., Mokeev, A.V., Popov, A.I.: Extending functionality and application scope of synchronised phasor measurement technology. In: International Youth Scientific and Technical Conference Relay Protection and Automation (RPA), 27–28 September 2018, pp. 1–15 (2018). https://doi.org/10.1109/rpa.2018.8537229
8. Kokorin, A.O., Korppoo, A.: Russia's Greenhouse Gas Target 2020: Projections, Trends and Risks. Friedrich-Ebert-Stiftung (FES), Berlin (2014)
9. Mathew, V., Sitaraman, R.K., Shenoy, P.J.: Energy-aware load balancing in content delivery networks. CoRR, vol. Abs/1109.5641 (2011). https://doi.org/10.1109/infcom.2012.6195846
10. dsplab – PyPI. https://pypi.python.org/pypi/dsplab. Accessed 16 Feb 2019
11. Bovykin, V.N., Mokeev, A.V., Popov, A.I., Rodionov, A.V. Expansion of the field of application of the technology of synchronized vector measurements. Automatizatciya i IT v energetike **12**(113), 44–50 (2018). ISSN: 2410-4043
12. NumPy. http://www.numpy.org/. Accessed 16 Feb 2019
13. Multiprocessing—Process-based parallelism. https://docs.python.org/3/library/multiprocessing.html. Accessed 16 Feb 2019
14. Malladi, K.T., et al.: Towards energy-proportional datacenter memory with mobile DRAM. In: Proceedings of the 39th Annual ISCA, pp. 37–48. (2012). https://doi.org/10.1109/isca.2012.6237004
15. Lewis, A.W., Tzeng, N.-F., Ghosh, S.: Runtime energy consumption estimation for server workloads based on chaotic time-series approximation. ACM Trans. Archit. Code Optim. **9**(3), 15:1–15:26 (2012). https://doi.org/10.1145/2355585.2355588
16. HPC NArFU. http://fujitsu-hpc-02.narfu.ru/. Accessed 16 Feb 2019

A Comparison of MPI/OpenMP and Coarray Fortran for Digital Rock Physics Application

Galina Reshetova[1]([✉]), Vladimir Cheverda[2], and Tatyana Khachkova[2]

[1] The Institute of Computational Mathematics and Mathematical
Geophysics SB RAS, Novosibirsk 630090, Russia
kgv@nmsf.sscc.ru
[2] The Trofimuk Institute of Petroleum Geology and Geophysics SB RAS,
Novosibirsk 630090, Russia

Abstract. A new parallel numerical technique to estimate the effective elastic parameters of a rock core sample from the three-dimensional Computed Tomography images is presented. The method is based on the energy equivalence principle and a new approach to solving the 3D static elasticity problem by the iterative relaxation technique.

The method in the three-dimensional case requires the obligatory parallel implementation. The most commonly used strategy of parallelization is MPI and OpenMP. The latest Fortran extension offers the new Coarray Fortran (CAF) features, which can potentially compete with the MPI due to its efficiency and simple implementation. We compare three parallel approaches based on the MPI, MPI+OpenMP and CAF to solve the problem. Comparison of these methods has shown that the CAF brings about a sufficiently compact parallel code with a simple syntax, thus making the parallelism easier to understand. The results presented demonstrate that the CAF implementation provides comparable performance to an equivalent MPI version.

Keywords: Effective parameters · Elastic moduli ·
3D Tomographic images · Coarrays · Fortran · PGAS languages · MPI

1 Introduction

The technologies for the core research based on the computer-aided simulation are a new direction of Digital Rock Physics. The tomographic core images make it possible to obtain the digital three-dimensional reconstruction of a core sample material and to perform numerical experiments using modern computer technologies. Digital Rock Physics includes the whole complex of the research into digital tomographic core images and their processing.

In this paper, we present a parallel numerical algorithm for estimating the elastic properties of a rock sample from their three-dimensional tomographic

© Springer Nature Switzerland AG 2019
V. Malyshkin (Ed.): PaCT 2019, LNCS 11657, pp. 232–244, 2019.
https://doi.org/10.1007/978-3-030-25636-4_19

images. Within the framework of these studies, there are a number of approaches to determining the effective heterogeneous elastic moduli, including the methods based on the analysis of inclusions [4,14,15], the use of a wide range of homogenization methods [1] and some other approaches.

Currently, to solve this type of problems, the methods that, as a rule, are being developed for studying wave fields with surface acquisitions are used [6,7]. Such methods are well studied both from the mathematical and from mechanical points of view, and in terms of organizing parallel computing for providing maximal scalability [2,11]. The current experience in solving the problems demonstrates the key role of optimizing exchanges among the processes organized for parallel computing. Therefore the main attention in this paper is concentrated on a comparative analysis of the well-known parallelization schemes based on MPI/OpenMP and the recently introduced approach on the base of CoArray Fortran (CAF). In the published works we have not found any other comparisons of MPI and CAF technologies in the field of associated studies.

For solving the problem of estimating the elastic properties of a rock sample from their three-dimensional tomographic images, the method based on the energy equivalence principle proposed in [16] to study the properties of composite materials and a new approach to solving 3D static elasticity problem by the iterative relaxation technique [12] were chosen. This choice is made because of the possibility to carry out the numerical parallelization of the original problem with linear acceleration. The elastic moduli are determined by the parallel computation of potential energy of the elastic deformations in a sample under static homogeneous stresses applied to the boundary, thus simulating the effects occurring in laboratory measurements.

2   Statement of the Problem

The effective elastic properties of a sample are determined based on the generalized Hooke's law, which expresses the relationship between the averaged deformations and stresses over a representative volume:

$$\bar{\sigma}_{ij} = c^*_{ijkl}\bar{\varepsilon}_{kl} \quad or \quad \bar{\varepsilon}_{ij} = s^*_{ijkl}\bar{\sigma}_{kl}. \tag{1}$$

The components of the stiffness tensor c^*_{ijkl} and of the compliance tensor s^*_{ijkl} form the fourth rank tensors which, by definition, are the effective stiffness C^* and the compliance S^* tensors. The average stresses and strains are determined by the formulas:

$$\bar{\sigma}_{ij} = \frac{1}{V}\int_V \sigma_{ij}dV, \quad \bar{\varepsilon}_{ij} = \frac{1}{V}\int_V \varepsilon_{ij}dV, \tag{2}$$

where σ_{ij} and ε_{ij} are the components of the stress and the strain tensors describing the stress-strain state of a sample in the representative volume V and satisfying the equilibrium equations and the Saint-Venant compatibility equations.

3   Method

To find the effective stiffness C^* and the compliance S^* tensors we use the energy equivalence principle method [16]. To this end, we introduce the notion of the homogeneous boundary conditions [1]. Such conditions can be either kinematic or static and are defined in such a way that when applied to the boundary S of a homogeneous elastic body of volume V, they cause within it the uniform (constant) stresses and displacements. In particular, the homogeneous static (3a) and kinematic (3b) boundary conditions are boundary conditions with stresses (3a) and displacements (3b) specified on the boundary in the form of the linear functions

$$(a)\ t_i(S) = \sigma_{ij}^0 n_j\ ,\quad (b)\ u_i(S) = \varepsilon_{ij}^0 x_j, \tag{3}$$

where $\sigma_{ij}^0, \varepsilon_{ij}^0$ are some constant symmetric stress and strain tensors, respectively, and n is the vector of the outer normal to the boundary S.

The energy equivalence principle method is based on the theorem [1] asserting that the homogeneous static (kinematic) boundary conditions applied to the boundary S of a non-homogeneous representative volume V generate such a stress field σ_{ij} (strain ε_{ij}) that its averaging over volume (2) is equal to the value of the constant stress σ_{ij}^0 (strain ε_{ij}^0) applied to boundary (3):

$$\bar{\sigma}_{ij} = \sigma_{ij}^0, \quad \bar{\varepsilon}_{ij} = \varepsilon_{ij}^0. \tag{4}$$

The potential energy of deformations in the heterogeneous elastic body V is expressed by the formula:

$$U = \frac{1}{2} \int_V \sigma_{ij}\varepsilon_{ij}dV. \tag{5}$$

We calculate the energy of deformations when the homogeneous static boundary conditions are applied to a heterogeneous elastic body:

$$\begin{aligned} U &= \tfrac{1}{2}\int_V \sigma_{ij}\varepsilon_{ij}dV = \tfrac{1}{2}\int_S \sigma_{ij}u_i n_j dS = \tfrac{1}{2}\sigma_{ij}^0\int_S u_i n_j dS \\ &= \tfrac{1}{2}\sigma_{ij}^0\int_V u_{i,j}dV = \tfrac{1}{2}\sigma_{ij}^0\int_V \varepsilon_{ij}dV \\ &= \tfrac{1}{2}\bar{\sigma}_{ij}\bar{\varepsilon}_{ij}V = \tfrac{1}{2}s_{ijkl}^*\sigma_{kl}^0\sigma_{ij}^0 V. \end{aligned} \tag{6}$$

It follows that the potential energy of a heterogeneous elastic body in the stress-strain state is represented in the following form:

$$U = \frac{1}{2}s_{ijkl}^*\sigma_{kl}^0\sigma_{ij}^0 V. \tag{7}$$

Thus, if the value of the potential energy U of the stress-strain state of the elastic body in which it has been transferred under the homogeneous boundary conditions (static stresses) σ_{ij}^0 is known, then Eq. (7) can be used to find the components of the effective compliance tensor s_{ijkl}^*. If we calculate the potential energy U_0 with elastic parameters corresponding to the effective stiffness tensor C^*, then we obtain the expression:

$$U_0 = \frac{1}{2} s^*_{ijkl} \sigma^0_{kl} \sigma^0_{ij} V. \tag{8}$$

Hence, it follows from formulas (7) and (8) that the energy method can be regarded as a method based on the equivalence principle of the potential energies for heterogeneous and homogeneous samples:

$$U_0 = U. \tag{9}$$

4   The Algorithm for Determining the Components of the Tensor S^*

We suppose the volume V to be fixed in space by a rectangular (parallelepiped in 3D) region with the sides (edges) parallel to the coordinate axes. To find the components s^*_{ijkl}, we seek the solution of the boundary value problem of the static linear elasticity theory

$$\sigma_{ij,j} = 0, \tag{10}$$

$$\sigma_{ij} = c_{ijkl} \varepsilon_{kl} = c_{ijkl} u_{k,l}, \quad i,j = 1,2 \tag{11}$$

with the corresponding homogeneous static boundary conditions applied to the faces of the sample.

4.1   Two-Dimensional Case

In the case of a 2D sample, the tensor S^* is written down in the form:

$$\begin{bmatrix} \varepsilon_{11} \\ \varepsilon_{22} \\ 2\varepsilon_{12} \end{bmatrix} = S^* \begin{bmatrix} \sigma_{11} \\ \sigma_{22} \\ \sigma_{12} \end{bmatrix}, \quad S^* = \begin{pmatrix} s^*_{1111} & s^*_{1122} & s^*_{1112} \\ & s^*_{2222} & s^*_{2212} \\ sym & & s^*_{1212} \end{pmatrix}. \tag{12}$$

When calculating s^*_{1111}, s^*_{2222} and s^*_{1212} with the static boundary conditions (Table 1) according to (7), we obtain:

$$U^{(1)} = \frac{1}{2} s^*_{1111} V, \quad s^*_{1111} = 2U^{(1)}/V. \tag{13}$$

Here and below, the superscript in the notation indicates to number of the case under consideration. Cases 1–3 are presented in Table 1. To determine the remaining components s^*_{1122}, s^*_{2212} and s^*_{1112}, we use the linearity property of the elasticity problem and define them by the formula presented in Table 1.

Table 1. The boundary conditions and the formula for finding the components of S^*.

Case	U	Faces a	Faces b	Faces a, b	Value s^*_{ijkl}
1	$U^{(1)}$	$\sigma_{11} = 1$	$\sigma_{22} = 0$	$\sigma_{12} = 0$	$s^*_{1111} = 2U^{(1)}/V$
2	$U^{(2)}$	$\sigma_{11} = 0$	$\sigma_{22} = 1$	$\sigma_{12} = 0$	$s^*_{2222} = 2U^{(2)}/V$
3	$U^{(3)}$	$\sigma_{11} = 0$	$\sigma_{22} = 0$	$\sigma_{12} = 1$	$s^*_{1212} = 2U^{(3)}/V$
4	$U^{(4)} = U^{(1)} + U^{(2)} + U^{(1,2)}$	$\sigma_{11} = 1$	$\sigma_{22} = 1$	$\sigma_{12} = 0$	$s^*_{1122} = U^{(1,2)}/V$
5	$U^{(5)} = U^{(2)} + U^{(3)} + U^{(2,3)}$	$\sigma_{11} = 0$	$\sigma_{22} = 1$	$\sigma_{12} = 1$	$s^*_{2212} = U^{(2,3)}/V$
6	$U^{(6)} = U^{(1)} + U^{(3)} + U^{(1,3)}$	$\sigma_{11} = 1$	$\sigma_{22} = 0$	$\sigma_{12} = 1$	$s^*_{1112} = U^{(1,3)}/V$

Table 2. The boundary conditions for finding the components s^*_{ijkl}.

U	Faces a	Faces b	Faces c	Value s^*_{ijkl}
1	$\sigma_{33} = \sigma_{13} = \sigma_{23} = 0$	$\sigma_{22} = \sigma_{12} = \sigma_{23} = 0$	$\sigma_{11} = 1, \sigma_{12} = \sigma_{13} = 0$	$s^*_{1111} = 2U^{(1)}/V$
2	$\sigma_{33} = \sigma_{13} = \sigma_{23} = 0$	$\sigma_{22} = 1, \sigma_{12} = \sigma_{23} = 0$	$\sigma_{11} = \sigma_{12} = \sigma_{13} = 0$	$s^*_{2222} = 2U^{(2)}/V$
3	$\sigma_{33} = 1, \sigma_{13} = \sigma_{23} = 0$	$\sigma_{22} = \sigma_{12} = \sigma_{23} = 0$	$\sigma_{11} = \sigma_{12} = \sigma_{13} = 0$	$s^*_{3333} = 2U^{(3)}/V$
4	$\sigma_{23} = 1, \sigma_{33} = \sigma_{13} = 0$	$\sigma_{23} = 1, \sigma_{22} = \sigma_{12} = 0$	$\sigma_{11} = \sigma_{12} = \sigma_{13} = 0$	$s^*_{2323} = 2U^{(4)}/V$
5	$\sigma_{13} = 1, \sigma_{33} = \sigma_{23} = 0$	$\sigma_{22} = \sigma_{12} = \sigma_{23} = 0$	$\sigma_{13} = 1, \sigma_{11} = \sigma_{12} = 0$	$s^*_{1313} = 2U^{(5)}/V$
6	$\sigma_{33} = \sigma_{13} = \sigma_{23} = 0$	$\sigma_{12} = 1, \sigma_{22} = \sigma_{23} = 0$	$\sigma_{12} = 1, \sigma_{11} = \sigma_{13} = 0$	$s^*_{1212} = 2U^{(6)}/V$

4.2   Three-Dimensional Case

In the three-dimensional case, the algorithm for finding the components of the compliance tensor is analogous to the two-dimensional one. The tensor S^* is written down in the form:

$$
\begin{bmatrix} \varepsilon_{11} \\ \varepsilon_{22} \\ \varepsilon_{33} \\ 2\varepsilon_{23} \\ 2\varepsilon_{13} \\ 2\varepsilon_{12} \end{bmatrix} = S^* \begin{bmatrix} \sigma_{11} \\ \sigma_{22} \\ \sigma_{33} \\ \sigma_{23} \\ \sigma_{13} \\ \sigma_{12} \end{bmatrix}, \quad S^* = \begin{pmatrix} s^*_{1111} & s^*_{1122} & s^*_{1133} & s^*_{1123} & s^*_{1113} & s^*_{1112} \\ & s^*_{2222} & s^*_{2233} & s^*_{2223} & s^*_{2213} & s^*_{2212} \\ & & s^*_{3333} & s^*_{3323} & s^*_{3313} & s^*_{3312} \\ & & & s^*_{2323} & s^*_{2313} & s^*_{2312} \\ & sym & & & s^*_{1313} & s^*_{1312} \\ & & & & & s^*_{1212} \end{pmatrix}. \quad (14)
$$

Similar to the two-dimensional case, we calculate $U^{(1)}$-$U^{(6)}$ (Table 2), and then use the linearity property to compute the remaining components of s^*_{ijkl} (Table 3), where the values of $U^{(k,l)}$ are calculated by the formula:

$$
U^{(k,l)} = \frac{1}{2}\int_V (\sigma^{(k)}_{ij}\varepsilon^{(l)}_{ij} + \sigma^{(k)}_{ij}\varepsilon^{(l)}_{ij})dV = \int_V \frac{1}{E}(\sigma^{(k)}_{11}\sigma^{(l)}_{11} + \sigma^{(k)}_{22}\sigma^{(l)}_{22} + \sigma^{(k)}_{33}\sigma^{(l)}_{33})dV
$$
$$
- \int_V \frac{\nu}{E}(\sigma^{(k)}_{11}\sigma^{(l)}_{22} + \sigma^{(l)}_{11}\sigma^{(k)}_{22} + \sigma^{(k)}_{22}\sigma^{(l)}_{33} + \sigma^{(l)}_{22}\sigma^{(k)}_{33} + \sigma^{(k)}_{11}\sigma^{(l)}_{33} + \sigma^{(l)}_{11}\sigma^{(k)}_{33})dV
$$
$$
+ \int_V \frac{2(\nu+1)}{E}(\sigma^{(k)}_{12}\sigma^{(l)}_{12} + \sigma^{(k)}_{23}\sigma^{(l)}_{23} + \sigma^{(k)}_{13}\sigma^{(l)}_{13})dV.
$$

$$(15)$$

Table 3. The formulas for computing s^*_{ijkl}.

$s^*_{1122} = U^{(1,2)}/V$	$s^*_{1123} = U^{(1,4)}/V$	$s^*_{1113} = U^{(1,5)}/V$	$s^*_{2313} = U^{(4,5)}/V$	$s^*_{3312} = U^{(3,6)}/V$
$s^*_{1133} = U^{(1,3)}/V$	$s^*_{2223} = U^{(2,4)}/V$	$s^*_{2213} = U^{(2,5)}/V$	$s^*_{1112} = U^{(1,6)}/V$	$s^*_{2312} = U^{(4,6)}/V$
$s^*_{2233} = U^{(2,3)}/V$	$s^*_{3323} = U^{(3,4)}/V$	$s^*_{3313} = U^{(3,5)}/V$	$s^*_{2212} = U^{(2,6)}/V$	$s^*_{1312} = U^{(5,6)}/V$

5   Numerical Solution to a Static Elasticity Problem

The most time-consuming computations are associated with the solution of a series of static problems in the elasticity theory with external stresses given at the boundaries. In the final analysis, these problems are reduced to systems of linear algebraic equations, for which ·it is possible to apply both direct and iterative methods. The fact is, the direct methods, having certain advantages, in this case are not suitable for solving three-dimensional problems due to excessive demands for computer resources. Therefore, for determining effective parameters we have chosen iterative methods.

We propose to find a solution of static problem (10), (11) with the static boundary conditions (3a) by finding the steady-state solution of the dynamic problem of the elasticity theory in the formulation of the stress/displacement velocity with additional dissipative terms to equations of motion (16):

$$\rho \dot{v}_i + \alpha v_i = \sigma_{ij,j} \tag{16}$$

$$\dot{\sigma}_{ij} = C_{ijkl} \dot{\varepsilon}_{kl} = C_{ijkl} v_{k,l} \tag{17}$$

with zero initial conditions for $t = 0$:

$$v_i = 0, \quad \sigma_{ij} = 0 \tag{18}$$

and constant in time boundary conditions on the boundary S (3a). Here $v_i = \dot{u}_i$ is the·displacement velocity of the i-th component of the displacement vector.

In order to show the convergence of problem (16)–(18) to the static problem (10), (11), we use the virial theorem ([10], §10) asserting that the kinetic energy of the mechanical system T averaged over an infinite time interval is equal to the virial averaged over the same time interval. If the potential energy U is a homogeneous function of the first degree of inverse values of the radius vectors, then the relation

$$2T = -U \tag{19}$$

is satisfied.

Therefore it follows that if the kinetic energy of the system is reduced through an artificially introduced damping mechanism, then the rigid connection between the kinetic and the potential energies provided by this theorem leads to a decrease in the potential energy up to its minimum. Then, based on the Lagrange-Dirichlet principle, for a statically stressed body (*of all possible stress-strain states of a deformable solid, the actual stress state corresponds to a minimum of the total deformation energy*), we can conclude that the solution of the dynamic problem (16)–(18) converges to the solution of the stationary problem (10), (11). For a numerical solution of the initial boundary value problem (16)–(18), we apply a finite difference scheme on staggered grids [18], whose coefficients are modified to provide approximation in heterogeneous media [8,17].

6   Parallel Implementation

6.1   MPI/OpenMP Parallelization

The most time-consuming part of the algorithm, which in the three-dimensional case requires the obligatory parallel implementation, consists in solving six stress-strain linear elasticity problems for calculating the potential energy of the elastic deformations in the sample under boundary static stresses. As these problems can be solved independently, the most natural way is to use the MPI parallelization to split the calculations to individual tasks. Further, the solution of an individual task can be parallelized with the MPI or the OpenMP, depending on the number of nodes and cores the problem is to be solved. This is the commonly used strategy of parallelization. Briefly, let us consider these versions.

MPI. Parallelization has two stages. At the first stage, using the MPI group constructor, the solution of the problem is divided into six independent tasks, as is mentioned above. Each task is assigned to its independent MPI group. Each group solves the local problem (16)–(18) with the help of finite difference time domain staggered grid scheme combined with the domain decomposition method. The domain decomposition is applied in order to decompose the original computational domain to multiple elementary subdomains of lower dimensions, each one being handled by its individual Processor Unit (PU) thus solving the system of equations within the subdomain (Fig. 1). Updating unknown data while moving from a time layer to the next one requires the exchange of values in the grid nodes along the interface between the adjacent subdomains. The message passing library MPI is used to communicate data between neighboring PU. The necessity of this exchange negatively impacts the scalability of the method. However, the impact is less visible on the 3D domain decomposition than on one- and two-dimensional ones [9]. In this implementation, we choose the 3D domain decomposition. In order to reduce the idle time, the asynchronous computations based on the non-blocking MPI procedures are used. The non-blocking MPI functions Isend()/MPI Irecv() allow us to overlap communications and computations, thus hiding communication latencies and improving the performance of an MPI application.

MPI+OpenMP. The choice of a specific method of parallelization depends on the number of resources allocated to solve the problem. If, for example, we are limited by only six nodes, a possible way to numerically solve the problem may be the use of a combination of the MPI with the OpenMP. In this case, parallelization is also performed in two stages. At the first stage, the MPI is

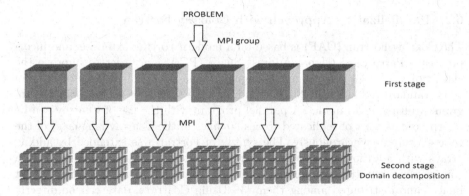

Fig. 1. MPI parallelization scheme.

used in order to split the calculations to individual tasks. Then the solution of an individual task is parallelized with the OpenMP, using the threads with shared memory on the node (Fig. 2). This approach is simpler in terms of writing a code, but has a limitation on the number of nodes used. In both cases, the MPI or the MPI+OpenMP, after solving six stress-strain elasticity problems, each process sending the calculated values to the zero process, which saves them into a disk as binary files, containing the values of the stress components in the representative volume. After this, the zero process produces a sequential reading of the information from files and calculates the result using the formulas from Table 2. The time required for the zero process for this operation is negligible as compared to the time needed for solving the static elasticity problems. From the MPI+OpenMP parallelization scheme, it follows, that the best architecture for calculating the problem: the choice of six nodes with a maximum number of cores per node.

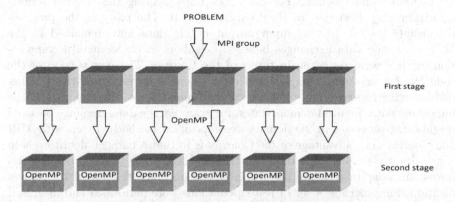

Fig. 2. The MPI+OpenMP parallelization scheme.

6.2   Parallelization Approach with Coarray Fortran

The Coarray Fortran (CAF) is based on a modern Fortran extension and incorporates a Partitioned Global Address Space (PGAS) in order to improve the clarity of a parallel programming language. The CAF is a feature of Fortran 2008 standard published in 2010 [5] and, like the MPI is based on a Single Program, Multiple Data model. A parallel program with the use of Coarray can be interpreted as a set of replicated copies (*images* in the Coarray language) of the code executed asynchronously. The syntax of Fortran was extended by adding arrays with additional trailing subscripts in square brackets, which provide a concise representation of references to data that can be accessed from other images and distributed among them [3]. Using Coarrays, data can be directly accessed in the neighbor memory without sending and receiving functions. Since the MPI uses the same SPMD model, the Fortran features allow the MPI and Coarray live together in a program. This fact is very convenient for a gradual conversion of the MPI program to a Coarray language.

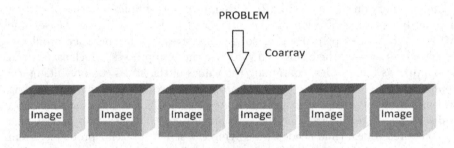

Fig. 3. Coarray parallelization scheme.

The MPI and OpenMP Fortran codes implementing the above-described algorithm were rewritten in the Coarray (Fig. 3). The parts of the program responsible for the parallel input/output of big data have remained in the MPI, while the data exchanges between neighbors in the domain decomposition method were rewritten in terms of the Coarray. The new version of the code has become more compact and clear, there is no need to write sending and receiving messages and to check the correspondence of the packing and the unpacking data. Figure 4 demonstrates, for example, the data exchange between neighboring processes in 2D domain decomposition method written with MPI and Coarray. The advantage of the Coarray is in that a parallel algorithm is in a significantly simpler style than the MPI and less prone to the programmer's errors. In order to estimate the real acceleration and efficiency of the above-mentioned approaches, a set of test calculations were performed and discussed below.

MPI	Coarray
real :: u(0:N+1, 0:M+1) ... call mpi_isend(u(1,1:M), M, mpi_real, top(myid), tag1,...) call mpi_irecv (u(N+1,1:M), M, mpi_real,bottom(myid), tag1,...) call mpi_isend(u(N,1:M), M, mpi_real,bottom(myid), tag2,...) call mpi_irecv (u(0,1:M), M, mpi_real, top(myid), tag2,...) call mpi_isend(u(1:N,M), N, mpi_real, right(myid), tag3,...) call mpi_irecv (u(1:N,0), N, mpi_real, left(myid), tag3,...) call mpi_isend(u(1:N,1), N, mpi_real, left(myid), tag4,...) call mpi_irecv (u(M+1,1:N), N, mpi_real, right(myid), tag4,...) call mpi_waitall(...)	real :: u(0:N+1, 0:M+1)[pN,*] ... u(N+1,1)[top(1),top(2)] = u(1,1:M) u(0,1:M)[bottom(1),bottom(2)] = u(N,1:M) u(1:N,0)[right(1),right(2)] = u(1:N,M) u(1:N,M+1)[left(1),left(2)] = u(1:N,1) sync all

Fig. 4. Data exchange between neighboring processes in 2D domain decomposition method written with MPI and Coarray.

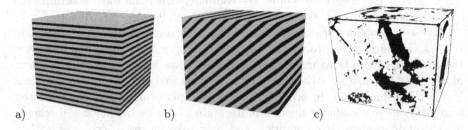

a)   b)   c)

Fig. 5. Different models of layered media (a–b) and the segmented digital model of a core sample (c).

7   Numerical Experiments

7.1   Validation of the Numerical Algorithm

To validate the algorithm proposed for estimating the effective elastic parameters of a rock core sample, a representative series of numerical experiments has been carried out.

First, the homogeneous isotropic materials samples (plexiglas, copper and steel) were considered, and the calculated effective parameters were compared with the elastic moduli of a material itself (the difference is 10^{-6}).

Second, the elastic moduli for the samples of layered materials were calculated (Fig. 5). The size of the models varied along the interlayers, across them, the number of layers and their incline being changed. The results of the method proposed were compared with the Schoenberg averaging method [13]. The difference decreased with increasing the size along the interlayer. For the model of 500*500*30 size along the interlayer this difference was about 4%.

Finally, calculations were done for a three-dimensional segmented digital 500*500*500 model of a carbonate core (Fig. 5c). The seismic velocities were estimated and compared with the results of laboratory measurements. The difference makes up less than 3%.

7.2 Comparison of MPI, MPI+OpenMP and Coarray Fortran

To compare the performance of the MPI, MPI+OpenMP and the Coarray Fortran communications in terms of the speedup, a three-dimensional segmented
digital model of a carbonate core from the previous section (Fig. 5c) was considered as a validation test. To assess a strong scaling, the problem of 500*500*500
size remains the same as the number of processors (cores) increases. This test
was performed on the Siberian Supercomputer Center cluster, Novosibirsk,
Russia, that includes 27 CPU Intel Xeon E5-2697A v4 with 16*2 logical cores
per node (32 threads). The Intel Fortran Compiler 2019.1.144 was used to create an executable file. We have chosen the performance on 96 cores as a baseline
for comparison. The CPU number is scaled from 3 to 16. Figure 6 (on the left)
presents the strong scaling results measured for the MPI, MPI+OpenMP and
the Coarray. The measured values are compared with the ideal speedup. We
observe the speedup of about 4.5 when scaling from 96 cores to 576 cores.

To estimate weak a scaling, we have to increase the problem size at the same
rate as the number of processors, keeping the amount of work per processor
the same. To be able to make this comparison, we have chosen the problem
of 500*500*100 size as a baseline and have increased the size of the problem
in the third dimension. In order to analyze the performance, we have limited
computations to a constant number of iterations, because the numerical scheme
is subject to the stability condition and may take longer to converge with a
denser grid. Figure 6 (on the right) presents the week scaling efficiency results
measured for the MPI, MPI+OpenMP and the Coarray Fortran versions of the
code. The results presented demonstrate that with an increase in the number of
cores, the CAF implementation provides comparable performance to an equivalent MPI version, while the MPI+OpenMP hybrid model is worse than the pure
MPI model and CAF. We are aware of the fact it is quite possible that the best
implementation of the MPI+OpenMP among many possible implementations
was not chosen. However we did not intend to compare all alternative task distributions among the MPI processes and the OpenMP threads within them for

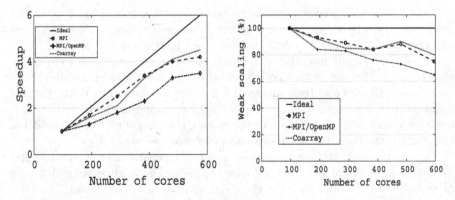

Fig. 6. The strong scaling speedup (on the left) and the week scaling efficiency (on the
right) on CPU Intel Xeon E5-2697A v4.

choosing the best one. We were aimed at revealing whether the Coarray can be considered to be an alternative to the conventional parallel parallelizations.

8   Conclusion

We have implemented the three parallel algorithms, written in Fortran 90 language, to solve the problem to determine the effective elastic moduli of a rock core sample from their tomographic images. This new method is based on the energy equivalence principle and the new approach to solving the 3D static elasticity problem by the iterative relaxation technique. The elastic moduli are determined by the parallel computation of potential energy of the elastic deformations arising in a sample under a certain homogeneous stress applied to the boundary, thus simulating the effects occurring in laboratory measurements. The numerical experiments have shown a high accuracy of the method proposed.

The three parallel implementation strategies of this method were compared in terms of performance and ease of programming. From this comparison it follows that Coarray Fortran can be considered as an alternative to the MPI and MPI+OpenMP. The analysis of acceleration and efficiency shows that Coarrays start to manifest superiority when using rather a large number of processes.

9   Author's Contribution and Funding

Galina Reshetova has developed the 3D parallel implementations for static virtual experiments with MPI, OpenMP and Coarray Fortran. Vladimir Cheverda has justified convergence of the relaxation processes used for the 3D elastic static solution. Tatyana Khachkova has prepared digital core samples and performed numerical experiments. Galina Reshetova has been supported by the Russian Science Foundation project 19-77-20004. Vladimir Cheverda and Tatyana Khachkova have been supported by RFBR project 19-01-00347.

Acknowledgements. The research has been carried out using the equipment of the shared research facilities of HPC computing resources at the Joint Supercomputer Center of RAS, the Siberian Supercomputer Center of SB RAS and the Irkutsk Supercomputer Center of SB RAS.

References

1. Aboudi, J.: Mechanics of Composite Materials: A Unified Micromechanical Approach. Elsevier Science, Amsterdam (1991)
2. Belonosov, M.A., Kostov, C., Reshetova, G.V., Soloviev, S.A., Tcheverda, V.A.: Parallel numerical simulation of seismic waves propagation with intel math Kernel library. In: Manninen, P., Öster, P. (eds.) PARA 2012. LNCS, vol. 7782, pp. 153–167. Springer, Heidelberg (2013). https://doi.org/10.1007/978-3-642-36803-5_11
3. Chivers, I., Sleightholme, J.: Introduction to Programming with Fortran. Springer, Cham (2015). https://doi.org/10.1007/978-3-319-17701-4

4. Christensen, R.: Introduction to Mechanics of Composite Materials, 1st edn. Wiley, New York (1979)
5. ISO/IEC 1539–1:2010, Fortran – Part 1: Base language, International Standard (2010)
6. Dupuy, B., Garambois, S., Virieux, J.: Estimation of rock physics properties from seismic attributes - Part 1: Strategy and sensitivity analysis. Geophysics **81**(3), M35–M53 (2016)
7. Dupuy, B., et al.: Estimation of rock physics properties from seismic attributes – Part 2: Applications. Geophysics **81**(4), M55–M69 (2016)
8. Kostin, V., Lisitsa, V., Reshetova, G., Tcheverda, V.: Parallel algorithm with modulus structure for simulation of seismic wave propagation in 3D multiscale multiphysics media. In: Malyshkin, V. (ed.) PaCT 2017. LNCS, vol. 10421, pp. 42–57. Springer, Cham (2017). https://doi.org/10.1007/978-3-319-62932-2_4
9. Kostin, V., Lisitsa, V., Reshetova, G., Tcheverda, V.: Simulation of seismic waves propagation in multiscale media: impact of cavernous/fractured reservoirs. In: Jónasson, K. (ed.) PARA 2010. LNCS, vol. 7133, pp. 54–64. Springer, Heidelberg (2012). https://doi.org/10.1007/978-3-642-28151-8_6
10. Landau, L.D., Lifshitz, E.M.: Mechanics. Nauka, Moscow (1988)
11. Pevzner, R., et al.: Feasibility of time-lapse seismic methodology for monitoring the injection of small quantities of $CO2$ into a saline formation, CO2CRC Otway Project. Energy Procedia **37**, 4336–4343 (2013)
12. Reshetova, G., Khachkova, T.: Parallel numerical method to estimate the effective elastic moduli of rock core samples from 3D tomographic images. In: Dimov, I., Faragó, I., Vulkov, L. (eds.) FDM 2018. LNCS, vol. 11386, pp. 452–460. Springer, Cham (2019). https://doi.org/10.1007/978-3-030-11539-5_52
13. Schoenberg, M., Muir, F.: A calculus for finely layered anisotropic media. Geophysics **54**(5), 581–589 (1989)
14. Sendetski, J. (ed.): Composition Materials, vol. 2. Mechanics of Composition Materials [Russian translation]. Mir, Moscow (1978)
15. Shermergor, T.: The Theory of Elasticity of Microinhomogeneous Media [Russian translation]. Nauka, Moscow (1977)
16. Zhang, W., Dai, G., Wang, F., Sun, S., Bassir, H.: Using strain energy-based prediction of effective elastic properties in topology optimization of material microstructures. Acta. Mech. Sin. **23**(1), 77–89 (2007)
17. Vishnevsky, D., Lisitsa, V., Tcheverda, V., Reshetova, G.: Numerical study of the interface errors of finite-difference simulations of seismic waves. Geophysics **79**, T219–T232 (2014)
18. Virieux, J.: P-SV wave propagation in heterogeneous media: velocity-stress finite-difference method. Geophysics **51**, 889–901 (1986)

Computational Issues in Construction of 4-D Projective Spaces with Perfect Access Patterns for Higher Primes

Shreeniwas N. Sapre[1]([✉]) [iD], Sachin B. Patkar[2], and Supratim Biswas[1]

[1] Department of Computer Science and Engineering, IIT Bombay, Mumbai, India
{sapre,sb}@cse.iitb.ac.in
[2] Department of Electrical Engineering, IIT Bombay, Mumbai, India
patkar@ee.iitb.ac.in

Abstract. Matrix operations are some of the important computations in scientific and engineering domains. Parallelization approaches for such operations have been a common topic of research. One of the novel approach proposed during the 90s is architectures based on finite projective spaces. A key benefit of this approach is the communication efficiency that can be achieved by exploiting perfect access patterns in the architecture. Such spaces of dimension 4 appear suitable for matrix-matrix multiplication and are amenable for distributions with good performance potential. The construction of such 4-dimensional spaces with perfect access patterns, however, has been reported only for the smallest space – the one corresponding to prime 2. In this paper, we explore the construction for primes greater than 2. We compare two alternative methods for computational construction of such spaces, based on their efficiency. We present the successful construction of such a space for prime 3 and indicate directions for future work.

Keywords: Projective spaces · Parallel computations

1 Introduction

Matrix operations are common to many applications, from fluid dynamics equations to algorithms used to rank the result of web searches. Linear system solvers often use LU or Cholesky factorization. LU decomposition of a matrix $\mathbf{A}$ decomposes the matrix into two - a lower triangular matrix $\mathbf{L}$ and an upper triangular matrix $\mathbf{U}$. Cholesky decomposition decomposes a *symmetric positive definite matrix* $\mathbf{A}$ as $\mathbf{A} = \mathbf{L} \cdot \mathbf{L}^{\mathbf{T}}$. Multiplication of two matrices has often been used as an indicator of the performance of a single computer, or a cluster. As the matrix sizes involved in engineering applications tend to be large, parallelizing such algorithms has been an actively researched topic [2].

Parallelization of matrix computations often requires splitting the matrix across memory blocks and distributing the actual computations across processing

© Springer Nature Switzerland AG 2019
V. Malyshkin (Ed.): PaCT 2019, LNCS 11657, pp. 245–259, 2019.
https://doi.org/10.1007/978-3-030-25636-4_20

units [3]. To address these, Karmarkar used principles of *finite projective geometries* to propose a novel interconnection scheme [4]. Finite projective geometries are parameterized by their dimension d and a prime power p^k. The elements of these geometries correspond to the processors and memories. The incidence relations of the geometry elements correspond to the interconnections. The allocation of data to memory blocks and computations to processors is governed by the geometry and its incidence relations. The definition of the scheme ensures that the communication of data is carried out without any conflicts, and involves all the processors and memories, by exploiting perfect-access patterns and sequences based on the properties of the geometry. These characteristics can be used to develop algorithms for various problems.

In an earlier, as yet unpublished write-up [7] we investigated the properties of Sparse Matrix-Vector multiplication with a 2 dimensional (or 2-D for short) projective space *for any prime*, and LU Decomposition with a 4 dimensional (4-D) projective space corresponding to prime 2, with promising results. We are extending these results in our ongoing research. We are interested in understanding the potential of 4-D finite projective spaces for higher primes, currently in the context of matrix-matrix multiplication.

Performance comparisons in such studies are usually addressed as an asymptotic analysis of the resource appetite - memory footprint, amount of computations or communications - of the scheme. However, an asymptotic analysis may not be the best choice for this *first study of 4-D projective spaces*. The sizes of the 4-D projective spaces (*this corresponds to the number of computers in the network*) increase rapidly for higher primes. Moreover, there are large gaps in the sizes corresponding to consecutive primes. For instance, the number of computers in the space for prime 2 is 155, for prime 3 is 1210, for prime 5 is 20306 and for prime 7 is 140050 (Table 1). Hence the number of distinct practical 4-D projective spaces we can consider in our *first* comparison is a discrete few. In light of this, we aim first at an exact comparison of the communication appetites. Once we get a concrete handle on the exact comparisons, we plan to address an asymptotic comparison in a subsequent study.

Table 1. Characteristics of subspaces of 4-D spaces for some primes

Prime	#Points	#Lines	#Planes	#Lines incident on every plane
2	31	155	155	7
3	121	1,210	1,210	13
5	781	20,306	20,306	31
7	2,801	140,050	140,050	57
11	16,105	1,964,810	1,964,810	133
13	30,941	5,259,970	5,259,970	183
17	88,741	25,734,890	25,734,890	307
19	137,561	49,797,082	49,797,082	381

The construction of 2-D projective spaces is well-understood and is computationally simple as well. Construction of the smallest 4-D projective space corresponding to prime 2 has been documented in the literature [7]. Constructing 4-D spaces for higher primes turns out to be non-trivial *if targeting the desirable property of perfect access patterns*. Constructing these spaces for higher primes becomes an essential step since we are aiming for an exact comparison of the communication appetite.

In this paper, we document our study of two approaches for computationally constructing 4-D spaces for higher primes with perfect access patterns, their performance comparison at a high level, as'well as the implementation of the selected approach. We report the successful construction of such a space for prime 3, for what we believe is the first time in literature. We conclude the paper by summarizing the results, as well as spelling out future work directions enabled by this research.

2   Projective Spaces Based Interconnect Topologies

We describe the basic concepts of projective spaces first. Projective spaces are usually constructed from Galois fields [6]. Consider a finite field $\mathbb{F} = GF(p^k)$ with p^k elements for a prime p.

An example *Finite Field* can be generated as follows. For each value of s in $GF(s)$, one needs to first find a *primitive polynomial* for the field. Such primitive polynomials are well-known and have been documented in the past. The smallest geometry, a 2-D geometry for prime 2 is generated using $GF(2^3)$. One primitive polynomial for this finite field is $(x^3 + x + 1)$. Powers of the root of this polynomial, x, are then successively taken, $2^3 - 1$ times, modulo this polynomial - and during this operation, the coefficients are treated modulo 2. This means, x^3 is substituted with $(x + 1)$, wherever required, since over base field $GF(2)$, $-1 = 1$. A *sequence* of such evaluations leads to the generation of the sequence of $(s - 1)$ finite field elements, **other than** 0. Thus, the sequence of 2^3 elements for $GF(2^3)$ is **0** (by default), $\alpha^0 = 1$, $\alpha^1 = \alpha$, $\alpha^2 = \alpha^2$, $\alpha^3 = \alpha + 1$, $\alpha^4 = \alpha^2 + \alpha$ (see note[1]), $\alpha^5 = \alpha^2 + \alpha + 1$, $\alpha^6 = \alpha^2 + 1$. In the multiplicative group of these non-zero elements, which is used to construct the subspaces of the geometry, the point α^i is often denoted by its index i.

To avoid mixing up the two notations, we term the α^i notation describing the point as the **point notation**, and i notation describing the index of the point as the **index notation**. Since $\alpha^i \times \alpha^j = \alpha^{i+j}$, multiplications in the point notation correspond to additions in the index notation. Similarly, $(\alpha^i)^j = \alpha^{i \times j}$, and exponentiation in point notation corresponds to multiplications in the index notation.

A projective space of dimension d is denoted by $\mathbb{P}(d, \mathbb{F})$, and consists of one-dimensional subspaces of the $(d+1)$-dimensional vector space over $\mathbb{F}$. The *points* in this space correspond to the zero-dimensional subspaces. The total number of

---

[1] $\alpha^4 = \alpha * \alpha^3 = \alpha * (\alpha + 1)$.

points in $\mathbb{P}(d, \mathbb{F})$ are $P(d) = \dfrac{(s^{d+1} - 1)}{s - 1}$. Let us denote the collection of all the l-dimensional projective subspaces by Ω_l. To count the number of elements in each of these sets, we define the function ϕ [4].

$$\phi(n, l, s) = \frac{(s^{n+1} - 1)(s^n - 1) \cdots (s^{n-l+1} - 1)}{(s - 1)(s^2 - 1) \cdots (s^{l+1} - 1)} \qquad (1)$$

The number of m-dimensional subspaces of $\mathbb{P}(d, \mathbb{F})$ is $\phi(d, m, s)$. Hence, the number of l-dimensional subspaces contained in an m-dimensional subspace (where $0 \leq l < m \leq d$) is $\phi(m, l, s)$, while the number of m-dimensional subspaces containing a particular l-dimensional subspace is $\phi(d-l-1, m-l-1, s)$. For details on projective space construction, refer [6]. We summarize subspaces of 2-D and 4-D spaces in Table 2.

Table 2. Characteristics of subspaces of 2-D and 4-D spaces for prime p

Dimension	#Points	#Lines	#Points/Line	#Planes	#Lines/Plane
2	$p^2 + p + 1$	$p^2 + p + 1$	$p + 1$	-	-
4	$\frac{p^5 - 1}{p - 1}$	$\frac{(p^5 - 1)(p^2 + 1)}{p - 1}$	$p + 1$	$\frac{(p^5 - 1)(p^2 + 1)}{p - 1}$	$p^2 + p + 1$

For example, to generate *Projective Geometry* corresponding to above Galois Field example ($\mathbb{GF}(2^3)$), the 2-d projective plane, we treat each of the *non-zero* element as a point of the geometry. Further, we pick subfields (vector subspaces) of $\mathbb{GF}(2^3)$, and label them as lines. Thus, the 7 lines of the projective plane are $\{1, \alpha, \alpha^3 = 1 + \alpha\}$, $\{1, \alpha^2, \alpha^6 = 1 + \alpha^2\}$, $\{\alpha, \alpha^2, \alpha^4 = \alpha^2 + \alpha\}$, $\{1, \alpha^4 = \alpha^2 + \alpha, \alpha^5 = \alpha^2 + \alpha + 1\}$, $\{\alpha, \alpha^5 = \alpha^2 + \alpha + 1, \alpha^6 = \alpha^2 + 1\}$, $\{\alpha^2, \alpha^3 = \alpha + 1, \alpha^5 = \alpha^2 + \alpha + 1\}$ and $\{\alpha^3 = 1 + \alpha, \alpha^4 = \alpha + \alpha^2, \alpha^6 = 1 + \alpha^2\}$. The corresponding geometry can be seen in Fig. 1.

Automorphisms: Automorphisms on the points of a projective space retain the incidence relations, mapping lines to lines, and planes to planes. Thus, if f is an automorphism, the line between points i and j will be mapped to the line between points $f(i)$ and $f(j)$. Similarly, the plane containing i, j, and k will be mapped to the plane containing $f(i)$, $f(j)$ and $f(k)$. The **Frobenius** and **shift** automorphisms are of particular interest. The Frobenius automorphism is $\Phi(x) = x^p$, p being the characteristic of $\mathbb{F}$. This automorphism corresponds to multiplying index of each point by p modulo the number of points. Similarly, the shift automorphism corresponds to *adding 1* to each point index modulo the number of points. We will revisit the automorphisms and their important role in the construction in subsequent sections.

Order of Frobenius Automorphism: Note that $x^{p(d+1)} \equiv x \; mod(p)$. Also, $x^{p(d+1)}$ corresponds to multiplying p and $(d+1)$ in the index notation, which is the same as raising x^p to $(d+1)^{\text{th}}$ power in the point notation. This operation is the same as $(d+1)$ repeated applications of the Frobenius automorphism. Thus, $(d+1)$, or 5 repeated applications of this automorphism cycle back to original point.

$$\text{Frobenius automorphism is cyclic with order 5} \tag{2}$$

Order of Shift Automorphism: The shift automorphism maps each point index to the next point index, and thus shifts the points cyclically. Hence, repeated application of this automorphism by the number of points cycles back to the original point.

$$\text{Shift automorphism is cyclic with order equal to number of points, i.e. } \frac{p^5 - 1}{p - 1} \tag{3}$$

Since Frobenius and Shift automorphisms are cyclic, each generates a subgroup.

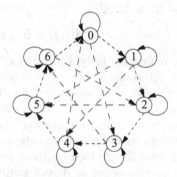

Fig. 1. 2-D projective geometry

For projective spaces, Karmarkar had evolved an architecture [4] for parallel computing. There are problems which have been found to be amenable for parallel computation using processing units and memories, connected using subgraph of an instance of projective geometry. For such problems, we choose a pair of dimensions d_m and d_p. The subspaces of dimension d_p and d_m are associated with the processing units and memories respectively. If a particular subspace of dimension d_m and d_p have a non-empty intersection, the corresponding processing unit and memory are connected together [4].

For a 2-D space, a *perfect access pattern* is defined to be a collection of tuples $\{(a_i, b_i), i = 1, \ldots n\}$, where n denote the number of points in the space, satisfying:

1. Each of the collections of points $\{a_i\}$ and $\{b_i\}$ forms a permutation of all points in the geometry
2. The collection of lines $l_i = \langle a_i, b_i \rangle$ forms a permutation of all lines in the geometry

Once the problem is broken down to parallelizable atomic computations, and corresponding memory blocks for storing data, the computations and memory blocks can be allocated to processors and memory ensuring that the computation takes place on a processor directly connected to the memory holding the requiring data. Thus, data required for computation is brought in parallelly, computations on each processor are carried out parallelly, leading to efficient and conflict-free use of resources. A weak point of projective geometry based schemes may be the fact that the number of processes is determined by the projective space, so it is likely that some modern hardware may be used inefficiently.

In the context of 4-D spaces, the scheme proposed by Karmarkar for allocating and sequencing data/computations to the memories and processors respectively, for the problem of LU Decomposition, is applicable for matrix multiplication as well. In this scheme, memories correspond to the lines of a 4-D space, and processors to the planes of a 4-D space. The block row and column indices of the matrices are mapped to the points of the same space. A typical operation in matrix multiplication is of the form

$$C(i,k) \leftarrow C(i,k) + A(i,j) \times B(j,k)$$

In general case, every two points uniquely determine a line (memory), and two lines/three points uniquely determine a plane (processor). When a triplet (i, j, k) is in a general position, the points i, j, k uniquely determine a plane, say, $\mathcal{P}$. The computation of the partial sum $A(i,j) \times B(j,k)$ is allocated to the processor corresponding to $\mathcal{P}$. To carry out this computation, the processor $\mathcal{P}$ needs to communicate with the memory modules corresponding to the point pairs (i, j), (j, k) and (i, j). Since the lines determined by these pairs are incident on the plane $\mathcal{P}$, the necessary connections exist as **direct connections**.

A *perfect access pattern* for a 4-D space is defined to be a collection of triplets $\{(a_i, b_i, c_i), i = 1, \ldots n\}$, where n is the number of lines in the space, satisfying:

1. Each of the collections of lines $\{\langle a_i, b_i \rangle\}$, $\{\langle b_i, c_i \rangle\}$ and $\{\langle c_i, a_i \rangle\}$ forms a permutation of all lines in the geometry
2. The collection of planes $\langle a_i, b_i, c_i \rangle$ forms a permutation of all planes in the geometry

It is easy to see that in a perfect access pattern, the *sequence* of the points in the triplet is critical. Each triplet is a 3-tuple, and not an unordered set of 3 points. In an analogous way, the construction of 2-D spaces also relies on the sequence of points in a line to ensure perfect access patterns.

We now see the detailed methods to construct projective spaces.

3   Construction of Projective Space

3.1   2-D Space Construction

The method described in Sect. 2, viz. identifying a primitive polynomial, then identifying points (0-dimensional subspaces), and then lines (1-dimensional subspaces), can be used to construct a 2-D projective space for any prime power p^k.

As a naive method, one can try all possible linear combinations of distinct points, and enumerate all the lines. For instance, with the geometry for $p = 2$ described in the earlier section, if the lines are identified as *sets* of points, the lines will get enumerated as in the columns marked ¶ in Table 3. In this enumeration, it is easy to see that none of the columns of points is a *permutation* of all the points.

Table 3. Lines in 2-D space as *sets* vis-a-vis *tuples*

Line	¶ Lines as sets			§ Lines as tuples		
	Point 1	Point 2	Point 3	Point 1	Point 2	Point 3
0	0	1	3	0	1	3
1	1	2	4	1	2	4
2	2	3	5	2	3	5
3	3	4	6	3	4	6
4	0	4	5	4	5	0
5	1	5	6	5	6	1
6	0	2	6	6	0	2
	Does not produce permutations			Produces permutations		

Obviously, this method does not retain the sequence of points. It is possible to construct the space by exploiting the structure for the 2-D spaces, by using the **Shift** automorphisms. Here, it is important to note that the repeated applications of the shift automorphism give rise to n distinct automorphisms where n is the number of points in the space. Thus the repeated applications of shift automorphism on a point result in *cycling* through all the points. We see that the tuple of points $(0, 1, 3)$ corresponding to line number 0 in columns marked § in Table 3, yields each subsequent line, when each of the point index is incremented by 1 modulo 7 - the number of points. Thus, by repeated application of the shift automorphism on the tuple of points corresponding to first line, we can construct the incidence relation of lines to points, *while also generating the perfect access pattern*. Also note that applying the shift automorphisms to a *tuple* naturally results in another tuple. This can be seen clearly in the column marked § in Table 3. We also note that the order of the shift automorphism is the number of points, and that for a 2-D space, the number of points and lines is exactly the same. We see that *shift automorphism is adequate to generate all the lines in a 2-D space.*

The construction of 2-D spaces is summarized in Algorithm 1.

3.2 Construction of 4-D Projective Space - Permuting Orbit Representatives

In existing literature [1,7], the generation of 4-D projective spaces *with perfect access patterns* has been detailed out only for prime $p = 2$. The generation for higher primes has not been reported in the literature yet.

Algorithm 1. Identify lines in 2-D space $\mathbb{F}$

---

1. Consider points $x = 0$ and $y = 1$
2. Identify new line w as the tuple consisting of maximal set of linear combinations of x and y.
    - Let $X = \{0, 1\}$, the set of points *identified till now* on the line w
    - Iterate while no new points can be added to X
        - Find $X_{next} = \{c_1 x_1 + c_2 x_2, c_i \in Z_p, x_i \in \mathbb{F}\}$
        - Stop iteration if $X_{next} = X$
        - $X \leftarrow X_{next}$
1. $w \leftarrow \text{tuple}(X)$
2. Apply shift transformation on w successively to identify additional lines.

We saw that the shift automorphism can only generate n elements (points, lines, planes) from a given element, since it has order n (number of points in the space). From Table 2, we see that the number of lines or planes in a 4-D space is much higher than the number of points, and hence we conclude that shift automorphism alone will not be adequate to generate the entire space.

Will using a combination of Frobenius and shift automorphisms be adequate? It turns out that adding the Frobenius automorphism for generation increases the number of elements that can be generated, but still can generate the **entire** space only for $p = 2$ and for no higher primes. The theorem below clearly shows the reasons behind this inadequacy, and overcoming this limitation by identifying ways to **generate the entire space** is the main focus of the work reported in this paper.

Theorem 1. *Frobenius and shift automorphisms together can generate the entire 4-D projective space for $p = 2$ and for no higher primes*

Proof. 1. Number of lines in the space is $\dfrac{(p^{5-1})(p^2 + 1)}{(p - 1)}$

2. Frobenius automorphism has order $d + 1 = 5$

3. Shift automorphism has order $\dfrac{(p^{5-1})}{(p - 1)}$

4. Repeated application of *only* Frobenius and Shift automorphisms on a single line can generate $5 \times \dfrac{(p^{5-1})}{(p - 1)}$ other lines.

5. $p^2 + 1 = 5$ for $p = 2$, and $p^2 + 1 > 5$ for $p > 2$ □

These special automorphisms can be considered for generation of *parts* of the entire space. We first state and prove a few results on these parts that are thus generated.

Definition 1. *For $\mathbb{F} = \mathbb{P}(4, GF(p))$, we define the **F-subgroup** as the set of repeated applications of the Frobenius automorphism; **S-subgroup** as the set of repeated applications of the Shift automorphism, and **SF** as the union of these two subgroups.*

Since both the Frobenius and Shift automorphisms are cyclic, their repeated applications would result in a subgroup. In our current research work, we do not investigate the algebraic structure **SF** in further details, since we only look at its adequacy (*or lack of it*) in generating the entire space. We denote by n, the number of points in $\mathbb{F}$, i.e. $\dfrac{p^{4+1} - 1}{p - 1}$.

We state some results and prove the not so obvious ones about the orbits with respect to these subgroups.

Theorem 2. *1. The orbit of any point under the F-subgroup contains exactly 5 distinct elements.*

2. The orbit of any point under the S-subgroup contains exactly n distinct elements.

3. The orbit of any line or plane under the S-subgroup contains exactly n distinct elements.

4. The orbit of any line or plane under the F-subgroup contains either 1 or exactly 5 distinct elements.

Proof. Most of the statements in this theorem are trivial. We provide an example for (3) where the F-subgroup application can result in a single line.

For $\mathbb{P}(4, GF(5))$, the first line identified through points 0 and 1 has points $\{0, 1, 5, 25, 125, 625\}$. When each point is multiplied by 5 modulo 781 (number of points in the space), the resultant set of points is identical to the same set. $\square$

These results allow us to identify the parts that can be generated by **SF**.

Theorem 3. *The orbit of any line/plane under **SF** has either n or $5 \times n$ elements.*

Theorem 4. *There are $\left\lfloor \dfrac{p^2 + 1}{5} \right\rfloor + ((p^2 + 1)\%5)$ orbits of **SF**, where % denotes the integer remainder.*

Proof. 1. There are $n \times (p^2 + 1)$ lines/planes, and each orbit contains either n or $5 \times (p^2 + 1)$ lines/planes.

2. Thus, there are $\left\lfloor \dfrac{p^2 + 1}{5} \right\rfloor$ orbits each containing $5 \times (p^2 + 1)$ lines/planes.

3. Remaining $(p^2 + 1)\%5$ orbits each has n lines/planes

4. For instance, $p = 5$ results in $\left\lfloor \frac{26}{5} \right\rfloor$ i.e. 5 orbits each with 781×5 i.e. 3905 lines, and 1 orbit with 781 lines, to result in the total of 20306 lines.

$\square$

We define the function Ξ as this number of orbits: $\Xi(p) = \left\lfloor \dfrac{p^2 + 1}{5} \right\rfloor + ((p^2 + 1)\%5)$

With this background, we now present two different approaches for generating the entire space, and consider the characteristics of these approaches. The first approach attempts to generate one orbit at a time, and then permutes

---

Algorithm 2. 4-D Space Generation by permuting orbit representatives

---

1. **Identify a new Line**
   (a) Identify two points p_1 and p_2, such that the line joining these two points ($\{c_1 p_1 + c_2 p_2, \forall c_i \in Z_p\}$) has not yet been identified
2. **Identify a new Plane**
   (a) Identify two lines l_1 and l_2, such that the line joining l_1 and l_2 has not yet been identified.
   (b) Find new plane P as all linear combinations of points on these two lines $\{c_1 p_1 + c_2 p_2, \forall c_1 \in l_1, c_2 \in l_2\}$
3. **Generate new Lines/Planes**
   (a) Repeat
       i. Let l be a new line identified by line-identification step
       ii. Find the set of lines in the orbit of l by repeated application of Frobenius and Shift automorphisms
       iii. Let P be a new plane identified by plane-identification step
       iv. Find the set of planes in the orbit of P by repeated application of Frobenius and Shift automorphisms
   (b) until no new lines/planes can be added
4. **Permute the orbits**
   (a) Permute the sequence of lines in **each** of the planes P identified in 3(a)iii above, until the collection of their orbits is a perfect access pattern

We term the lines identified by 3(a)i and planes identified by 3(a)iii as **orbit representatives**.

---

the sequences across the tuples to get perfect access patterns. This approach is summarized in Algorithm 2.

We now identify ways to permute the results and the complexity of doing that. With the approach in Algorithm 2 *but without the permutation step 4*, consider two lines identified in two distinct invocations of 3(a)i or two planes identified in distinct invocations of 3(a)iii in Algorithm 2. Since the two lines (or two planes) are selected, not by considering the structure of the space, but by virtue of the incident points (or lines) not having been combined together earlier, the sequence of the points in the two lines (or sequence of lines in the two planes) may not lead to a permutation. Hence, this approach can not directly result in the space definition *with perfect access patterns*. Though the resulting incidence relations between lines and planes will be correct, the sequence of lines on the planes would not lead to perfect access patterns.

The permutation step 4 in Algorithm 2 tries different permutations of the sequence of lines in the orbit *representative*, till a combination of permutations across the orbits results in a perfect access pattern.

Theorem 5. *The number of different sequences required by the permutation step is* $(\Xi(p) \times (p^2 + p + 1))!$

Proof. There are $\Xi(p)$ orbits, and each orbit representative has $p^2 + p + 1$ lines   □

The number of these sequences is so large, that we have tabulated (Table 4) *not the number itself*, but the logarithm to base 10 of these numbers. We can see that even for small primes, the number of trials would be very high for the permutation approach. The next approach described aims to reduce this complexity by addressing the problem in a different direction.

3.3 Construction of 4-D Projective Space - Using Non-singular Matrices

In the earlier approach, the complexity increases because the orbits are identified first, and then permuted. However, if orbits can be generated in a way that does not require permutations, then it may be possible to control the complexity. This approach aims at such a generation. Remember that for 2-D spaces, and 4-D space for p = 2, the generation effort is in identifying one line/plane, and the remaining space is generated by simple arithmetic transformations on the line/plane. If we are able to identify an automorphism, which together with Frobenius and Shift automorphisms, can generate the space, then we can do away with the permutation complexity.

Table 4. Number of operations required

p	Log_{10} (Number of operations)	
	Permutations	Non-singular matrices
3	26.6	11.9
5	342.8	17.5
7	1325.0	21.2

General automorphisms on our field of interest are represented by $(d+1) \times (d+1)$ non-singular matrices over Z_p [5]. For 4-D spaces, this search space has 5×5 non-singular matrices over Z_p. An analytical structure of these general automorphisms is in progress, and will be reported in a separate paper. In this paper, we focus on a computational approach. The total number of such non-singular matrices is bounded above by the total number of 5×5 matrices over Z_p, i.e. by p^{25}. The logarithm to base 10 of this number for some prime values is tabulated in Table 4, and it is clear that trying to search suitable non-singular matrices is a simpler problem compared to permuting the orbits.

Generating non-singular matrices, particularly over Z_p can be done in a more predictable manner, avoiding trial-and-error with singularity checks. Instead, we use the following observations to systematically generate *only* the non-singular matrices.

1. The first row can be any row other than the zero row, with $p^5 - 1$ possibilities.
2. The second row can be any row other than a multiple of the first row, with $p^5 - p$ possibilities

3. The third row can be any row other than a linear combination of the first two rows, with $p^5 - p^2$ possibilities
4. In general, m'th row can be any row other than a linear combination of the first $m - 1$ rows, with $p^5 - p^{m-1}$ possibilities.

These observations allow us to generate the non-singular matrices in a *generative* way, rather than exhaustively generating *all* the matrices and checking for singularity. These observations also clearly indicate the number of non-singular matrices to be

$$(p^5 - 1) \times (p^5 - p) \times (p^5 - p^2) \times (p^5 - p^3) \times (p^5 - p^4) \approx O(p^{25}).$$

Suitability of Non-singular Matrices: Ideally, a candidate non-singular matrix to help us jointly generate the entire space, should complement the Frobenius and Shift automorphisms. That is, it should generate *precisely* the complete space along with Frobenius and Shift automorphisms, and nothing more. Since we do not have any existing characterizations of the general automorphisms, we use a weaker condition - the non-singular matrix should generate *at least* the entire space. Thus, if a non-singular matrix (its associated automorphism) has an order *larger* than required to complement Frobenius and Shift automorphisms, we do not discard the matrix. As we refine the criteria after studying the characteristics of the non-singular matrices in the planned study, we will aim to strengthen this criterion further, aiming to arrive at an optimal one.

With this background, we are now ready to define the algorithm for this approach, as detailed in Algorithm 3. Since the number of non-singular 5×5 matrices over Z_p is $O(p^{25})$, that is also an upper bound for the complexity of the algorithm in the non-singular matrix based approach.

3.4   Implementation

Even though the complexity of the non-singular matrix based approach is comparatively lower, it is still large at p^{25}. The implementation will therefore have to test these many candidate non-singular matrices when looking for a solution.

Implementing the generation algorithm as a sequential implementation is not practical. This problem turns out to be embarrassingly parallel, since a candidate non-singular matrix can be tested for suitability, independent of the matrices.

We have implemented this algorithm using Python 3, and are running it on the SpaceTime parallel cluster at computer center, IIT Bombay. The cluster has 216 CPU-compute nodes, with each compute node having processor 2xIntel Skylake 6148 2.4 GHZ and 192 GB Ram, for a total peak performance of 663.5 TFlops.

Since the problem is embarrassingly parallel, we choose a queue (the *small* queue) on the cluster which has a higher throughput. This queue provides 60 processes per user. We have been using 30 processes for finding perfect access patterns for prime 3, and 30 processes for prime 5. While one solution will be adequate per prime, we are aiming to enumerate multiple solutions, to be able to

---

Algorithm 3. 4-D Space Generation using Non-Singular Matrices

---

1. **Identify a new Line**
   (a) Identify two points p_1 and p_2, such that the line joining these two points $(\{c_1p_1 + c_2p_2, \forall c_i \in Z_p\})$ has not yet been identified
2. **Identify a new Plane**
   (a) Identify two lines l_1 and l_2, such that the line joining l_1 and l_2 has not yet been identified.
   (b) Find new plane P as all linear combinations of points on these two lines $\{c_1p_1 + c_2p_2, \forall p_1 \in l_1, p_2 \in l_2$ and $c_1, c_2 \in Z_p\}$
3. **Generate new Lines / Planes**
   (a) Let l_0 be a new line identified by line-identification step
   (b) Determine the orbit of l_0 under repeated application of Frobenius and Shift automorphisms
   (c) Let P_0 be a new plane identified by plane-identification step
   (d) Determine the orbit of P_0 under repeated application of Frobenius and Shift automorphisms
   (e) For each non-singular matrix M
       i. let $i \leftarrow 0$
       ii. For $1 < i < \#$ of orbits:
           A. let l_i be the map of line l_{i-1} under automorphism corresponding to M. Similarly let P_i be the map of plane P_{i-1} under M
           B. Determine the orbits of l_i and P_i.
           C. If either of the orbits contain lines / planes already identified in earlier orbits, continue the outermost iteration with the next non-singular matrix
       iii. If all the orbits identified in the iteration above are distinct, then we have generated orbits resulting in perfect access pattern. The automorphism corresponding to non-singular matrix M successfully generates the entire space.
4. **Generate the non-singular matrices**
   (a) 5-deep nested loop, iterating over all the possible combinations of rows
   (b) For each choice of row at a particular row number, identify the discard list for next depth of the loop
   (c) After choosing 5 linearly independent rows, produce this as a candidate matrix
   (d) Keep producing candidate matrices until all combinations are produced

---

get insights into the structure of the non-singular matrices that turn out to be suitable. When enumerating multiple solutions, distinct non-singular matrices can and do end up in identical perfect-access-patterns.

Code Overview: The code implements step 4 of Algorithm 3 as a python generator, yielding one non-singular matrix on every invocation. The rows of the matrix are also constructed using another generator. Since the code is run on multiple processes, the entire search space is divided across the number of processes. The code also implements an in-built timeout feature, with each process saveing its internal state after a specific time. A subsequent execution of the process with *the same MPI rank* resume rather than repeat the run. Thus, the search can span multiple runs.

To simplify the work division and resumable execution, we assign each matrix a unique number, based on the sequence in which the matrices are enumerated. This numbering allows quick mapping from sequence number to the matrix. This numbering scheme enables each process to efficiently identify the resumption point.

With this non-singular matrix generator, the code determines the first orbit, and then tries the generated matrices one by one. For each matrix, the code applies it repeatedly on the orbit representative identified in the beginning. On each result of the matrix application, the Frobenius and Shift automorphisms are applied. If the resulting line tuples do *not* form a permutation, the matrix is discarded at the earliest opportunity. If all the tuples turn out to be permutations, for the entire space, then the matrix is a success. In such a case, the orbit representative and its images under the successful matrix together is the signature for the candidate matrix. Solutions are tested for uniqueness with respect to this signature.

With this setup, we have been able to successfully generate several alternative solutions for prime 3. The summary of the results **so far** appears in Table 5.

For prime 5, the ongoing runs have not identified any solution so far.

Table 5. Results of generation

Prime	3	5
# non-singular matrices	475,566,474,240	226,614,960,000,000,000
Searched so far	5,806,126,476	225,286,319
Percentage	1.22%	9.94e-8%
Success (suitable matrices)	4,156,797	0
Unique patterns	4834	0

4   Conclusions and Future Work

We have summarized our work on the construction of projective spaces of dimension 4 *with perfect access patterns*, for higher primes $(p > 2)$. We have identified the analytical and computational issues in constructing such spaces. We have compared two methods for such construction, and implemented the one with computationally lower complexity. With this implementation, we have, for the first time reported the successful construction of a 4-D projective space *with perfect access patterns* for prime $p = 3$.

For our larger research goal, to characterize the performance of matrix-matrix multiplications over parallel computers connected in such perfect access patterns *for higher primes - $p > 3$ as well*, we plan to extend this work in the following directions:

1. Explore feasibility of *analytically* identifying a set of generators, preferably a minimal set, for the automorphism group. Also, study the suitable non-singular matrices for $p = 3$ to gain insights into the structure of such matrices, aiming to reduce the search space further.
2. Explore computational/implementation optimizations for efficient construction.
3. Study the performance characteristics of matrix-matrix multiplications using the spaces so constructed, for at least two or three primes.

Acknowledgements. The authors would like to thank Dr. B. S. Adiga for the many discussions on the concepts of projective geometry, and Prof Milind Sohani for the concepts of algebraic structure of the projective spaces.

References

1. Amrutur, B.S., Joshi, R., Karmarkar, N.K.: A projective geometry architecture for scientific computation. In: Proceedings of the International Conference on Application Specific Array Processors, pp. 64–80, August 1992. https://doi.org/10.1109/ASAP.1992.218581
2. D'Azevedo, E.F., Dongarra, J.: The design and implementation of the parallel out-of-core ScaLAPACK LU, QR and cholesky factorization routines. University of Tennessee, Knoxville. Technical report (1997)
3. Grama, A., Gupta, A., Karypis, G., Kumar, V.: Introduction to Parallel Computing. Addison–Wesley (2003)
4. Karmarkar, N.: A new parallel architecture for sparse matrix computation based on finite projective geometries. In: Proceedings of Supercomputing, pp. 358–369 (1991)
5. Karmarkar, N.: Massively parallel systems and global optimization (2008). http://math.mit.edu/crib/08/Extended-abstract.pdf
6. Lin, S., Costello, D.J.: Error Control Coding, 2nd edn. Prentice Hall, Upper Saddle River (2004)
7. Sapre, S., Sharma, H., Patil, A., Adiga, B.S., Patkar, S.: Finite projective geometry based fast, conflict-free parallel matrix computations. https://arxiv.org/abs/1107.1127

Data Processing

Dimensional Reduction Using Conditional Entropy for Incomplete Information Systems

Mustafa Mat Deris[1,2,3,4]($\boxtimes$), Norhalina Senan[1], Zailani Abdullah[2], Rabiei Mamat[3], and Bana Handaga[4]

[1] Faculty of Computer Science and Information Technology, Universiti Tun Hussein Onn Malaysia, Parit Raja, Malaysia
{mmustafa,halina}@uthm.edu.my
[2] Faculty of Entrepreneurship and Business, Universiti Malaysia Kelantan, Pengkalan Chepa, 16100 Kota Bharu, Kelantan, Malaysia
zailania@umk.edu.my
[3] Faculty of Informatics and Applied Mathematics, University of Malaysia Terengganu, 21030 Kuala Terengganu, Terengganu, Malaysia
rab@umt.edu.my
[4] Program Studi Informatika, Universitas Muhammadiah Surakarta, 57162 Surakarta, Central Java, Indonesia
bana.handaga@ums.ac.id

Abstract. Dimension reduction approach is one of the main data reduction approaches in order to reduce the storage and processing time while maintaining the integrity of the original data. A wide range of dimension reduction approaches are based on classical approaches such as PCA and Bayer's, and machine learning approaches such as clustering, and feature selection techniques. However, many of the approaches do not consider the incomplete information systems where some attribute values are missing or incomplete. Only few studies were proposed for the problem in incomplete information systems due to its complexities, specifically on attribute selection. The most popular approaches is based on probability theory to replace missing values with the most common values, or remove the missing objects from the information systems. However, it needs to know the probability distribution of data in advance. To overcome these issues, we propose a new approach based on conditional entropy to reduce dimensionality. The results show that the proposed approach achieves better data reduction with higher accuracy for objects and dimensionality reduction in incomplete information systems.

Keywords: Dimension reduction · Conditional entropy · Incomplete information system

1 Introduction

With the massive data generated daily to computer systems, it is difficult to manage and do analysis on it. The massive volume of data not only causes the data heterogeneity but also the diverse of dimensionalities in the datasets. For example social data

V. Malyshkin (Ed.): PaCT 2019, LNCS 11657, pp. 263–272, 2019.
https://doi.org/10.1007/978-3-030-25636-4_21

aggregators, scientific experimental systems, the profiles of internet users, etc., are sparse with high dimensionalities [1]. Thus, it is imperative to reduce the data while retaining the most important and useful data. Data reduction is a process to reduce the volume/size of data to make effective data analysis. It is mainly based on the dimension reduction to reduce the number of features in a dataset without having to lose much information. Dimension reduction techniques are useful to handle the heterogeneity and massiveness of data by reducing variables data into manageable size.

A wide range of dimension reduction approaches are based on classical approaches such as PCA and Bayer's, and machine learning approaches such as clustering, and feature selection techniques. However, many of the approaches do not consider the incomplete information systems where some attribute values are missing or incomplete. Only few studies were proposed for the problem in incomplete information systems due to its complexities, specifically on attribute selection.

Rough set theory [2] proposed by Pawlak was successful in the study of soft computing characterized by uncertainty of information, especially in rule extraction [3], uncertainty reasoning [4], granular computing [5, 6], data clustering [7–9], and data classification [10]. It has been proven to be an efficient mathematical tool as compared with PCA, neural networks and support vector machine [11, 12] methods. Unlike those methods, rough set theory allows knowledge discovering process to be conducted automatically by the data themselves without any dependence on the prior knowledge [13]. The rough set theory however, only be used to solve complete information systems where all available objects in information system have attribute values. It is basically based on the indiscernibility relation that conforms with the reflexive, symmetric and transitive properties. A problem arises when certain attribute values in information systems are missing that cause imprecise answer to some queries, which sometimes happens in the real world. This information system is called incomplete information system (IIS). Because some attribute values are missing in incomplete information systems, such relational properties are difficult to generate, and it is hard to process the incomplete information systems with the indiscernibility relation. There have been many efforts in studying incomplete information systems, including the works of [14–20].

To some fields, such as data mining, bio-informatics, and machine learning, data sets have huge number of dimensions/attributes that often be encountered. Some attributes are irrelevant or redundant that can complicate the problem and subsequently, degrade the performance and solution accuracy. Thus, some redundant or irrelevant attributes need to be removed which is the main objective of attribute selection.

Some approaches on attribute selection for IIS have been proposed: tolerance relation approach [15], and tolerance relation using conditional entropy approach [13]. However, tolerance relation approach leads to poor results in terms of approximation. Consequently, Stefanowski and Tsoukias [17, 18] introduced similarity relation to refine the results obtained using tolerance relation approach. However, Wang [19] and Yang et al. [20] prove that similarity relation will lost some information and proposed limited tolerance relation. Nevertheless, some information may also loss because the limited tolerance relation does not consider the similarity precision between objects. Nguyen et al. [21] improve the tolerance relation by considering the probability matching between two objects. However, the probability distribution should be known in advance. Consequently, we proposed a new limited tolerance relation based on similarity precision between the two objects [22]. In this paper, the similarity precision proposed in [22] will be adopted to reduce the similarity objects, and the conditional entropy will be used for dimensions/attributes reduction. The main aim of this paper is to construct a precise uncertainty measure evaluating the accuracy of knowledge to find attribute selection in IIS. Comparative analysis and experiment results between the proposed approach with the limited tolerance relation approach in terms of accuracy are presented. We found that, the proposed approach is more precise and better in terms of attribute selection.

The rest of the paper is organized as follows. Section 2 discusses the theoretical background of information system, rough set theory and limited tolerance relation based on similarity precision in IIS. Section 3 describes the proposed new approach based on conditional entropy for incomplete information systems. Conclusions of this work are presented in Sect. 4.

2 Theoretical Background

The basic concepts of information systems and the similarity precision for limited tolerance relation will be explained in this section.

2.1 Information Systems

An *information system* is a 4-tuple (quadruple) $S = (U, A, V, f)$, where $U = \{u_1, u_2, \cdots, u_{|U|}\}$ is a non-empty finite set of objects, $A = \{a_1, a_2, \cdots, a_{|A|}\}$ is a non-empty finite set of attributes, $V = \bigcup_{a \in A} V_a$, V_a is the domain (value set) of attribute a, $f : U \times A \to V$ is an information function such that $f(u, a) \in V_a$, for every $(u, a) \in U \times A$, called information function [7]. If U in $S = (U, A, V, f)$ contains at least one object with an unknown or missing value, the S is *called incomplete information system (IIS)*. The unknown value is denoted as "*" in incomplete information system. In this paper, we use the quadruple $S^{\#} = (U, A, V_{\#}, f)$ to denote an incomplete information system. From the notion of an information system above, in the following sub-section we recall the notion of a tolerance relation as an approach for incomplete information system.

2.2 Rough Set Theory

The fundamental concept of Rough Set Theory proposed by Pawlak [2] is the approximation of lower and upper spaces of a set, where it is based on indiscernibility relation. The indiscernibility relation is the starting point to form the partition. Two elements $x, y \in U$ in $S = (U, A, V, f)$ is said to be *B-indiscernible* (indiscernible by the set of attribute $B \subseteq A$ in S) if and only if $f(x, a) = f(y, a)$, for every $a \in B$. An indiscernible relation induced by the set of attribute B, denoted by $IND(B)$, is an equivalence relation. It is well-known that an equivalence relation can induce a unique partition. The partition of U induced by $IND(B)$ in $S = (U, A, V, f)$ denoted by U/B and the equivalence class in the partition U/B contains $x \in U$ and denotes by $[x]_B$.

Let B be any subset of A in S and let X be any subset of U, the B-*lower approximation* of X, denoted by $\underline{B}(X)$ and B-*upper approximation* of X, denoted by $\overline{B}(X)$ respectively, are defined by

$$\underline{B}(X) = \{x \in U | [x]_B \subseteq X\} \text{ and } \overline{B}(X) = \{x \in U | [x]_B \cap X \neq \phi\}.$$

2.3 The Similarity Precision for Limited Tolerance Relations

Given an incomplete information system $S^{\#} = (U, A, V_{\#}, f)$, *where* $A = C \cup \{d\}$, C *is a set of condition attributes and d the decision attribute, such that* $f : U \times A \to V_*$. *For any* $a \in A$, *where* V_a *is called domain of an attribute* a *and a subset* $B \subseteq C$, *the similarity precision is defined as follows.*

Definition 1. *Let* $P_B(x) = \{b \mid b \in B \wedge b(x) \neq^*\}$, *the similarity precision* δ, *is defined as*

$$\delta(x, y) = \frac{|P_B(x) \cap P_B(y)|}{|C|}, \tag{1}$$

where $|\cdot|$ *represents the cardinality of the set.*

From (1), it is clear that $0 < \delta(x, y) \leq 1$. From Definition 1, the limited tolerance relation with similarity precision is given as follow:

Definition 2. *Let an given IIS,* $S^{\#} = (U, A, V_{\#}, f)$. *The limited tolerance relation with similarity precision* $L\delta$ *is defined as follows*

$$\forall_{x,y \in U \times U} \ (L\delta_B(x, y) \Leftrightarrow \forall_{b \in B}(b(x) = b(y) =^*) \vee ((\delta(x, y)) \geq \alpha) \wedge$$
$$\forall_{b \in B} (((b(x) \neq^*) \wedge (b(y) \neq^*)) \to (b(x) = b(y)))$$

where $\alpha \in (0, 1]$ *is a threshold value.*

Since $\alpha \in (0, 1]$, then $0 < \delta(x, y) \leq 1$ which implies that $P_B(x) \cap P_B(y) \neq \phi$ holds, but not vice versa if certain threshold value of the similarity is given.

Now, the similarity precision for limited tolerance with a threshold value will be defined as follows.

Definition 3. *Let given an IIS,* $S^{\#} = (U, A, V_{\#}, f)$, *a subset* $B \subseteq C$ *and a threshold* α. *The limited tolerance relation with similarity precision is defined as;* $L\delta_B(x, y) \Leftrightarrow \delta_B(x, y) \geq \alpha$.

The above relation is reflexive and symmetric but not necessarily transitive. The concept of similarity precision between objects x and y in order to determine both objects are tolerant will be adopted [22].

Table 1. An incomplete information (for scholarship-application)

Students	C_1	C_2	C_3	C_4	Decision (d)
s_1	Good	Good	Fluent	*	Accept
s_2	Poor	*	Fluent	Good	Accept
s_3	*	*	Not fluent	Good	Reject
s_4	Good	*	Fluent	Good	Accept
s_5	*	Good	Fluent	Good	Accept
s_6	Poor	Good	Fluent	*	Accept
s_7	Poor	Good	Fluent	Good	Accept
s_8	*	Good	*	*	Reject

We can illustrate the above concepts with an IIS (for scholarship-application) below:

Table 1, is a list of students $S = \{s_i | i = 1, 2, \ldots, 9\}$ who apply for the scholarship sponsored by a Malaysian company. The decision is based on four criteria or condition attributes; the ability to do analysis (C_1), Studying BSc in Mathematics (C_2), the communication skills (C_3), and the ability to speak in Malay language (C_4). The table is an incomplete information system, where some values are not available, stated as '*'.

The decision (d), where its domain values are $Accept = \{s_1, s_2, s_4, s_5, s_6, s_7, s_9\}$ and $Reject = \{s_3, s_8\}$.

To clearly depict the limited tolerance with similarity precision as defined above, we illustrate through an example from Table 1.

Example 1. From Table 1, two objects s_1 and s_8 are not tolerant if $\alpha = 0.4$. However, two objects s_4 and s_5 are tolerant due to $\delta(s_4, s_5) \geq 0.4$.

Properties and Correctness of Proof

Proposition 1. *Let given an IIS, $S^{\#} = (U, A, V_{\#}, f)$, a subset $B \subseteq C$ and $x \in U$. If $\delta > 0$, then a. For any x and y, $L\delta_B(x, y) \Rightarrow L_B(x, y)$*

b. $L\delta_B(x, y) \Leftarrow L_B(x, y)$ except the case when

$$P_B(x) \cap P_B(y) = \phi$$

Proof

a. When $\delta > 0$, then
$$L\delta_B(x, y) \Leftrightarrow \alpha_B(x, y) > 0$$
$$\Leftrightarrow P_B(x) \cap P_B(y) \neq \phi$$
$$\wedge \forall a \in P_B(x) \cap P_B(y), f_a(x) = f_a(y)$$
$$\Rightarrow L_B(x, y)$$

b. It is clear that $L_B(x, y) \Rightarrow L\delta_B(x, y)$ except the case when $P_B(x) \cap P_B(y) = \phi$.

Definition 4. *Let an given IIS, $S^* = (U, A, V_*, f)$ and $B \subseteq C$. The limited tolerance class with similarity precision is defined as $I_B^{L\delta}(x) = \{y \mid y \in U \wedge L\delta_B(x, y)\}$.*

To clearly depict the new limited tolerance class as defined above, we illustrate through an example from Table 1.

Example 2. From Table 1, and let $\delta \geq 0.5$, we have the new tolerance classes as follows

$I_C^{L\delta}(s_1) = \{s_1\}, I_C^{L\delta}(s_2) = I_C^{L\delta}(s_6) = I_C^{L\delta}(s_7) = \{s_2, s_6, s_7\}, I_C^{L\delta}(s_3) = \{s_3\},$
$I_C^{L\delta}(s_4) = I_C^{L\delta}(s_5) = \{s_4, s_5\}, I_C^{L\delta}(s_8) = \{s_8\},$ and $Accept_C^{L\delta} = \{s_1, s_2, s_4, s_5, s_6, s_7\},$
$Reject_C^{L\delta} = \{s_3, s_8\},$

From the above analysis, s_1 and s_8 are divided into different class.

Definition 5. *Let an given IIS, $S^{\#} = (U, A, V_{\#}, f)$. The lower approximation and the upper approximation of an object x based on the limited tolerance class with similarity precision $I_B^{L\delta}(x)$ denoted as $D_{L\delta}^B(x)$ and $D_B^{L\delta}(x)$ respectively are defined as*

$$D_B^{L\delta} = \{x \mid x \in U \wedge I_B^{L\delta}(x) \subseteq D\} \text{ and } D_{L\delta}^B = \{x \mid x \in U \wedge I_B^{L\delta}(x) \cap D \neq \phi\}.$$

From Definition 2, we can generalize Proposition 1 as describe in the following proposition.

Proposition 2. *Let given an incomplete information system $S^* = (U, A, V_*, f)$, a subset $B \subseteq A$ and $x \in U$. If $0 \leq \delta_1 < \delta_2 \leq 1$, then $I_B^{L\delta_2} \subseteq I_B^{L\delta_1}$.*

Proof. For every $s \in I_B^{L\delta_2}(x)$, we have $\alpha_B(x, y) \geq \delta_2$. Since $\delta_2 > \delta_1$, then $\alpha_B(x, y) \geq \delta_1$, that is $\forall s \in I_B^{L\delta_1}(x)$ which implies $I_B^{L\delta_2}(x) = I_B^{L\delta_1}(x)$. However, if $\alpha_B(x, y) \geq \delta_1$ then it does not necessarily $\alpha_B(x, y) \geq \delta_2$. Hence $I_B^{L\delta_2} \subseteq I_B^{L\delta_1}$.

To clearly depict the property of generalized tolerance class in Proposition 2, we illustrate through an example from Table 1.

Example 3. From Table 1, we have $I_C^{L\delta_1}(s_2) = I_C^{L\delta_2}(s_6) = \{s_2, s_6, s_7\}$ for $\delta_1 = 0.5$. However,

for $\delta_2 = 0.75$, we have $I_C^{L\delta_2}(s_6) = \{s_6, s_7\}$ and thus, $I_C^{L\delta_2}(s_6) \neq I_C^{L\delta_1}(s_6)$.

Reduction Based on Similarity Precision

Let an given IIS, $S^{\#} = (U, A, V_{\#}, f)$. The similarity precision for limited tolerance relation $L\delta$ is defined as in Definition 5. The reduction based on similarity precision can be defined as follows,

Definition 6. The reduction of $L\delta$ is given as; $R^{L\delta}(x) = \{x \mid x \in L\delta_B(x, y) \wedge |b(x)| > |b(y)| \text{ for } b(x), b(y) \neq '^*'\}$ and the reduction of U can be defined as;

$R^U = |\{x \in U : R^{L\delta}(x)\}|$

For example, for the case of $\delta = 0.75, I_C^{L\delta_2}(s_6) = \{s_6, s_7\}$. Since $|b(s_6)| = |\{poor, good, fluent, *\}| = 3$, and $|b(s_7)| = |\{poor, good, fluent, good\}| = 4$, then $R^{L\delta}(s_6) = \{s_7\}$.

Relative Reduct

Definition 7. Let $S^{\#} = (U, A, V_{\#}, f)$ and $* \in V_C$ be an IIS, and $A = C \cup \{d\}$, then the generalized decision $\psi_B = U \rightarrow V_d$ is defined as follows:

$$\psi_B(x) = \{u \mid y \in L\delta_B(x) \cap u = d(y)\}$$

And, if $|\psi_B(x)| = 1$ for any $x \in U$, then the incomplete information system is consistent.

Definition 8. Let $S^{\#} = (U, A, V_{\#}, f)$ be an IIS and $A = C \cup \{d\}$ with $B \subseteq C$. The attribute set B is a relative reduct of *IIS*, if and only if

(a) $\psi_B(x) = \psi_A(x)$ for all $x \in U$,
(b) $b \in B, \psi_{B-\{b\}} \neq \psi_C$

are satisfied.

3   Conditional Entropy for Incomplete Information Systems

In this sub-section we will introduce conditional entropy on Similarity Precision Tolerance Relation approach to measure the uncertainty of knowledge in IIS.

Definition 9. Given an *IIS* = $(U, C \cup \{d\})$ and $B \subseteq C$. Let $U/I_B = \{I_B(x_1), I_B(x_2), \ldots, I_B(x_{|U|})\}$, $U/d = \{d_1, d_2, \ldots, d_m\}$. The conditional entropy of B to d is defined as follows:

$$EN(d \mid B) = -\sum_{i=1}^{|U|} p(I_B(x_i)) x \sum_{j=1}^{|U/d|} p(d_j \mid I_B(x_i)) \log p(d_j \mid I_B(x_i)) \tag{2}$$

where, $p(I_B(x_i)) = \frac{|I_B(x_i)|}{|U|}, i = 1, 2, \ldots |U|$, and $p(d_j \mid I_B(x_i)) = \frac{|p(I_B(x_i) \cap d_j)|}{|I_B(x_i)|}, i = 1, 2, \ldots$
$|U|, j = 1, 2, \ldots, m$
From Eq. 2, it is obvious that $EN(d|B) = 0$ when $I_B(x_i) \cap d_j = 0$.

Proposition 3. Let *IIS* = $(U, C \cup \{d\})$ be a consistent incomplete information system. Then we have $EN(d|C) = 0$.

Proof. Since IIS is consistent, then $|\psi_B(x)| = 1$, for $x_i \in U$. This means that $L\delta_C(x_i) \subseteq d_j, d_j \in U/d$. Hence, we have $L\delta_C(x_i) \cap d_k = \phi, d_k \neq d_j \in U/d$,
Consequently,

$$EN(d|C) = -\sum_{i=1}^{|U|} p(L\delta_C(x_i)) \sum_{j=1}^{|U/d|} p(d_j|L\delta_C(x_i)) \log p(d_j|L\delta_C(x_i))$$

$$= -\sum_{i=1}^{|U|} \frac{|L\delta_C(x_i)|}{|U|} \sum_{j=1}^{|U/d|} \frac{|d_j \cap L\delta_C(x_i)|}{|L\delta_C(x_i)|} \log \frac{|d_j \cap L\delta_C(x_i)|}{|L\delta_C(x_i)|}$$

$$= -\sum_{i=1}^{|U|} \frac{|L\delta_C(x_i)|}{|U|} (0 + 0 + \ldots + 1 \log 1 + 0 + \ldots 0) = 0.$$

The Algorithm

The algorithm that we impose in this paper is based on breath-first search algorithm in order to find the minimal attribute selection for incomplete information system. The algorithm is given as follows:

> Input: An IIS $(U, A \cup \{d\})$.
> Output: A minimal attribute selection result,M
> a. For all sizes = 0 to $|A|$
> b. For every subset Attributeselection with Attributeselection = size
> c. If EN(d|Attributeselection) $\neq$ EN(d|A), go to step b, otherwise return M=Attributeselection
> d. End
> e. End

Alg. 1: Breath-first search for attribute reduction

Example 4. Given the IIS shown in Table 1. We obtain, $C = \{C_1, C_2, C_3, C_4\}$ $\frac{U}{d} = \{\{s_1, s_2, s_4, s_5, s_6, s_7\}, \{s_3, s_8\}\}$

Let $\alpha \geq 0.5$, then we have,

$I_C^{L\delta}(s_1) = \{s_1\}, I_C^{L\delta}(s_2) = I_C^{L\delta}(s_6) = I_C^{L\delta}(s_7) = \{s_2, s_6, s_7\}, I_C^{L\delta}(s_3) = \{s_3\},$
$I_C^{L\delta}(s_4) = I_C^{L\delta}(s_4) = \{s_4, s_5\}, I_C^{L\delta}(s_8) = \{s_8\}$

Based on the Definition 9, for the conditional entropy, we obtain

$$EN(d|C) = -[\tfrac{1}{8}(\tfrac{1}{1}\log\tfrac{1}{1} + \tfrac{0}{1}\log\tfrac{0}{1})] - [\tfrac{1}{8}(\tfrac{1}{1}\log\tfrac{1}{1} + \tfrac{0}{1}\log\tfrac{0}{1})] - [\tfrac{1}{8}(\tfrac{0}{1}\log\tfrac{0}{1} + \tfrac{1}{1}\log\tfrac{1}{1})]$$
$$-2[\tfrac{2}{8}(\tfrac{2}{2}\log\tfrac{2}{2} + \tfrac{0}{2}\log\tfrac{0}{2})] - 2[\tfrac{3}{8}(\tfrac{3}{3}\log\tfrac{3}{3} + \tfrac{0}{3}\log\tfrac{0}{3})] - [\tfrac{1}{8}(\tfrac{0}{1}\log\tfrac{0}{1} + \tfrac{1}{1}\log\tfrac{1}{1})].$$
$$= 0$$

The other conditional entropy for different conditional attributes such as $EN(d| \{C_1\})$, $EN(d|\{C_1, C_2\})$,..., $EN(d|\{C_1, C_2, C_3\})$, $EN(d|\{C_2, C_3, C_4\})$, can be deduced in the same way. To make it short, we would like to show the $EN(d|\{C_1, C_3\}) = 0$, which is a relative reduct of the whole attribute set in C. It can be calculated as;

$I_{C1,C3}^{L\delta}(s_1) = I_{C1,C3}^{L\delta}(s_4) = I_{C1,C3}^{L\delta}(s_5) = \{s_1, s_4, s_5\}, I_{C1,C3}^{L\delta}(s_3) = \{s_3\}$
$I_{C1,C3}^{L\delta}(s_2) = I_{C1,C3}^{L\delta}(s_5) = I_{C1,C3}^{L\delta}(s_6) = I_{C1,C3}^{L\delta}(s_7) = \{s_{21}, s_5, s_6, s_7\}, I_{C1,C3}^{L\delta}(s_8) = \{s_8\},$

$$EN(d|\{C_1, C_2\}) = -3[\tfrac{3}{8}(\tfrac{3}{3}\log\tfrac{3}{3} + \tfrac{0}{3}\log\tfrac{0}{3})] - 4[\tfrac{4}{8}(\tfrac{4}{4}\log\tfrac{4}{4} + \tfrac{0}{4}\log\tfrac{0}{4})]$$
$$-[\tfrac{1}{8}(\tfrac{0}{1}\log\tfrac{0}{1} + \tfrac{1}{1}\log\tfrac{1}{1})] - [\tfrac{1}{8}(\tfrac{0}{1}\log\tfrac{0}{1} + \tfrac{1}{1}\log\tfrac{1}{1})]$$
$$= -(3(0) + 4(0) + 0 + 0) = 0$$

Thus, the attribute selection is $\{C_1, C_3\}$.

Table 2 is the reduction information system from Table 1 using similarity precision and conditional entropy. The objects have been reduced to 4 objects only, and the dimensions/attributes have been reduced to 2 dimensions/attributes only, which is of 50% reduction on both number of objects and dimensions/attributes.

Table 2. Reduction Information from Table 1

Students	C_1	C_3	Decision (d)
s1, s4, s5	Good	Fluent	*Accept*
s2, s5, s6, s7	Poor	Fluent	Accept
s3	*	Not-fluent	Reject
s8	*	*	Reject

4 Conclusion

Dimension reduction approach is one of the main data reduction approaches in order to reduce the storage and processing time while maintaining the integrity of the original data. This paper establishes a new approach based on conditional entropy for dimension/attribute reduction in incomplete information systems, besides objects/records reduction using limited tolerance relation with similarity precision. From the example, we manage to make a dimensional reduction up to 50% from the original data set which subsequently reduce the processing time as well as storage usage. The practical applications including feature selection on large databases will be used in the near future.

Acknowledgment. The research was supported from Ministry of Higher Education through Fundamental Research Grant Scheme (FRGS) vote number 1643.

References

1. Chandramouli, B., Goldstein, J., Duan, S.: Temporal analytics on Big Data for web advertizing. In: 28th IEEE International Conference on Data Engineering, pp. 90–101 (2012)
2. Pawlak, Z.: Rough sets. Int. J. Comput. Inform. Sci. **11**(5), 341–356 (1982)
3. Lu, Z., Qin, Z.: Rule extraction from incomplete decision system based on novel dominance relation. In: Proceedings of the 4th International Conference on Intelligent Networks and Intelligent Systems, pp. 149–152 (2011)
4. Dai, J., Wang, W., Xu, Q., Tian, H.: Uncertainty measurement for interval-valued decision systems based on extended conditional entropy. Knowl. Based Syst. **27**, 443–450 (2012)
5. Skowron, A., Wasilewski, P.: Toward interactive Rough-Granular Computing. Control Cybern. **40**(2), 213–235 (2011)
6. Skowron, A., Stepaniuk, J., Swiniarski, R.: Approximation spaces in Rough-Granular Computing. Fundamentae Informaticae **100**(1–4), 141–157 (2010)
7. Yanto, I.T.R., Vitasari, H.T., Deris, M.M.: Applying variable precision rough set model for clustering suffering student's anxiety. Expert Syst. Appl. **39**(1), 452–459 (2012)
8. Herawan, T., Deris, M.M., Abawajy, J.H.: A rough set approach for selecting clustering attributes. Knowl. Based Syst. **23**(3), 220–231 (2010)
9. Parmar, D., Wu, T., Blackhurst, J.: MMR: an algorithm for clustering categorical data using rough set theory. Data Knowl. Eng. **63**(3), 879–893 (2007)
10. Kim, D.: Data classification based on tolerant rough set. Pattern Recogn. **34**(8), 1613–1624 (2001)
11. Trabelsi, S., Elouedi, Z., Lingras, P.: Classification systems based on rough sets under the belief function network. Int. J. Approximate Reasoning **52**(9), 1409–1432 (2011)

12. Kaneiwa, K.: A rough set approach to multiple dataset analysis. J. Appl. Soft Comput. **11**(2), 2538–2547 (2011)
13. Yan, T., Han, C.: A novel approach of rough conditional entropy-based attribute selection for incomplete decision system. Math. Probl. Eng. **2014**, 1–15 (2014)
14. Grzymala-Busse, J.W.: Rough set strategies to data with missing attribute values. In: Proceedings of the workshop on Foundation and New Directions in Data Mining, associated with the 3rd IEEE International Conference on Data Mining, pp. 56–63 (2003)
15. Kryszkiewicz, M.: Rough set approach to incomplete information systems. Inf. Sci. **112**(1–4), 39–49 (1998)
16. Kryszkiewicz, M.: Rules in incomplete information systems. Inf. Sci. **113**(3–4), 271–292 (1999)
17. Stefanowski, J., Tsoukiàs, A.: On the extension of rough sets under incomplete information. In: Zhong, N., Skowron, A., Ohsuga, S. (eds.) RSFDGrC 1999. LNCS (LNAI), vol. 1711, pp. 73–81. Springer, Heidelberg (1999). https://doi.org/10.1007/978-3-540-48061-7_11
18. Stefanowski, J., Tsoukias, A.: Incomplete information table and rough classification. Comput. Intell. **17**(3), 545–566 (2001)
19. Wang, G.Y.: Extension of rough set under incomplete system. In: IEEE International Conference on Fuzzy Systems, pp. 1098–1103 (2002)
20. Yang, X., Song, X., Hu, X.: Generalization of rough set for rule induction in incomplete system. Int. J. Granular Comput. Rough Sets Intell. Syst. **2**(1), 37–50 (2011)
21. Nguyen, D.V., Yamada, K., Unehara, M.: Extended tolerance relation to define a new rough set model in incomplete information systems. Adv. Fuzzy Syst. **2013**, 1–11 (2013)
22. Deris, M.M., Abdullah, Z., Mamat, R., Yuan, Y.: A new limited tolerance relation for attribute selection in incomplete information systems. In: IEEE International Conference on Fuzzy Systems and Knowledge Discovery, pp. 964–969 (2015)

Data-Parallel Computational Model
for Next Generation Sequencing
on Commodity Clusters

Majid Hajibaba[1,2]([✉]), Mohsen Sharifi[2], and Saeid Gorgin[1]

[1] Department of Electrical and Information Technology,
Iranian Research Organization for Science and Technology, Tehran, Iran
hajibaba.m@irost.ir
[2] Department of Computer Engineering,
Iran University of Science and Technology, Tehran, Iran

Abstract. It is obvious that the next generation sequencing (NGS) technologies, are poised to be the next big revolution in personalized healthcare, and caused the amount of available sequencing data growing exponentially. While NGS data processing has become a major challenge for individual genomic research, commodity computers as a cost-effective platform for distributed and parallel processing in laboratories can help processing such huge volume of data. To deploy sequence-processing methods on these platforms, in this paper we present a parallel computational model for BLAST on commodity clusters that works in a data parallel manner. The suggested model has a master-worker paradigm. The master stores temporarily incoming requests and splits the database to chunks according to the number of available workers. Each worker pulls, formats, and searches queries against a unique chunk of the database. To show that our model works well, we used queries with different lengths to search against a small database (i.e. UniProtKB/SWISS-PROT) and a large database (i.e. UniProtKB/TrEMBL). The results were equal with the output of the golden method (i.e. NCBI BLAST) and the performance of our model outperformed the most popular distributed form of BLAST (i.e. mpi-BLAST) with 25% higher performance.

Keywords: Distributed systems · Next generation sequencing ·
Parallel computational models · Parallel programming paradigm ·
Commodity clusters

1 Introduction

The amount of available sequencing data is growing exponentially [1] by the advent of high-throughput next-generation sequencing (NGS) technologies. As a result, the processing and storing such big data has become a major challenge for modern genomic research. Bioinformatics algorithms, like finding genes in DNA sequences and aligning similar proteins are complex and time-consuming. Therefore, there is high demand for faster methods to speed up the processing these data [1].

© Springer Nature Switzerland AG 2019
V. Malyshkin (Ed.): PaCT 2019, LNCS 11657, pp. 273–288, 2019.
https://doi.org/10.1007/978-3-030-25636-4_22

Since the use of dedicated supercomputers is often very expensive, most research groups have only limited access to such computers. In contrast, the use of computer clusters has proved more affordable [2]. Therefore, parallel and distributed computing on multiple computers has long been recognized as an attractive approach to parallelization of database searches. Accordingly, to deal with the massive growth in the quantity of data that is generated by NGS, researchers have become interested to use new parallelization techniques over distributed systems to accelerate search of such data.

Sequence alignment is one of the most important research topics in Bioinformatics that focuses on developing methods and tools for the discovery of similar biological sequences and comparing their similarities [3]. BLAST [4] tool is extensively used for searching similarities between biological sequences on a databases of known sequences [5]. The standard BLAST is insufficient to handle the increasing demands for sequence alignments on big databases [6]. Therefore, parallel distributed processing of BLAST on multi computers has proved attractive to achieve faster execution times [7].

Two main approaches to parallelization of BLAST are query segmentation and database segmentation. Our work is founded on distributed BLAST that use database segmentation to parallelize BLAST, not just query segmentation such as SparkBLAST [8]. Moreover, we concentrate on both distributed and parallel computing beyond parallel applications such as H-BLAST [9]. The main research related to our work is mpiBLAST [10], which adopts MPI (Message Passing Interface) for distributing BLAST. Nevertheless, MPI suffers from scalability on big data. In addition, it cannot hide system level details from user for parallel programming. The research on distributed parallel processing using BLAST extends to some other works too [11–14], including ScalaBLAST [15] that uses a software implementation of shared memory or G-BLAST [16] that can operate in heterogeneous distributed environments as well as works reported in [17, 18]. There are also other researches [19, 20] that considered other factors such as accelerating the rate of BLAST calculations in proportion to available compute resources.

In BLAST, raw databases need to be indexed to becoming searchable which is done in a separate process called format. Since BLAST databases are updated daily [21], for short and medium length queries, the formatting database takes longer time than the search on the formatted databases, which is quite a challenge, especially in commodity computers. The above-mentioned works try to improve search time, which is acceptable for a batch of queries or a long query. However, we try to improve search time and format time by distributing these tasks among workers to gain higher performance.

BLAST as an embarrassingly parallel problem can be divided into a number of independent subtasks to be executed by separate processors, which do not need to pass messages between each other. It is ideally suited to master-worker programming paradigm. Therefore, in this paper, we present a data-parallel model for running BLAST application in a distributed manner using the master-worker paradigm. In this model, a master process executes the sequential part of a parallel program that takes requests, splits data and dispatches tasks to workers, and multiple workers do the BLAST processing. Socket data transfer as the means for message passing is responsible for data communication and coordination between the master and the workers.

The main contribution of the paper can be summarized as a new parallel computational model that addresses BLAST calculations using a data-parallel programming model to distribute BLAST tasks adapted for commodity clusters.

2   Methods

2.1   Parallel Computational Model

A parallel computational model, also called parallel programming paradigm [22], is a coherent collection of mechanisms for communication, synchronization, partitioning, placement, and scheduling of tasks [23]. Typical parallel computational models include master-worker, pipeline, divide-and-conquer, and domain decomposition [24–26]. To develop applications, these paradigms must be explicitly programmed in the source code via a programming model using special language constructs, complex directives, or libraries. Three dominant parallel programming models in common use are shared memory, distributed memory (message passing) and data-parallel models.

We adopt a simple data-parallel programming model to develop distributed BLAST. It uses a master-worker programming paradigm, where the master works as the manager and receives the request, split the data on a coarse-grained manner, and sets up tasks; in the coarse-grain case, sequences are partitioned equally among the processing elements [27]. It then transfers the data and tasks to his workers. The overall view of the computational model is illustrated in Fig. 1. We have used the following mechanisms in our parallel computational model to distribute BLAST:

- *Partitioning*, which is based on data decomposition that is database segmentation in the problem domain.
- *Communication*, which uses sockets to transfer data or logically access data in workers.
- *Scheduling*, which is based on First-in-First-out scheduling.
- *Task placement*, which is based on a simple peer selection.

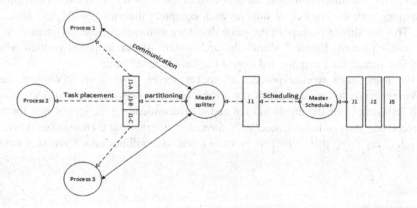

Fig. 1. An overview of our parallel computational model to distribute BLAST

It should be noted that in contrast to some other computational models, our computational model in proposed method does not use any synchronization mechanism since it is based on a data-parallel model which no synchronizations required between subtasks.

2.2 Architecture and Implementation

To design our data-parallel computational model for distributing BLAST, by a master-worker model for scheduling tasks, we used socket data transfer to exchange data between master and worker nodes. Socket data transfer based methods are less dependent on platform than system based methods (like NFS), and thus can be installed and operated more easily [16]. We employ a fine-grained resource-sharing model, wherein nodes are subdivided into slots (corresponding to CPU cores) and a BLAST job binds to each slot. Indeed, for each slot in each node in the cluster, we have one BLAST worker. The master process is the coordinator that pumps the BLAST input information to workers. The way the master works is outlined in Algorithm 1.

Algorithm 1: MASTER algorithm for the BLAST master process

$Q \leftarrow$ query, $D \leftarrow$ database, $W=\{w_1, w_2, ..., w_n\} \leftarrow$ WorkerList
$N_W \leftarrow$ size of W
$D=\{D1, D2, ..., Dn\} \leftarrow$ split D in N_W chunks
For each w_i in W
    Send D_i and Q paths to worker w_i
    $R_i \leftarrow$ wait (assync) for BLAST Result from w_i
Result $\leftarrow$ merge R_1 to R_n
Give Result to user

In Algorithm 1, in the partitioning phase, the database is split into equal sized chunks based on the total number of slots in the cluster. To make the sequence aware of the splitting files, the master uses *gt splitfasta* from genome tools package[1]. In the task placement phase, the master assigns chunks of the database to slots from a worker list by using a simple peer selection strategy. The worker list is read from an input file called *hosts*. To launch a cluster, the user creates the *hosts* file, which must contain the hostnames with the number of slots on each computer that runs the BLAST, one per line. This file should be kept in the same directory as the executable program running the master process. Figure 2 shows the architecture of our proposed method where solid line means data transfer and dotted means command issue.

We name each worker process as socket server process or SSWorker. Each SSWorker binds to one slot and waits for the master process to send BLAST inputs. Each SSWorker gets the inputs via *ssh* and after downloading the inputs, first formats the received chunk of the database using *formatdb* program and then searches it against a given query using the *blastall* program. Algorithm 2 outlines how SSWorker works.

---

[1] GenomeTools. Available at: http://genometools.org

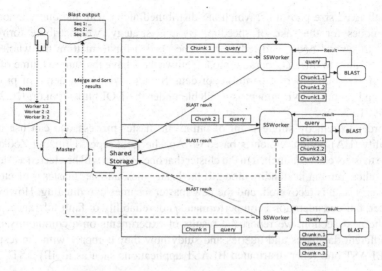

Fig. 2. The architecture of our proposed method

Algorithm 2: *SSWORKER* algorithm for the BLAST worker process

$Q \leftarrow$ query, $D_i \leftarrow$ Database Chunk i

Pull Q and D_i from shared storage

$C_i = \{c_{i1}, c_{i2}, ..., c_{im}\} \leftarrow$ Format D_i

For each c_{ij} in C_i

　　$R_{ij} \leftarrow$ BLAST Q against c_{ij}

$R_i \leftarrow$ Merge $\{R_{i1}, R_{i2}, ..., R_{ij}\}$

Push R_i to shared Storage

The BLAST results are saved in a file using the *blastall* program and the SSWorker uploads the final output file to shared storage. The master process waits for all results from workers and upon receiving the results from all SSWorkers, it merges and sorts them using an extended version of *blastmerge* [5], called *sblastmerge*. *sblastmerge* sorts results according to e-values, the most significant hits appearing at the top, and implemented by authors which is available in Code Ocean DOI https://doi.org/10. 24433/co.0216508.v2. At last, master delivers the final results to the user as a file similar to the BLAST output.

We use database segmentation to parallelize BLAST in a coarse-grained manner, but due to the data-parallel processing, we not only distribute the BLAST task but also distribute the format database task. In most cases in sequence analysis, there is a big database and a query that is commonly not long. BLAST databases are updated daily with no established incremental update scheme and it is recommended that databases be downloaded at regular intervals to keep the content of local copy current [21]. In this case, the formatting of the database is a very time-consuming process taking longer time to execute than the search process. This time can be much more important for running BLAST on commodity computers. Thus, in our model, the database is split into small fixed size partitions, which are distributed along with the query among the worker nodes for the sake of speedup. By this strategy, we distribute formatting database to several nodes and reduce the time needed for formatting the whole database. According to this data-parallel model, all workers have the same volume of work to be performed on different partitions of data. Source code and a demo of proposed method and running environment is available under the DOI https://doi.org/10.24433/co.6424991.v2.

In order to remedy single point of failure in master process, we can use a high availability (HA) option, which is based on Apache Zookeeper. Utilizing Zookeeper, multiple masters can be launched in the cluster that one of them will be elected as "leader" and the others remain in standby mode. If the leader dies, another master is elected, the old master's state is recovered, and the new master resumes coordinating. However, in this paper, our consideration is on performance not reliability or fault-tolerance.

In the next section, we report the results of experiments on a commodity cluster using different databases and queries and study how they compare with the results of NCBI BLAST and other distributed BLAST applications such as mpiBLAST.

3   Results

To evaluate our proposed method, we performed different evaluations and experiments, each with a different goal. Firstly, we evaluated the validity of the results of our proposed method and compared our results with NCBI BLAST. Then, we measured the performance of our method and compared them to the performance of other works. To compare our work with others, we selected mpiBLAST that is an open-source parallel and distributed BLAST tool and is highly similar to our work. It uses database segmentation by distributed and parallel methods not just parallel methods.

3.1   Query Sequences and Database Choices

For performance and validity tests, we used two publicly available protein databases with different sizes: SWISS-PROT as a small database that is manually annotated and reviewed, and UniProtKB/TrEMBL as a huge database that is automatically annotated and not reviewed. For validity evaluation, we used SWISS-PROT as the target database alongside *Surface protein gp120 (Accession P04578)* and *Amyloid beta A4 (Accession P05067)* as query sequences. We used the *Amyloid beta A4* protein because it hits sequences on all partitions (six partitions in our test-bed), but *Surface protein gp120* hits sequences just in one partition. To measure the performance, we used three queries with different lengths: *Fibrinogen beta chain (Accession P14472)* as a small query, *Amyloid beta A4* as a medium query, and *Titin homolog (Accession G4SLH0)* as a long query.

3.2   Cluster Testbed

Because of the evaluation on commodity clusters, available in individual laboratories, the testbed composed of a cluster with just three commodity computers, each with 4 GB memory and Intel processors with the same architecture, connected together with a conventional network switch. As well, we used Ubuntu 12.04LTS 32 bits for running with BLAST, and Ubuntu 14.04LTS 64 bits for running with BLAST+. In order to fairly compare our results with mpiBLAST, we used the latest version of mpiBLAST (i.e. v1.6.0) that is based on NCBI BLAST v2.2.20 released on 2010 and updated until 2012.

3.3   Validity Test

To investigate validity, we counted the number of differences in the top 100 sequences in the final output by (1).

$$diff = \sum \begin{cases} 0 & if\ ord_{res1}(s) = ord_{res2}(s) \\ 1 & if\ ord_{res1}(s) <> ord_{res2}(s) \end{cases} \tag{1}$$

Where $ord_{res1}(s)$ is the order of sequence s in output $res1$. We ignored the differences in the order of sequences with the same e-value and score. The similarity of the proposed method and mpiBLAST results to NCBI BLAST, all with the same version, are illustrated in Fig. 3.

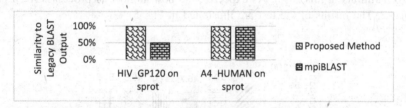

Fig. 3. Similarity of our method and mpiBLAST vs. NCBI blast.

Figure 3 shows that BLAST with partitioning the database and formatting it in distributed mode lead to the same results with BLAST on a single system.

3.4   Performance Test

For performance test, we compare the performances of NCBI BLAST with a shared memory model and mpiBLAST with a message-passing model to the performance of our method, which has a data-parallel model.

Optimization of Parameters for mpiBLAST. mpiBLAST formats and divides the database into many small fragments of approximately equal size. But, the user is responsible to determine the number of fragments of database and also the number of processes in the system to search queries on these fragments. With this static fragmentation and processing method, two parameters must be determined: Number of database fragments as Parameter1, Number of processes as Parameter2.

Since two processes are dedicated for task scheduling and coordinating the output in mpiBLAST, we need $n + 2$ processes to actually perform search tasks on n fragments. To determine the number of fragments, we consider Parameter2 as $n + 2$ where

n is the value of Parameter1, while evaluating it on SWISS-PROT and a simple query. We observed that the best execution time is when the number of fragments is equal to the number of processing cores, in this case 6 fragments. The execution times of mpiBLAST with different number of parts are illustrated in Fig. 4. When the number of fragments increases, both the search time and non-search time rise [28].

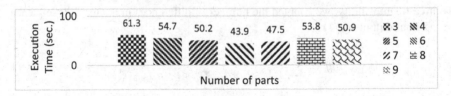

Fig. 4. Parameter setting for the Database Parts in mpiBLAST

To determine the number of processes that must be known when running mpi-BLAST, we tested different settings for Parameter2 by setting Parameter1 to the best case (6 partitions in this case). As expected, the best number of processes for n parts was $n + 2$. The results of the test are illustrated in Fig. 5.

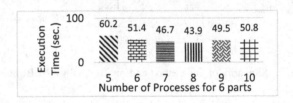

Fig. 5. Parameter setting for the number of processes in mpiBLAST

Moreover, in mpiBLAST, when the number of processes rose above $n + 2$, duplicated copy fragments were loaded on worker nodes. If the number of processes fell below n, some processors had nothing to do. Once the number of processes was set between $n - 1$ and $n + 2$, there were some processes doing more work.

We further evaluated these parameters with TrEMBL database, but during the test, we faced critical problems in mpiBLAST. When database size grows, system failed because of memory heap error. We tried with more fragments and larger memory size, but got some other memory errors. This problem arises because a single BLAST database can contain up to 4 billion letters [29]. We need to set Parameter1 such that each fragment size not be larger than 4 GB. This parameter must be a multiple of six to allow fair balancing of load among computing nodes. Therefore, we found the parameter setting of our environment experimentally as is shown in Table 1.

Table 1. Parameter setting for mpiBLAST

Database	# of partitions	# of processes
SWISS-PROT/TrEMBL	6/24	8/26

Comparison with mpiBLAST and NCBI BLAST. For comparing performance in a distributed system, we measure the total time as (2) [30].

$$T_{total} = T_{comp} + T_{comm} \qquad (2)$$

Where T_{comm} is the time involved in communication among master and workers, and T_{comp} is the computation time which is calculated as (3).

$$T_{comp} = T_{split} + T_{blast} + T_{merge} \qquad (3)$$

Where T_{split} is the time required to split database, T_{blast} is time taken to perform BLAST search (including format) and T_{merge} is the time relative to merge results.

The comparisons of the performances of our method with mpiBLAST and NCBI BLAST, for a small and a big database, are illustrated in Figs. 6 and 7, respectively. The times include executing BLAST, partitioning the database and communication and exclude the overheads of remotely launching programs or monitoring the progress.

Figures 6 and 7 show that for medium and short length queries, mpiBLAST is inefficient especially on a commodity cluster, while our method performs better in all cases.

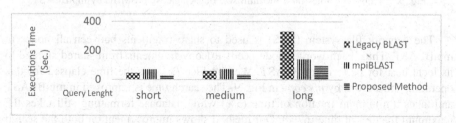

Fig. 6. Performance comparison of the proposed method, mpiBLAST and NCBI BLAST in a small database

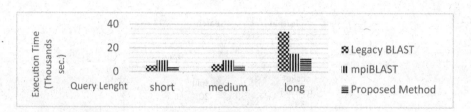

Fig. 7. Performance comparison of the proposed method with mpiBLAST and NCBI BLAST in a huge database

To highlight the outperformance of our method, we measured the time spent in each method. Figure 8 illustrates the fraction of times spent in medium and large queries on a big database in NCBI BLAST.

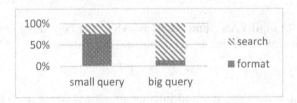

Fig. 8. Fraction of times on medium and large size queries against TrEMBL by NCBI blast

For a big query, a large fraction of time is attributed to the search time. mpiBLAST tries to improve the search time by distributing the search task. However, in a medium size query, database formatting takes a large fraction of the total spent time. The elapsed times for a medium query on a big database in mpiBLAST are illustrated in Fig. 9.

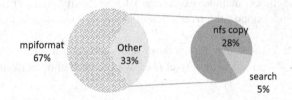

Fig. 9. Fraction of times on a medium size query against TrEMBL by mpiBLAST

The network file system (NFS) is used to share fragments between all nodes in mpiBLAST. Thus, each worker node needs to copy fragments from shared storage to its local host for performing BLAST. We consider T_{comm} as the time elapsed for this operation and show it by *nfs copy* in Fig. 9. The search time is improved in mpiBLAST and takes a minimum fraction of time (5%) while database formatting still takes the maximum fraction of time (67%). Our method shows improvement on both the search time and the database formatting time by distributing both of these two tasks. Figure 10 shows the elapsed times for a medium query on a big database.

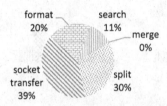

Fig. 10. Fraction of times on a medium size query against trembl by the proposed method

Besides, due to data-parallel operations, for the same database, T_{comm} is less than mpiBLAST for *blastp* program. Because mpiBLAST first formats database and then transfers fragments, while proposed method firsts transfer database chunks and then formats them. The formatted fragments contain files other than sequence file such as index file. T_{comm} for huge database is as Fig. 11 for mpiBLAST and proposed method.

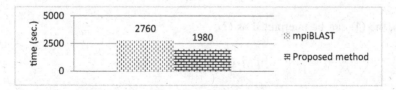

Fig. 11. The communication time for transferring TREMBL database chunks

However, for *blastn* program, the formatted chunks have less size than the original ones and hence mpiBLAST has less T_{comm} related to our method. In mpiBLAST, the main limitations are the formatting of databases and the copying of formatted fragments to local disks in NFS. By contrast, the main limitation of our method is the copying of chunks to computing nodes.

4 Discussion

In this section, we evaluate the load balance and estimate the scalability, which is important for each distributed and parallel model accompanied by performance.

4.1 Load Balance

To efficiently use of a parallel computer system, a balanced workload among the processors is required. To compare the workload balancing effectiveness of a computational method, the percentage of load imbalance (PLIB) [31] was defined as (4).

$$PLIB = \frac{MaximumLoad - MinimumLoad}{MaximumLoad} \times 100 \qquad (4)$$

PLIB is the percentage of the overall processing time that the first processor must wait for the last processor to finish his work. This number also indicates the degree of parallelism. For example, if PLIB is less than one, we achieve over a 99% degree of parallelism. Therefore, a parallel method with a lower PLIB is more efficient than another one with a higher PLIB. The workload is perfectly balanced if PLIB is equal to zero [31]. So, if in a cluster, we are given m workers for scheduling, indexed by the set $W = \{w_1, \ldots, w_m\}$ and there are furthermore given n tasks, indexed by the set $T = \{t_1, \ldots, t_n\}$, which T_i be the set of tasks scheduled on w_i, then the load of worker i (that is w_i) can be achieved from (5).

$$\ell_i = \sum_{t_j \in T_i} C_{i,j} \tag{5}$$

Where task t_j takes $C_{i,j}$ units of time if scheduled on w_i. The maximum load of scheduling is called the *makespan* of the schedule and is equal to (6).

$$\ell_{max} = max_{i \in \{1,\dots,m\}} \ell_i \tag{6}$$

So, the (4) can be interpreted as (7).

$$PLIB = \frac{\ell_{max} - \ell_{min}}{\ell_{max}} \times 100 \tag{7}$$

In the sequence alignment, the duration of the operation depends on the length of the sequence. If the sequence is longer, the search time will be higher. Of course, there are other factors like the similarity of the reference sequence with the query sequence. However, for ease of measurement, it is assumed that the duration of the operation depends only on the length of the sequence. Therefore, for the sequence S, we have (8).

$$C_S = v \times n + p \times n \tag{8}$$

Where v is considered as the processing time required for a base pair and is a constant value, which is the same in all workers. Also, p is the time related to transfer a base pair from shared storage to a local storage. Therefore, if the expected sequences 1 to s are scheduled to run on the worker i (i.e., $T_i = \{t_1.t_2.t_3.\cdots.t_s\}$), we will have (9).

$$\ell_i = \sum_{t_j \in T_i} C_{i,j} = C_{i,1} + C_{i,2} + \cdots + C_{i,n} = (p+v) \times \sum_{k=1}^{s} len(t_k) \tag{9}$$

Similarly, the total length of the sequences in a machine will be equal to the size of the chunk delivered to that machine for processing. In fact, we will have the (10).

$$\ell_i = \sum_{t_j \in T_i} C_{i,j} = (p+v) \times \sum_{k=1}^{s} len(t_k) = (p+v) \times S_{F_i} \tag{10}$$

Where F_i is the chunk assigned to the worker i and S_{F_i} is the size of the chunk.

Since the power of machines is considered equal (homogeneous), so the running time of this operation is the same in all machines and the (11) is resulted.

$$C_{1,a} = C_{2,a} = \cdots = C_{m,a} \tag{11}$$

To evaluate PLIB on all workers, we merge the (10) with (7) which is resulted (12).

$$PLIB = \frac{\ell_{max} - \ell_{min}}{\ell_{max}} \times 100 = \frac{S_{F_{max}} - S_{F_{min}}}{S_{F_{max}}} \times 100 \tag{12}$$

Therefore, the workload of workers is measured based on the amount of data assigned to it. In proposed data-parallel model, the data is divided equally according to the number of workers. Therefore, the sizes of all chunks are equal, which means $S_{F_{max}} = S_{F_{min}}$ and hence PLIB is zero.

4.2   Scalability

To estimate scalability, we measure size scalability by using an insensitive search of *Titin homolog* sequence on TrEMBL database. Although three nodes are not enough for scalability test, but since our target is commodity computers and more importantly, the master just issue commands with no synchronization between workers, so the master and synchronization will not prevent scalability for more nodes.

Size scalability indicates how well the performance will improve when the size of the cluster increased by additional processors. Add resources in this test, means to add more nodes to the cluster, or scale horizontally not vertically by adding more cores to a single node. Figure 12 shows the scalability of the proposed method running on 1, 2 and 3 nodes respectively.

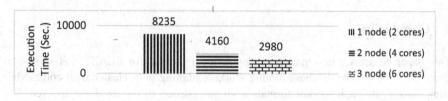

Fig. 12. The scalability of proposed method

To measure the relative performance of two systems processing the same problem, the speedup in parallel computing is often used. Speedup is defined by the (13).

$$S(N) = \frac{T_s}{T_p} \qquad (13)$$

Where $S(N)$ is the theoretical speedup of the parallel program; N is the number of processors in parallel mode; T_s is the execution time of program in serial and T_p is the execution time of program in parallel mode. Moreover, scalability is the ratio between the two efficiency estimates and can extended to (14).

$$\left\{ S = \frac{T_{p_m}}{T_{p_n}} \middle| n = k * m, k > 0 \right\} \qquad (14)$$

Where T_{p_m} is the execution time of program with m processors and T_{p_n} is the execution time of program with n processors, while k is an integer. Linear speedup is obtained when $S(N) = N$. When running a task with linear speedup, doubling the size of the processors, then linear speedup doubles the overall speedup. By using the (14), and the experiment results, we can conclude that the proposed method in commodity clusters has nearly linear scalability. With doubling the size of the nodes from 1 to 2 nodes we have 1.97 times speedup, while with tripling the size of nodes from 1 to 3 we have 2.76 times speedup.

$$S(2) = \frac{8235}{4160} = 1.97, S(3) = \frac{8235}{2980} = 2.76$$

According to Amdahl's Law [32], the sequential fraction of the code is fundamental limitation of the speedup of the system, and thus the maximum speedup that can be achieved using N number of processors is as (15).

$$S(N) = \frac{1}{(1 - P) + \frac{P}{N}} \tag{15}$$

Where P is the proportion of a program that can be made parallel and $1-P$ is the proportion that remains serial. Amdahl's Law yields our speedup is logarithmic and remains below the line $S(N) = N$.

5 Conclusion

This paper presents a new parallel computational method for BLAST that splits and distributes a sequence database among workers running on a cluster with commodity computers while each worker formats a unique portion of the database and searches query in it. The formatting of database takes a significant portion of the non-search fraction of the BLAST runtime. Therefore, we distribute this task as well search task.

The parallel computational model of the proposed method is similar to other works, such as mpiBLAST, but differs in some conceptual and implementation aspects. So, we compared our results with mpiBLAST to show the superiority of the proposed method. BLAST suffers from high execution time for big database while mpiBLAST attempts to reduce search time by distributing database's chunks using MPI. In this work we reduce both search and non-search time on a cluster of commodity computers by a data-parallel computational model. We first present the performance of NCBI BLAST on a single computer with 1 and 2 cores, to demonstrate its poor performance. Through distribution by mpiBLAST we achieved a better performance compared to NCBI BLAST with the same material. But our comparative study of the performances of the proposed method showed that there is a good improvement on execution time over mpiBLAST. We improved the performance 25% by proposed data-parallel model.

As we observed, the transfer time of chunks is high in our method and mpiBLAST. During the transfer, CPUs are idle. Therefore, if we split the database into smaller chunks and send them to workers one by one, we can do formatting and searching during the transfer. In addition, we can merge the BLAST time in transfer time. For future work, we decide to investigate these issues and compare them with this paper.

References

1. Fu, L., Niu, B., Zhu, Z., Wu, S., Li, W.: CD-HIT: accelerated for clustering the next-generation sequencing data. Bioinformatics **28**(23), 3150–3152 (2012)
2. Wilkinson, B., Allen, M.: Parallel Programming: Techniques and Applications Using Networked Workstations and Parallel Computers. Prentice-Hall Inc., Upper Saddle River (2004)
3. Petsko, G., Ringe, D.: From sequence to function: case studies in structural and functional genomics. In: Protein Structure and Function. New Science Press (2004)
4. Altschul, S.F., Gish, W., Miller, W., Myers, E.W., Lipman, D.J.: Basic local alignment search tool. J. Mol. Biol. **215**(3), 403–410 (1990)
5. Mathog, D.: Parallel BLAST on split databases. Bioinformatics **19**(14), 1865–1866 (2003)
6. Bjornson, R., Sherman, A., Weston, S., Willard, N., Wing, J.: TurboBLAST: a parallel implementation of BLAST built on the TurboHub. In: Proceedings of the 16th International Parallel and Distributed Processing Symposium, Washington, DC, USA, p. 325 (2002)
7. Matsunaga, A., Tsugawa, M., Fortes, J.: CloudBLAST: combining MapReduce and virtualization on distributed resources for bioinformatics applications. In: Proceedings of the 2008 Fourth IEEE International Conference on eScience, Indianapolis, IN, USA, pp. 222–229 (2008)
8. Castro, M., Tostes, C., Dávila, A., Senger, H., Silva, F.: SparkBLAST: scalable BLAST processing using in-memory operations. BMC Bioinformatics **18**(1), 318 (2017)
9. Ye, W., Chen, Y., Zhang, Y., Xu, Y.: H-BLAST: a fast protein sequence alignment toolkit on heterogeneous computers with GPUs. Bioinformatics **33**(8), 1130–1138 (2017)
10. Darling, A., Carey, L., Feng, W.: The design, implementation, and evaluation of mpiBLAST. In: 4th International Conference on Linux Clusters, San Jose, CA, USA, p. 14p (2003)
11. Zhang, L., Tang, B.: Parka: a parallel implementation of BLAST with MapReduce. In: Xhafa, F., Patnaik, S., Zomaya, A.Y. (eds.) IISA 2017. AISC, vol. 686, pp. 185–191. Springer, Cham (2018). https://doi.org/10.1007/978-3-319-69096-4_26
12. Dong, G., Fu, X., Li, H., Li, J.: An accurate algorithm for multiple sequence alignment in MapReduce. J. Comput. Methods Sci. Eng. **18**(1), 283–295 (2018)
13. Guo, R., Zhao, Y., Zou, Q., Fang, X., Peng, S.: Bioinformatics applications on Apache Spark. GigaScience **7**(8), giy098 (2018)
14. Mondal, S., Khatua, S.: Accelerating pairwise sequence alignment algorithm by MapReduce technique for Next-Generation Sequencing (NGS) data analysis. In: Abraham, A., Dutta, P., Mandal, J., Bhattacharya, A., Dutta, S. (eds.) Emerging Technologies in Data Mining and Information Security. Advances in Intelligent Systems and Computing, vol. 813, pp. 213–220. Springer, Singapore (2018). https://doi.org/10.1007/978-981-13-1498-8_19
15. Oehmen, C.S., Baxter, D.J.: ScalaBLAST 2.0: rapid and robust BLAST calculations on multiprocessor systems. Bioinformatics **29**(6), 797–798 (2013)
16. Kim, D.-W., et al.: G-BLAST: BLAST manager in an heterogeneous distributed environment. In: 2012 Sixth International Symposium on Theoretical Aspects of Software Engineering, Tianjin, China, pp. 315–316 (2009)
17. Braun, R.C., Pedretti, K.T., Casavant, T.L., Scheetz, T.E., Birkett, C.L., Roberts, C.A.: Parallelization of local BLAST service on workstation clusters. Future Gener. Comput. Syst. **17**, 745–754 (2001)
18. Xiao, S., Lin, H., Feng, W.-C.: Accelerating protein sequence search in a heterogeneous computing system. In: Proceedings of the 2011 IEEE International Parallel Distributed Processing Symposium (IPDPS), Washington, DC, USA, pp. 1212–1222 (2011)

19. Kim, H.-S., Kim, H.-J., Han, D.-S.: Hyper-BLAST: a parallelized BLAST on cluster system. In: Sloot, P.M.A., Abramson, D., Bogdanov, A.V., Gorbachev, Y.E., Dongarra, J.J., Zomaya, A.Y. (eds.) ICCS 2003. LNCS, vol. 2659, pp. 213–222. Springer, Heidelberg (2003). https://doi.org/10.1007/3-540-44863-2_22

20. Pinthong, W., Muangruen, P., Suriyaphol, P., Mairiang, D.: A simple grid implementation with Berkeley Open Infrastructure for Network Computing using BLAST as a model. PeerJ 4, e1388 (2016)

21. Tao, T., Madden, T., Christiam, C., Szilagyi, L.: BLAST® Help. https://www.ncbi.nlm.nih.gov/books/NBK62345/

22. Li, L., Malony, A.D.: Model-based performance diagnosis of master-worker parallel computations. In: Nagel, W.E., Walter, W.V., Lehner, W. (eds.) Euro-Par 2006. LNCS, vol. 4128, pp. 35–46. Springer, Heidelberg (2006). https://doi.org/10.1007/11823285_5

23. Agarwal, A.: Parallel Computational Models, Handout, Lecture02, Multicore Systems Laboratory. MIT (2010)

24. Hamilton, S.: An Introduction to Parallel Programming. CreateSpace Independent Publishing Platform, Scotts Valley (2014)

25. Muresano, R., Rexachs, D., Luque, E.: Learning parallel programming: a challenge for university students. Procedia Comput. Sci. 1(1), 875–883 (2010)

26. Massingill, B., Mattson, T., Sanders, B.: Patterns for parallel application programs. In: 6th Pattern Languages of Programs Workshop (1999)

27. Hughey, R.: Parallel hardware for sequence comparison and alignment. CABIOS 12(6), 473–479 (1996)

28. Lin, H., Ma, X., Chandramohan, P., Geist, A., Samatova, N.: Efficient data access for parallel BLAST. In: Proceedings of the 19th IEEE International Parallel and Distributed Processing Symposium, Denver, Colorado, US, p. 72b (2005)

29. Korf, I., Yandell, M., Bedell, J.: BLAST - An Essential Guide to the Basic Local Alignment Search Tool. O'Reilly & Associates, Sebastopol (2003)

30. Vidyarthi, D., Sarker, B., Tripathi, A., Yang, L.: Scheduling in Distributed Computing Systems. Springer, New York (2009). https://doi.org/10.1007/978-0-387-74483-4

31. Yap, T., Frieder, O., Martino, R.: Parallel computation in biological sequence analysis. IEEE Trans. Parallel Distrib. Syst. 9(3), 283–294 (1998)

32. Amdahl, G.: Validity of the single processor approach to achieving large scale computing capabilities. In: Proceedings of the April 18–20, 1967, Spring Joint Computer Conference, New York, NY, USA (1967)

Parallelization of Algorithms for Mining Data from Distributed Sources

Ivan Kholod[1]([⊠]), Andrey Shorov[1], Maria Efimova[1],
and Sergei Gorlatch[2]

[1] Saint Petersburg Electrotechnical University "LETI", Saint Petersburg, Russia
iiholod@mail.ru, ashxz@mail.ru,
maria.efimova@hotmail.com
[2] University of Muenster, Muenster, Germany
gorlatch@uni-muenster.de

Abstract. We suggest an approach to optimize data mining in modern applications that work on distributed data. We formally transform a high-level functional representation of a data-mining algorithm into a parallel implementation that performs as much as possible computations locally at the data sources, rather than accumulating all data for processing at a central location as in the traditional MapReduce approach. Our approach avoids the main disadvantages of the state-of-the-art MapReduce frameworks in the context of distributed data: increased run time, high network traffic, and an unauthorized access to data. We use the popular data-mining algorithm – Naive Bayes – for illustrating our approach and evaluating it experimentally. Our experiments confirm that the implementation of Naive Bayes developed by using our approach significantly outperforms the traditional MapReduce-based implementation regarding the run time and the network traffic.

Keywords: Parallel algorithms · Distributed algorithms · Data mining ·
Distributed data mining · MapReduce · Homomorphisms

1 Introduction

The development of information technologies, smart devices and their wide use in the Internet of Things (IoT) [1] lead to an increase in the number of distributed sources of information which provide streams of data in large volumes.

Figure 1 presents an example of a system with distributed data sources – Remote Monitoring System (RMS) that is designed to control objects of large and complex systems such as network, factories, airports and other.

A system in Fig. 1 receives data from sensors that can be connected with a large number of middleware data storage nodes. These nodes are often low-cost and have low computational power. Therefore, data processing is carried out on a powerful cluster of the monitoring center. For this, all data from the middleware data storage nodes are gathered into the single data warehouse of the monitoring center.

For high-performance processing of data, scalable data processing systems like Apache Hadoop [2] and Apache Spark [3] are usually running on the computational

V. Malyshkin (Ed.): PaCT 2019, LNCS 11657, pp. 289–303, 2019.
https://doi.org/10.1007/978-3-030-25636-4_23

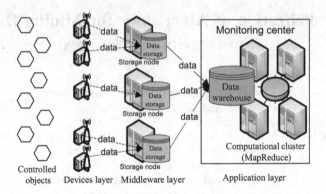

Fig. 1. Example of system with distributed data sources - Remote Monitoring System (RMS).

cluster. These systems are typically based on the MapReduce programming model [4] that performs distributed data processing using a single data storage, e.g., warehouse. Some components of Apache Hadoop use distributed replication [5] for reducing time of access to data and increasing data storage reliability; however, such storage is handled as single data storage in data processing.

The main feature of the RMS systems – collecting all data for processing in a single data warehouse – has several disadvantages: it leads to an increase in total processing time, network traffic, and a risk of unauthorized access to the data.

The goal of this paper to avoid these drawbacks by processing (mining) the data at storage nodes: we achieve this by moving computations to the data nodes. Our approach decomposes a data mining algorithm to perform its major parts locally on the storage nodes without transferring data by network. This helps to reduce the run time of the application and the network traffic. This paper extends our earlier approach to parallelising data mining algorithms on multi-core CPU [6].

2   Our Approach: From Monolithic to Distributed MapReduce

Figure 2(a) presents the traditional approach: the majority of frameworks used for distributed data mining are based on the MapReduce programming model [4].

The MapReduce model uses the abstraction inspired by the *map* and *reduce* primitives that are present in many functional programming languages and also actively exploited in the skeleton-based approach to parallel computing [7]. Parts of MapReduce can run in parallel on distributed nodes, thereby ensuring a high data mining performance. The most important feature of MapReduce in the context of this paper is that the distributed *map* functions take input data from a single data warehouse [4].

In the traditional approach shown in Fig. 2(a), for distributed execution of a data mining algorithm it is restructured on the *map* and *reduce* functions. There are several open-source data mining libraries and frameworks that contain implementations of data mining algorithms for distributed execution based on the MapReduce model, e.g.:

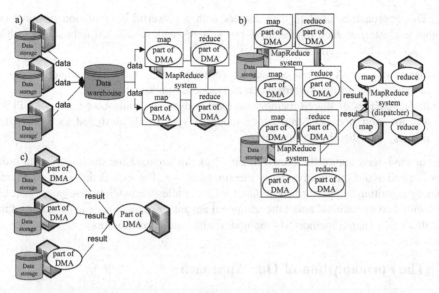

Fig. 2. Variants of data mining algorithms applied to distributed data: (a) traditional MapReduce; (b) geo-distributed batch processing MapReduce-based systems; (c) suggested approach.

Apache Spark MLlib [8], ML Grid [9] from Apache Ignite 2.0, Scalable Advanced Massive Online Analysis [10], Vowpal Wabbit (VW) [11].

There are two ways of distributed data mining when using MapReduce systems:

- collecting data in a single data warehouse;
- using a distributed file system that still presents distributed data as a single data warehouse.

In both cases, data are transferred from distributed sources to the location where data mining takes place. This variant shown in Fig. 2(a) implies the following problems:

- information transfer takes a comparatively long time which may be crucially disadvantageous, especially for real-time processing;
- an increase in network traffic limits the usability of low-capacity communication channels (satellites, wireless, etc.);
- information containing confidential data is transferred using public channels, which increases the risk of unauthorized access to the data;
- large volumes and types of data collected at a single location require enhanced protection to ensure data security and reliability.

Figure 2(b) shows an alternative approach that attempts to overcome these problems using geo-distributed batch processing MapReduce systems for pre-located distributed data. Examples are G-Hadoop [12], G-MR [13], Nebula [14]. These systems assume that there are local MapReduce systems (Hadoop or Spark) located at data sources and responsible for local data processing in Fig. 1(b). An additional MapReduce system dispatcher is used to manage such systems and produce a final result.

This approach is used in data centers with a powerful computation resource (a compute cluster) at each data source. For systems with distributed data sources, this approach still has several disadvantages:

- high requirements are put on the storage nodes – they must have enough computational power to run systems such as Hadoop, Spark, etc. locally;
- the advantages of shared memory on local nodes (e.g., multi-core CPUs or GPUs) are not used because MapReduce model is specifically designed to work with memory in a distributed manner.

In our envisaged approach, shown in Fig. 2(c), we improve the distributed data processing and avoid the disadvantages mentioned above. The idea is that parts of a data mining algorithm are performed at data sources, while intermediate results are sent to the central computational node; the additional advantage is that they can be executed in parallel using shared memory on the nodes with parallel processors.

3   The Formalization of Our Approach

We develop our approach using a general formalism, in order to cover a broad class of data mining algorithms. Capital letters are used for types, and lower-case letters are used for variables of these types and functions.

3.1   Data Mining Algorithm as a Composition of Functions

A data mining algorithm is represented in our approach as a function that takes a data set $d \in D$ as input and creates a mining model $\mu \in M$ from it as output:

$$\text{dma}: D \to M \tag{1}$$

We use capital letters to denote types and lower case letters for variables of these types. In (1), a data mining algorithm creates a mining model, without changing the input data.

A data set D usually contains characteristics (such as temperature, sound level, vibration, pressure etc.) of objects (e.g., elements of complex system, vehicles and other). We represent a data set as a 2-dimensional array (data matrix), e.g., for μ objects that are described by p characteristics [15]:

$$d = \begin{pmatrix} x_{1.1} & \cdots & x_{1.k} & \cdots & x_{1.p} \\ \cdots & & \cdots & & \cdots \\ x_{j.1} & \cdots & x_{j.k} & \cdots & x_{j.p} \\ \cdots & & \cdots & & \cdots \\ x_{m.1} & \cdots & x_{m.k} & \cdots & x_{m.p} \end{pmatrix} \tag{2}$$

where $x_{j.k}$ is the value of the kth characteristic of the jth object.

Figure 3 represents a distributed storage: data matrix d is split between s nodes:

$$d = d_1 \cup \ldots \cup d_s,$$

where data submatrix d_h is located at the storage node number h.

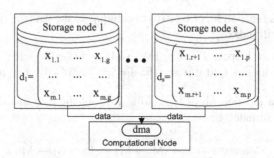

Fig. 3. Traditional processing distribution of data set.

Traditionally, data processing is performed by a data mining algorithm on a particular computational node or computational cluster as shown in Fig. 2(a). For this, all data from the storage nodes are transferred to it.

A mining model comprises elements that describe a knowledge extracted by a data mining algorithm from a data set. These elements can be, e.g., classification or association rules, cluster centers, decision tree nodes, etc. Thus, the mining model $\mu \in M$, $M = [E]$, can be represented as an array of elements e_i, $i = 0 \ldots u$:

$$\mu = [e_0,\ e_1,\ \ldots,\ e_u]. \tag{3}$$

We represent a data mining algorithm as a sequence of steps, formally expressed as a sequential composition of functions, e.g.:

$$dma = f_n^{\circ} f_{n-1}{}^{\circ} \ldots {}^{\circ} f_1^{\circ} f_0. \tag{4}$$

where $^\circ$ is composition operator that is applied from right to left.

In (4), function f_0: D$\rightarrow$M takes a data set $d \in D$ as an argument and returns a mining model $\mu_0 \in M$. Function f_t, $t = 1..n$ in (4) takes the mining model $\mu_{t-1} \in M$ created by the previous function f_{t-1} and returns the changed mining model $\mu_t \in M$:

$$f_t : M \rightarrow M. \tag{5}$$

The functions f_t, $t = 1..n$ of type (5) are called Functional Mining Blocks (FMB).

The functions that process data matrix d take an additional argument d:

$$fd_t : D \rightarrow M \rightarrow M. \tag{6}$$

Partial function application with the fixed first argument $f_t = fd_t\, d$ (i.e. parameter d is constant for the function f_t) allows us to use such functions in the composition (4).

To apply some function fd_t to each element of the data matrix, we invoke it in a loop. We introduce loops over columns and rows for iterative processing of the data matrix. We use an asterisk to refer to whole row (for example, $d[j,*]$ refers to the jth row) or whole column (for example, $d[*, k]$ refers to the kth column) in a data matrix:

- *loopc* applies a function fd_t of type (6) to columns of the data matrix $d \in D$ starting from index i_s till index i_e:

$$loopc :\ I \to I \to (M \to M) \to D \to M \to M$$
$$loopc\, i_s\, i_e\, fd_t\, d\, \mu\ =\ ((fd_t\, d[*, i_e])^\circ \ldots {}^\circ (fd_t\, d[*, i_s]))\, \mu; \tag{7}$$

- *loopr* applies a function fd_t of type (6) to rows of the data matrix $d \in D$ starting from index i_s till index i_e:

$$loopr :\ I \to I \to (M \to M) \to D \to M \to M$$
$$loopr\, i_s\, i_e\, f_t\, d\, \mu = (fd_t\, d[i_e, *])^\circ \ldots {}^\circ (fd_t\, d[i_s, *])\, \mu. \tag{8}$$

The first four arguments are fixed to use *loopc* and *loopr* in the composition (4).

3.2 Illustration for the Naive Bayes Algorithm

We use the Naive Bayes [16] algorithm to illustrate the parallelization of algorithms represented as (4) for distributed data: Naive Bayes belongs to Top 10 data mining algorithms [17]: it solves the classification task by analyzing data where attributes have type that is a finite set of values:

$$T_k = \{v_{1.k}, \ldots, v_{l.k}\}, \text{ where } k = 1..p.$$

Figure 4 shows the pseudocode of the Naive Bayes algorithm. It calculates:

- the number of vectors with the value $d[j, p] = v_{q.p}$, for each value $v_{q.p}$ ($v_{q.p} \in T_p$) of a p-th attribute (line 4 in Fig. 4);
- the number of vectors with value $d[j, k] = v_{i.k}$ of the k-th attribute and with value $d[j, p] = v_{q.p}$ of the p-th attribute, for each value $v_{i.k}$ of each k-th attribute ($v_{i.k} \in T_k$) (line 6 in Fig. 4).

```
1.  for q = 1... s // loop for each mining model's element
2.      μ[q] = 0; // initialization of mining model's elements
3.  for j = 1 ... μ // loop for each vector
4.      μ[d[j,p]]++; //increment count of vectors for value x_{j.p} of vector x_j;
5.      for k = 1 ...p-1 // loop for each attribute
6.          μ[φ(k-1)+(d[j, k]-1)·φ(0)+ d[j, p]]++; // increment count of vectors with
                                              // value x_{j.k} and value x_{j.p}
7.      end for;
8. end for;
```

Fig. 4. The Naive Bayes algorithm: pseudocode.

Using these values and Bayes' theorem, we can calculate an unknown value of the p-th attribute of a new object with some probability based on only the k-th ($k = 1..p - 1$) attributes describing the object.

We interpret the value of the data matrix as the index number of value in the set:

$$d[j, k] = v_{i.k} = i, \text{ where } i : 1 .. l.$$

Thus the mining model for Naive Bayes algorithm comprises:

- mining model's elements $\mu[1],\dots, \mu[s]$ are the counters ($\mu[q] = 0, 1 \le q \le s$) of vectors with value $v_{q.p}$ of the p-th attribute ($v_{q.p} \in T_p, s = |T_p|$);
- mining model's elements $\mu[s+1],\dots, \mu[u]$ are the counters ($\mu[g] = 0, s < g \le u$) of vectors with value $v_{i.k}$ of k-th attribute ($v_{i.k} \in T_k$) and value $v_{q.p}$ of the p-th attribute:

$$g = \varphi(k - 1) + (d[j, k] - 1) \cdot \varphi(0) + d[j, p], \text{ where } \varphi(k) = s + \sum_{q=1}^{k} (|T_q| \cdot s).$$

The function f_0 initializes the array of the mining model's elements (line 1–2 in Fig. 4). All other FMBs are formed for each line of the Naive Bayes algorithm:

- f_1 is the loop for the data set's vectors (line 3 in Fig. 4):

$$f_1 \, d\,\mu = \text{loopr } 1 \, \mu \, (f_3^{\circ} f_2) \, d\,\mu;$$

- f_2 increments the counter ($\mu[q], 0 \le q \le s$) of the vectors with the value $q = d[j, p]$ of the p-th attribute (line 4 in Fig. 4):

$$f_2 \, d\,\mu = \mu[d[j, p]] ++ ;$$

- f_3 is the loop for the data set's attributes (line 5 in Fig. 4):

$$f_3 \, d\,\mu = \text{loopc } 1 \, p - 1 \, f_4 \, d\,\mu;$$

- f_4 increments the counter ($\mu[\varphi(k - 1) + (d[j, k] - 1) \cdot \varphi(0) + d[j, p]], s < g \le v$) of the vectors with the value $d[j, k]$ of the k-th attribute and value $d[j, p]$ of the p-th attribute (line 6 in Fig. 4):

$$f_4 \, d\,\mu = \mu[\varphi(k - 1) + (d[j, k] - 1) \cdot \varphi(0) + d[j, p]] ++ .$$

Composition of these functions represents the Naive Bayes algorithm:

$$NB = f_1^{\circ} f_0 = (\text{loopr } 1 \, z \, (f_3^{\circ} f_2) \, d)^{\circ} f_0 = (\text{loopr } 1 \, z \, ((\text{loopc } 1 \, p - 1 f_4 \, d)^{\circ} f_2) \, d)^{\circ} f_0. \quad (9)$$

Summarizing, Fig. 5 represents the interaction of the FMBs in the Naive Bayes algorithm with distributed data.

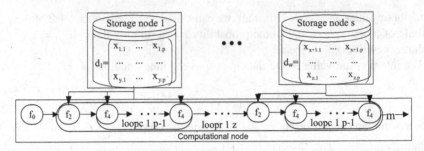

Fig. 5. Naive Bayes algorithm as a composition of functions on distributed data.

3.3 Functions for Parallelization

In case of distributed data, FMBs of type (6) can be performed on the storage nodes. In addition they can be executed by in parallel. This corresponds to the parallel execution with distributed memory. At the same time parallel execution on shared memory can be performed on a storage node using a multi-core processor. Our approach aims at parallelizing a data mining algorithm both for shared and distributed memory.

The higher-order function *parallel* expresses the parallel execution of FMBs by a system with shared memory:

$$\text{parallel} : [(M \rightarrow M)] \rightarrow M \rightarrow M$$
$$\text{parallel} [f_r, \ldots, f_s] \, \mu = \text{head fork} [f_r, \ldots, f_s] \, \mu, \tag{10}$$

where function *fork* invokes FMBs in parallel:

$$\text{fork} : [M \rightarrow M] \rightarrow M \rightarrow [M]$$
$$\text{fork} [f_r, \ldots, f_s] \, \mu = [f_r \, \mu, \ldots, f_s \, \mu]. \tag{11}$$

function *head* returns the first element of a mining model's list of elements.

The higher-order function *paralleld* for expresses the parallel execution of FMBs on a distributed memory:

$$\text{paralleld} : [(M \rightarrow M)] \rightarrow M \rightarrow M$$
$$\text{paralleld} [f_r, \ldots, f_s] \, \mu = \text{join} \, \mu \, (\text{forkd}[f_r, \ldots, f_s] \, \mu), \tag{12}$$

where function *forkd* allows to invoke FMBs in parallel on distributed memory:

$$\text{forkd} : [M \rightarrow M] \rightarrow M \rightarrow [M]$$
$$\text{forkd} [f_r, \ldots, f_s] \, \mu = [f_r \, \text{copy} \, \mu, \ldots, f_s \, \text{copy} \, \mu]; \tag{13}$$

function *copy* creates copies of the initial mining model in separate areas of the distributed memory for parallel processing by FMBs:

$$\text{copy} : M \rightarrow M$$
$$\text{copy} \, \mu = [\mu[0], \, \mu[1], \ldots, \mu[v]]; \tag{14}$$

function *join* joins the mining models that are built by parallel FMBs in separate areas of distributed memory:

$$\text{join} : M \rightarrow [M] \rightarrow M$$
$$\text{join } \mu \ [\mu_r, \ldots, \mu_s] = [\mu'[0], \ldots, \mu'[g], \ldots, \mu'[v]],$$
$$\text{where } \mu'[g] = \begin{cases} \mu[g] \text{ if } \mu_i[g] = \mu[g] \text{ for all } i = r..s \\ \text{union } \mu[g] \ [\mu_r[g], \ldots, \mu_s[g]] \text{ otherwise} \end{cases} \tag{15}$$

where function *union* merges elements of different mining models with the same index to a single mining model's element:

$$\text{union} : E \rightarrow [E] \rightarrow E. \tag{16}$$

The implementation of function *union* depends on the structure of the element.

Figure 6 shows the distributed execution of a data mining algorithm using *paralleld*.

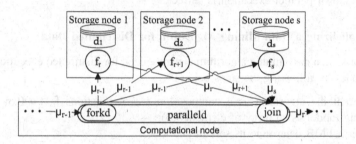

Fig. 6. Execution of a data mining algorithm on distributed data using distributed memory.

3.4   Conditions for Parallel Execution of FMBs

Not all FMBs can be correctly executed in parallel to each other, because of possible data dependencies between them. To check the correctness of the parallel execution using shared and distributed memory, we introduce the following conditions.

Bernstein's conditions [18] are traditionally used to verify parallel execution of the FMBs $f_1, \ldots, f_r$: two FMBs f_t and f_{t+1} can be executed in parallel on a shared memory if:

- there is no data anti-dependency: $\text{In}(f_t) \cap \text{Out}(f_{t+1}) = \varnothing$;
- there is no data flow dependency: $\text{Out}(f_t) \cap \text{In}(f_{t+1}) = \varnothing$;
- there is no output dependency: $\text{Out}(f_t) \cap \text{Out}(f_{t+1}) = \varnothing$;

where $\text{In}(f_t)$ is a subset of mining model elements used by FMB f_t; $\text{Out}(f_t)$ is a subset of mining model elements modified by FMB f_t.

Bernstein's conditions are sufficient, but not necessary. When representing an algorithm as (4), we can weaken these conditions for shared and distributed memory.

When executed in parallel using shared memory, FMBs f_t and f_{t+1} use and modify elements of the same mining model μ_{t-1}. Therefore, if following condition:

$$f_t^\circ f_{t+1} = f_{t+1}^\circ f_t \tag{17}$$

is true for them, then the result of parallel execution is correct.

During parallel execution on distributed memory, FMBs f_t and f_{t+1} receive a copy of the mining model μ_{t-1} constructed by FMB f_{t-1} and copied by the *copy* function (14) as an argument of the *forkd* function (13). Therefore, FMB f_t does not use elements that are modified by FMBs f_{t+1}, i.e. $\text{In}(f_t) \cap \text{Out}(f_{t+1}) = \varnothing$ always is true.

As a result, FMBs f_t and f_{t+1} create different instances of a mining model μ_t and their modified elements are merged by the *union* function (16). Therefore, if there exist the *union* functions for each modified elements $\mu_t[g] \in \text{Out}(f_t) \cap \text{Out}(f_{t+1})$ invoked by function *join* (15), such that the following condition is true:

$$(f_{t+1}^{\;\circ} f_t)\,\mu_{t-1} = \text{join}\,\mu_{t-1}\,[(f_t\,\text{copy}\,\mu_{t-1}),\,(f_{t+1}\,\text{copy}\,\mu_{t-1})], \tag{18}$$

then the result of parallel execution is correct.

3.5   Parallelizing a Data Mining Algorithm for Distributed Data

In our approach, a data mining algorithm is parallelized for distributed execution on the storage nodes by following the steps below:

(1) represent the algorithm as a composition (4) of functions f_t, $t = 0..n$ and the mining model as an array of elements (3);
(2) for each FMB determine its sets In and Out;
(3) for each pair of the functions of type (6) verify condition (18) for parallel execution on storage nodes using distributed memory;
(4) for all other FMBs verify condition (17) for parallel execution using shared memory on nodes with multi-core processors;
(5) convert the sequential execution of FMBs, which can be performed in parallel with distributed and shared memory, into parallel execution by using functions *paralleld* and *parallel*, respectively.

Applying function *paralleld* to FMBs that interact with the data set (i.e., they are of type (6)) allows their execution on the nodes where this data are stored, which makes it possible to perform local processing without transferring data to other nodes (Fig. 6). When the mining model of a data mining algorithm is significantly smaller than the volume of the input data set, such distributed execution allows to reduce:

- network traffic by sending over a network a mining model that is smaller than large volumes of data;
- run time of the algorithm by reducing the time that is necessary to transmit large volumes of data.

An important feature of our approach is that we retain the capability to execute FMBs in parallel on nodes with multi-core processors with shared memory. This enables using computational resources at the nodes more efficiently, by exploiting two different kinds of parallelism – on distributed and on shared memory.

3.6 Illustration of Approach: The Naive Bayes Algorithm

As a real-world use case, we apply our approach to parallelising the Naive Bayes algorithm. The first step is described in Sect. 3.2; its result is the algorithm represented as a composition of FMBs (9).The second step defines sets In and Out for each FMB of the algorithm. First, the sets are defined for the simple FMB f_2 and f_4 are determined based on the lines 4 and 6 of the pseudocode in Fig. 4 as follows:

$$In(f_2) = \{\mu[d[j,p]]\}, \; Out(f_2) = \{\mu[d[j,p]]\},$$
$$In(f_4) = \{\mu[\varphi(k-1) + (d[j,k]-1) \cdot \varphi(0) + d[j,p]]\}, \; Out(f_4) = \{\mu[\varphi(k-1) + (d[j,k]-1) \cdot \varphi(0) + d[j,p]]\}.$$

The In and Out sets of the loops are a union of the corresponding sets of iterative FMB that are called at each loop iteration. For it need to determinate the In and Out sets for one iteration and extends it for whole loop. For example, the loop f_3 invokes the FMB f_4 for attributes from 1 till p, thus, the In and Out sets are follows:

$$In(f_3) = \{\mu[d[j,1]\cdot\varphi(0) + d[j,p]], \ldots, \mu[\varphi(p-1) + (d[j,p]-1) \cdot \varphi(0) + d[j,p]]\},$$
$$Out(f_3) = \{\mu[d[j,1]\cdot\varphi(0) + d[j,p]], \ldots, \mu[\varphi(p-1) + (d[j,p]-1) \cdot \varphi(0) + d[j,p]]\}.$$

Table 1 presents the In and Out sets for all FMBs of the Naive Bayes algorithm.

Table 1. Sets In and Out for the FMBs of the Naive Bayes algorithm

FMB	In	Out
$f_1 = loopr \; 1 \; z$ $(f_3{}^\circ f_2)$ d	$\mu[1], \ldots, \mu[\varphi(p-1)]$	$\mu[1], \ldots, \mu[\varphi(p-1)]$
f_2	$\mu[d[j,p]]$	$\mu[d[j,p]]$
$f_3 = loopc \; 1 \; p$ f_4 d	$\mu[d[j,1]\cdot\varphi(0) + d[j,p]], \ldots, \mu[\varphi(p-1) + (d[j,p]-1)\cdot\varphi(0) + d[j,p]]$	$d[j,1]\cdot\varphi(0) + d[j,p]], \ldots, \mu[\varphi(p-1) + (d[j,p]-1)\cdot\varphi(0) + d[j,p]]$
f_4	$\mu[\varphi(k-1) + (d[j,k]-1)\cdot\varphi(0) + d[j,p]]$	$\mu[\varphi(k-1) + (d[j,k]-1)\cdot\varphi(0) + d[j,p]]$

In the 3rd step, we verify condition (18) for the loop $f_1 = loopr \; 1 \; z \; (f_3{}^\circ f_2)$ that is executed in parallel on storage nodes using distributed memory. The verification is carried out for composition $f_3{}^\circ f_2$ that is invoked on adjacent iterations as follows:

$$In((f_3^\circ f_2) \; d[j,*]) \cap Out((f_3^\circ f_2) \; d[j+1,*]) =$$
$$Out((f_3^\circ f_2) \; d[j,*]) \cap In((f_3^\circ f_2) \; d[j+1,*]) =$$
$$Out((f_3^\circ f_2) \; d[j,*]) \cap Out((f_3^\circ f_2) \; d[j+1,*]) =$$
$$\{\mu[d[j,p]], \; \mu[d[j,1]\cdot\varphi(0) + d[j,p]], \ldots, \mu[\varphi(p-1) + (d[j,p]-1) \cdot \varphi(0) + d[j,p]]\} \cap$$
$$\{\mu[d[j+1,p]], \; \mu[d[j+1,1]\cdot\varphi(0) + d[j+1,p]], \ldots, \mu[\varphi(p-1) + (d[j+1,p]-1) \cdot \varphi(0) + d[j+1,p]]\} =$$
$$\{\mu[d[j,p]], \; \mu[\varphi(k-1) + (d[j,k]-1) \cdot \varphi(0) + d[j,p]]\} \neq \emptyset,$$

when $d[j, p] = d[j + 1, p]$ and $d[j, k] = d[j + 1, k]$ for $k = 1 \ldots p$.

However, for these elements a *union* function exists that sums up vector counters has been defined for these elements:

$$\text{union} : E \rightarrow [E] \rightarrow E$$
$$\text{union } \mu[q] \, [m_1[q], \ldots, m_r[q]] = \mu[q] + (m_1[q] - \mu[q]) + \ldots + (m_r[q] - \mu[q]).$$

Thus, loop f_1 in (9) can be executed in parallel on distributed memory (storage nodes).

In the 4th step, we verify the condition (17) for the loop f_3=loopc 1 p f_4 for the execution using shared memory. The verification is carried out for iterative FMB f_4 that is invoked on adjacent iterations for different attributes:

$$(f_4 \, d[*, k]) \text{ and } (f_4 \, d[*, k+1])),$$

consequently change different mining model's elements:

$$(\mu[\varphi(k-1) + (d[j, k] - 1) \cdot \varphi(0) + d[j, p]] ++) \text{ and } (\mu[\varphi(k) + (d[j, k+1] - 1) \cdot \varphi(0) + d[j, p]] ++)$$

Consequently, the condition (17) is true:

$$((f_4 \, d[*, k])^\circ (f_4 \, d[*, k+1])) = ((f_4 \, d[*, k+1])^\circ (f_4 \, d[*, k])).$$

Thus loop f_3 can be executed in parallel using shared memory.

Figure 7 represents the expression that is obtained at step 5 for data distribution:

$$\text{NBPar} = (\textbf{paralleld } [\text{loopr d 1 z } (\textbf{parallel } [(\text{loopc 1 p } f_4)]^\circ f_2])^\circ f_0. \qquad (19)$$

As a result, a parallel version of the Naive Bayes algorithm is obtained for processing the distributed data at the storage nodes. Note that loop *loopc* is parallelized on n cores at each storage node. This allows us to use the computational resources at the nodes more efficiently.

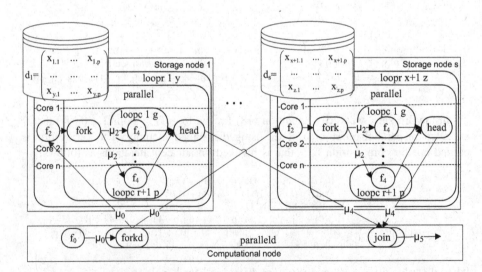

Fig. 7. Distributed execution of the Naive Bayes algorithm for distributed data

4   Experimental Evaluation

We implement the distributed variant of the Naive Bayes algorithm in the Java-based library DXelopes [19]. With it, we perform experiments described in the following.

In our experiments, we use the data set "Predict Outcome of Pregnancy" from the Kaggle Datasets [20]. This data set contains data from Annual Health Survey: Woman Schedule. The data set contains 68 attributes related to birth (birth history; type of medical attention at delivery; details of maternal health care (antenatal/natal/postnatal); immunization of children, etc.) and one attribute that contains information about the outcome of pregnancy (live birth/stillbirth/abortion). The data set comes as a text file (in CVS format) that is 2 Gb in size.

Table 2 describes how the data are partitioned for our experiments. The data set is split into 2 and then 4 parts and first distributed between two storage nodes and then between four storage nodes. A non-divided data set (for single storage node) is used for comparison.

Table 2. Distributed data sets.

Number of distributed data sets	Number of vectors in each data set	Number of attributes in each data set	Size of each data set (Mb)
4	3 402 670	68	500
2	6 805 350	68	1 000
1	14 461 451	68	2 000

The storage nodes are connected to the computational node by local network with bandwidth 1 Gbps. Each storage node has the following configuration: CPU Intel Xeon (4 physical cores), 2.90 GHz, 4 Gb. The computational node has: CPU Intel Xeon (12 physical cores), 2.90 GHz, 4 Gb. To imitate communication channel limitation, we use channel throttling with level 75 Mbps, that matches 4G wireless systems.

Figure 8(a) shows the run time for distributed data. We compare two approaches of processing distributed data: with gathering data into single data warehouse as in the usual MapReduce frameworks and without gathering data as in our approach. We observe that the run time when processing distributed data without gathering them is reduced when increasing number of storage nodes. For two storage nodes with 4 physical cores on each, the run time of processing without gathering data almost equals run time of processing without gathering data using12 physical cores. For four storage nodes, this run time is less by 30%. These results can be explained by processing a smaller volume of data on each storage node in parallel.

The difference between both approaches is increasing with limited bandwidth. With a limit of 75 Mbsp, the difference for four storage nodes is more than twice. It can be explained by the large time spent on data transfer. Obviously, with increasing volume of data, the run time of processing distributed data with gathering data will increase.

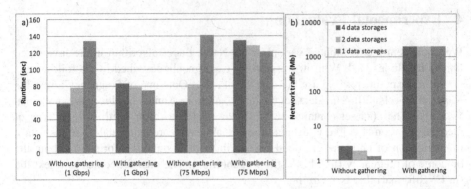

Fig. 8. Processing of distributed data by Naive Bayes algorithm: (a) run time; (b) network traffic.

Figure 8(b) shows a comparison of network traffic for both approaches. Distributed data mining with gathering data generates larger traffic (≥ 100 times), because for it all data are transferred by network. When processing distributed data in our approach without gathering data, only processing results are transferred over the network.

We observe that increasing the number of storage nodes leads to increasing the volume of network traffic, because a higher number of results are being sent over the network. The number of the results that are being transferred equals the number of distributed storage nodes (i.e., two mining models for two storage nodes and four models for four storage nodes). Therefore, the volume of the network traffic directly depends on the number of distributed storage nodes.

5   Conclusion

We suggest an approach to optimize data mining in the modern applications with distributed data sources. Our approach formally transforms a high-level functional representation of a data-mining algorithm into a parallel implementation that performs as much as possible computations locally at the data sources, rather than accumulating all data for processing at a central location as in the traditional MapReduce approach. Thereby our data mining implementation avoids the main disadvantages of the state-of-the-art MapReduce frameworks in the context of distributed data: increasing total processing time, high network traffic, and a risk of unauthorized access to the data.

We develop a functional formalism as a formal base of our approach: it allows proving the correctness of the formal program transformation and of the obtained parallel implementation. We use the popular data-mining algorithm – Naive Bayes – for illustrating our approach and for its experimental evaluation. Our experiments confirm that the distributed implementation of Naive Bayes developed by using our approach significantly outperforms the usual MapReduce-based implementation regarding the run time and the network traffic.

Acknowledgments. We thank the anonymous referees for very helpful remarks on the preliminary version of the paper. This work was supported by the Ministry of Education and Science of the Russian Federation in the framework of the state order "Organization of Scientific Research", task #2.6113.2017/BY, and by the German Ministry of Education and Research (BMBF) in the framework of project HPC2SE at the University of Muenster.

References

1. Santucci, G.: From internet to data to Internet of Things. In: Proceedings of the International Conference on Future Trends of the Internet (2009)
2. Apache Hadoop. http://hadoop.apache.org
3. Apache Spark. http://spark.apache.org/
4. Dean, J., Ghemawat, S.: MapReduce: simplified data processing on large clusters. In: Proceedings of Operating Systems Design and Implementation, San Francisco, CA, December 2004
5. Harshawardhan, S.B., et al.: A review paper on Big Data and Hadoop. Int. J. Sci. Res. Publ. **4**(10), 1–7 (2014)
6. Kholod, I., Shorov, A., Titkov, E., Gorlatch, S.: A formally-based parallelization of data mining algorithms for multi-core systems. J. Supercomputing (2018)
7. Gorlatch, S., Cole, M.: Parallel skeletons. In: Padua, D. (ed.) Encyclopedia of Parallel Computing. Springer, Boston (2011). https://doi.org/10.1007/978-0-387-09766-4
8. Introducing Apache Mahout. http://www.ibm.com/developerworks/java/library/j-mahout/
9. Apache Ignite. Documentation. Machine Learning. https://apacheignite.readme.io/docs/machine-learning
10. De Francisci, M.G., Bifet, A.: SAMOA scalable advanced massive online analysis. J. Mach. Learn. Res. **16**, 149–153 (2015)
11. Langford, J., Strehl, F., Li, L.: Vowpal wabbit (2007). http://hunch.net/~vw
12. Wang, L., et al.: G-Hadoop: MapReduce across distributed data centers for data-intensive computing. FGCS **29**(3), 739–750 (2013)
13. Jayalath, C., Stephen, J.J., Eugster, P.: From the cloud to the atmosphere: running MapReduce across data centers. IEEE Trans. Comput. **63**(1), 74–87 (2014)
14. Ryden, M., et al.: Nebula: distributed edge cloud for data intensive computing. In: IC2E, pp. 57–66 (2014)
15. Hastie, T., Tibshirani, R., Friedman, J.: The Elements of Statistical Learning. SSS. Springer, New York (2009). https://doi.org/10.1007/978-0-387-84858-7
16. George, H.J., Langley, P.: Estimating continuous distributions in Bayesian classifiers. In: Eleventh Conference on Uncertainty in Artificial Intelligence, pp. 338–345 (1995)
17. Xindong, W., et al.: Top 10 algorithms in data mining. Knowl. Inf. Syst. **14**(1), 1–37 (2007)
18. Bernstein, J.: Program analysis for parallel processing. IEEE Trans. Electron. Comput. **EC-15**, 757–762 (1966)
19. Prudsys Xelopes. https://prudsys.de/en/knowledge/technology/prudsys-xelopes/
20. Kaggle: Dataset: Predict Outcome of Pregnancy. https://www.kaggle.com/rajanand/ahs-woman-1

HaraliCU: GPU-Powered Haralick Feature Extraction on Medical Images Exploiting the Full Dynamics of Gray-Scale Levels

Leonardo Rundo[1,2,3], Andrea Tangherloni[3,4,5,6], Simone Galimberti[3],
Paolo Cazzaniga[7,8], Ramona Woitek[1,2,9], Evis Sala[1,2], Marco S. Nobile[3,8],
and Giancarlo Mauri[3,8(✉)]

[1] Department of Radiology, University of Cambridge, Cambridge, UK
[2] Cancer Research UK Cambridge Centre, Cambridge, UK
[3] Department of Informatics, Systems and Communication,
University of Milano-Bicocca, Milan, Italy
giancarlo.mauri@unimib.it
[4] Department of Haematology, University of Cambridge, Cambridge, UK
[5] Wellcome Trust Sanger Institute, Wellcome Trust Genome Campus, Hinxton, UK
[6] Wellcome Trust – Medical Research Council Cambridge Stem Cell Institute,
Cambridge, UK
[7] Department of Human and Social Sciences, University of Bergamo, Bergamo, Italy
[8] SYSBIO.IT Centre of Systems Biology, Milano, Italy
[9] Department of Biomedical Imaging and Image-guided Therapy,
Medical University Vienna, Vienna, Austria

Abstract. Image texture extraction and analysis are fundamental steps in Computer Vision. In particular, considering the biomedical field, quantitative imaging methods are increasingly gaining importance since they convey scientifically and clinically relevant information for prediction, prognosis, and treatment response assessment. In this context, radiomic approaches are fostering large-scale studies that can have a significant impact in the clinical practice. In this work, we focus on Haralick features, the most common and clinically relevant descriptors. These features are based on the Gray-Level Co-occurrence Matrix (GLCM), whose computation is considerably intensive on images characterized by a high bit-depth (e.g., 16 bits), as in the case of medical images that convey detailed visual information. We propose here HaraliCU, an efficient strategy for the computation of the GLCM and the extraction of an exhaustive set of the Haralick features. HaraliCU was conceived to exploit the parallel computation capabilities of modern Graphics Processing Units (GPUs), allowing us to achieve up to $\sim 20\times$ speed-up with respect to the corresponding C++ coded sequential version. Our GPU-powered solution highlights the promising capabilities of GPUs in the clinical research.

L. Rundo and A. Tangherloni—Contributed equally.

© Springer Nature Switzerland AG 2019
V. Malyshkin (Ed.): PaCT 2019, LNCS 11657, pp. 304–318, 2019.
https://doi.org/10.1007/978-3-030-25636-4_24

Keywords: Haralick features · GPU computing ·
Full gray-scale range · Medical imaging · Radiomics · CUDA

1   Introduction

Texture analysis has been effectively used in the classification and categorization of pictorial data in several Computer Vision tasks, such as object detection [1] and representation [2]. More specifically, texture features allow for quantitative analyses of the properties concerning scenes or objects of interest. Even though Deep Learning has recently gained ground, conventional Machine Learning models built on top of hand-engineered features remain fundamental in practical applications, especially thanks to the interpretability of the results [3]. With particular reference to biomedicine, quantitative imaging methods are increasingly gaining importance since they convey scientifically and clinically relevant information for prediction, prognosis, and treatment response assessment [4]. In this context, radiomic approaches are encouraging large-scale studies that can have a significant impact in the clinical practice [5]. Radiomics aims at extracting huge amounts of features from medical images and then mining them by means of cutting-edge computational techniques [6]. By so doing, radiomics exploits advanced imaging features to objectively and quantitatively describe tumor phenotypes [5]. Recently, radiomic studies have drawn considerable interest due to the potentialities for predicting treatment outcomes and cancer genetics, which may have important applications in personalized medicine [6,7]. Relying on the idea that radiomic features convey information about the different cancer phenotypes, they enable quantitative measurements for intra- and inter-tumoral heterogeneity.

The radiomic features can be essentially divided into four classes [8]. The first class comprises features related to region-based measurements (i.e., size, shape, diameter), while the other classes can be described as first-, second-, and higher-order statistical outputs, respectively. First-order statistical features concern the gray-level intensity histogram of a Region of Interest (ROI), such as mean, median, standard deviation, minimum, maximum, quartiles, kurtosis, and skewness. The second-order statistics consider texture analysis, which describes the texture of the ROI, by relying on the Gray-Level Co-occurrence Matrix (GLCM) that stores the co-occurrence frequency of similar intensity levels over the region (i.e., intensity value pairs). An alternative technique belonging to the second-order statistical outputs is fractal-based texture analysis, which examines the difference between pixels at different length scales (i.e., offset differences) [9]. Lastly, the higher-order methods extract repetitive or non-repetitive patterns by using kernel functional transformations, as in the case of the Gray-Level Run Length Matrix (GLRLM), which gives the size of homogeneous runs for each gray-level [10], and the Gray-Level Zone Length Matrix (GLZLM), which provides information on the size of homogeneous zones for each gray-level [11]. Moreover, some popularly used descriptors in transformed domains are Fourier transform, Wavelets, and Gabor filters [12,13].

Among the available radiomic descriptors, Haralick features are the most commonly used and clinically relevant [14,15], allowing radiologists to assess image regions characterized by heterogeneous/homogeneous areas or local intensity variations [16]. GLCM-based texture features have been extensively exploited in several medical image analysis tasks, such as breast Ultrasound (US) classification [17], brain tissue segmentation on Magnetic Resonance (MR) images [18], and volume-preserving non-rigid lung Computed Tomography (CT) image registration [19]. Unfortunately, the computation of these features is considerably intensive on images characterized by a high bit-depth (e.g., 16 bits), such as in the case of medical images that have to convey detailed visual information [20]. As a matter of fact, with the existing computational tools, the range of intensity values of an image must be reduced and limited to achieve an efficient radiomic feature computation [7].

In this work, we propose a novel strategy to compute the GLCM and extract an exhaustive set of the Haralick features. In particular, we aim at overcoming the limitations of the available feature extraction and radiomics tools that cannot effectively manage the full-dynamics of gray-scale levels. Our method, called HaraliCU, can offload the computations onto the GPU cores, thus allowing us to drastically reduce the running time required by the execution on Central Processing Units (CPUs).

This manuscript is organized as follows. Section 2 introduces the fundamental concepts regarding the GLCM-based textural features by presenting also the set of the extracted Haralick features. Section 3 introduces the Compute Unified Device Architecture (CUDA) and summarizes the state-of-the-art of the available software for Haralick feature extraction. Section 4 describes HaraliCU in details. The achieved results are shown and discussed in Sect. 5. Finally, some concluding remarks and future developments of this work are given in Sect. 6.

2 Haralick Features

Haralick features contain data about image textural characteristics, e.g., homogeneity, gray-tone linear dependencies, contrast, number and nature of boundaries present, along with indices of the inherent complexity of the image. All these features are calculated according to a GLCM.

2.1 GLCM: Basic Concepts

Formally, a GLCM with size $L \times L$, where L represents the maximum number of gray-levels according to the quantization scheme, denotes the second-order joint probability function $p(i, j)$ of an image region—where $i, j \in [0, 1, \ldots, L - 1]$ are gray-levels—defined as $\mathbf{P}(i, j)$. The GLCM considers the mapping of the initial full dynamics due to computational limitations.

In what follows, we will refer to two neighboring pixels, separated by a distance δ along an orientation θ, as the pair ⟨reference, neighbor⟩, where the reference pixel has gray-level equal to i, while the neighbor pixel is characterized by a gray-level j. More specifically, given a sliding window of size $\omega \times \omega$, the $\langle i, j \rangle$-th element of the matrix $\mathbf{P}(i, j | \delta, \theta, \omega)$ represents the number of times that the combination of the levels i and j occurs in two pixels $\langle \text{reference}, \text{neighbor} \rangle$ inside the sliding window, which are separated by a distance of δ pixels along the orientation θ. The distance δ is defined according to the infinity norm ℓ_∞. The undirected and directed distances denote the symmetric and non-symmetric GLCM, respectively, in terms of conditional co-occurrence probabilities. In some specific applications, valuable information could be lost in the symmetric approach [15]. Specifically:

- when computing the symmetric GLCM $\mathbf{P}_s$, since the pairs of gray-levels $\langle i, j \rangle$ and $\langle j, i \rangle$ are considered as the same element in $\mathbf{P}_s$, the frequency of both $\langle i, j \rangle$ and $\langle j, i \rangle$ is increased, so the resulting GLCM is symmetric across its main diagonal;
- when computing the non-symmetric GLCM $\mathbf{P}_{ns}$, the pairs of gray-levels $\langle i, j \rangle$ and $\langle j, i \rangle$ are considered separately in $\mathbf{P}_{ns}$.

In medical imaging, the selection of δ and θ used for the GLCM computation could depend on the specific application. For instance in breast US, the direction $\theta = 90°$ coincides with the direction of US propagation [17]. In order to obtain rotationally invariant features, it is common to average the GLCM-based statistics achieved over the four directions $\theta \in \{0°, 45°, 90°, 135°\}$.

2.2   Haralick Features in Medical Imaging

As a first step, we conducted an in-depth analysis of the literature to accurately define an exhaustive set of the Haralick features and avoid both ambiguities and redundancies. In the literature, some features exhibited potential in the characterization of the cancer imaging phenotype. For instance, entropy was shown to be a promising quantitative imaging biomarker for characterizing cancer heterogeneity, although it could be affected by acquisition protocols in multi-institutional studies [21]. With regard to the computation of the GLCM-based features, HaraliCU exploits the existing dependencies among Haralick features. Indeed, Gipp et al. [22] pointed out that some features can exploit some calculations pertaining to other features or intermediate results.

Considering the process of image digitalization, the compression of the initial intensity range is called quantization, which is generally irreversible and results in loss of information. For instance, in the case of texture features based on the Standardized Uptake Value (SUV) [23] within the tumor, a quantization phase is involved. Orlhac et al. [24] compared the different quantization strategies in metabolic activity pattern identification, by showing that they might significantly affect the texture values. In [25], the Positron Emission Tomography (PET)-derived texture features were calculated by quantizing the tumor voxel intensities with similar uptake to the same value. A similar study on Haralick features computed on the Apparent Diffusion Coefficient (ADC) MR images was presented in [16]. Even though the authors claimed that the impact of noise is reduced, this gray-scale compression could considerably decrease the discriminating power in feature-based classification tasks [26]. However, the main practical argument for the gray-scale compression is the computational cost. Therefore, to fully justify this choice, more advanced and adaptive quantization schemes should be devised [16]. With reference to the normalization of CT-based radiomics [27], the influence of gray-level quantification on radiomic feature stability for different CT scanners, tube currents and slice thickness was investigated in [28].

3   State-of-the-Art

The main limitation of the existing radiomics tools concerns their inability to deal with the full dynamics of 16-bit images, meaning that they are not capable of extracting the feature maps by preserving the initial gray-scale range. This drawback is emphasized when handling feature extraction tasks on the whole input image, especially for image classification purposes.

The aim of our approach is to tackle these issues, by effectively computing the feature maps for high-resolution images with their full dynamics. Since the calculation of Haralick features represents an embarrassingly parallel problem, several High-Performance Computing (HPC) technologies can be exploited. Among them, General-Purpose Computing on GPUs (GPGPU) is one of the most promising approaches. With reference to the existing Single Instruction Multiple Data (SIMD) architectures, NVIDIA CUDA is one of the most widespread and popular options [29]. CUDA is designed to exploit the parallelism provided by many-core GPUs for general-purpose scientific computing. Specifically, the idea is to offload intrinsically parallel calculations from the CPU, called the host, onto one or more devices (the GPUs) by means of kernels, that is, functions launched from the host and replicated in multiple threads running on the GPU cores.

CUDA threads are logically subdivided into thread blocks which, in turn, are organized in block grids. From a hardware standpoint, blocks are distributed over the GPU Streaming Multiprocessors (SMs) for their execution. When the blocks outnumber the available SMs, they are queued by the CUDA scheduler, transparently scaling the performance on different GPUs. Indeed, the higher the number of SMs, the higher the number of blocks running at the same time. The threads in execution on an SM are organized in tight groups of 32 threads named warps, which are executed in locksteps. Thus, blocks smaller than 32 threads imply a reduced occupancy of the GPU resources. In addition, due to this peculiar pattern of execution, any divergent path taken by some threads in a warp (e.g., the consequence of a conditional `if-then-else` statement) causes a serialization of the execution until re-convergence, affecting the overall performance. Therefore, in order to achieve optimal performance, CUDA code must be optimized to prevent any branch divergence in the execution.

CUDA has also a complex memory hierarchy, characterized by multiple memory types that provide different advantages and drawbacks. Notable examples are the global memory (large, visible by all threads, and affected by high access latency) and the shared memory (very small and used for intra-block communications with very slow access latency). A careful optimization of the data structures in these memories is mandatory to achieve the theoretical peak performance. Moreover, since any memory transfer between the host and device is very time consuming, they should be reduced as much as possible. Due to these peculiarities, CUDA programming could be challenging and generally requires the redesign of existing algorithms [30].

GPUs are representing an enabling factor for feasible computational solutions in medical image analysis [31,32]. Over the last years, GPUs proved to be fundamental for the practical use of computationally demanding algorithms [30], like the efficient training of deep neural networks [33]. Considering the GPU-accelerated Haralick feature extraction methods, Gipp *et al.* [22] proposed a packed representation of the symmetric GLCM, by storing only the rows and columns with non-zero elements. Afterwards, the Haralick features are computed by means of the lookup table that maps the index of the packed co-matrix. This clever solution reduces the accesses to global memory and, in turn, reduces the latencies due to memory reads, strongly improving the overall performances. The authors applied this implementation to cell images with 12-bit intensity depth. Tsai *et al.* [34] proposed an indirect encoding scheme for storing the GLCM, named the meta GLCM array, designed to fully exploit the GPU memory hierarchy. This approach was tested on brain MR images.

4   The Proposed GPU-Accelerated Method

Medical images convey a valuable amount of information, in terms of image resolution as well as pixel depth, which should be maintained for automated processing [20], since additional clinically useful pictorial details could be identified with respect to the naked eye perception. For these motivations, in the proposed approach, we aimed at keeping the whole initial information provided by the full dynamics of the gray-levels (i.e., 16 bits in the case of biomedical images), by efficiently managing the memory. As a matter of fact, the state-of-the-art methods exhaust the physical memory, such as in the case of the MatLab built-in function `graycomatrix`, even if running on machines equipped with 16 GB of RAM.

HaraliCU aims at supporting the user by providing low-level control. Indeed, the user can set the distance offset δ, the orientation θ, and the window size $\omega \times \omega$, while the neighborhood $\mathcal{N}$ is defined according to δ and θ. Therefore, the features can be computed for the four directions and then averaged to obtain a single aggregate value. The user can also set the padding conditions for the border pixels, either by choosing the zero padding or the symmetric padding. The number of quantized gray-levels Q can be also provided; HaraliCU linearly maps the initial minimum and maximum gray-levels onto 0 and $Q - 1$, respectively, in order to avoid the loss of a considerable amount of intensity bins.

The accuracy of the proposed efficient GLCM computation approach was evaluated against the built-in function `graycomatrix` provided by MatLab.

The computation of the Haralick features was carefully compared against the graycoprops function, which provides only the contrast, correlation, energy (i.e., angular second moment), and homogeneity features. For the other features, we relied also on a MatLab implementation publicly available on MatLab Central[1]. It is worth to note that our comparison was limited to the use of $L = 2^8$ gray-levels for the computation of the GLCM due to the computational limitations of the MatLab implementation. As a matter of fact, the graycomatrix function requires a double-precision $L \times L$ GLCM, by exceeding the main memory even in the case of 16 GB of RAM.

Since allocating a GLCM with 2^{16} rows and columns for each sliding window is memory demanding, and also considering that the size of each GLCM is strictly related to the number of different gray-levels inside the considered sliding window, we designed an effective and efficient encoding. More specifically, our novel encoding consists in storing each GLCM by using a list-based data structure in which every element of the list is a pair $\langle \mathsf{GrayPair}, \mathsf{freq} \rangle$, where $\mathsf{GrayPair}$ is a pair $\langle i, j \rangle$ of gray-levels and freq is the corresponding frequency (i.e., number of occurrences of the pair $\langle i, j \rangle$) inside the considered sliding window. The number of possible different elements composing the GLCM is given by the number of pairs $\langle \mathsf{reference}, \mathsf{neighbor} \rangle$ that can be identified inside the sliding window, taking into account the distance δ. The exact number of elements is provided by the following equation: $\#\mathsf{GrayPairs} = \omega^2 - \omega\delta$. The GLCM is dynamically computed by using the following procedure:

1. each pair $\langle \mathsf{reference}, \mathsf{neighbor} \rangle$, with gray-levels equal to $\langle i, j \rangle$, belonging to the sliding window is evaluated;
2. when a pair $\langle i, j \rangle$ is found, if the corresponding $\mathsf{GrayPair}$ element in the list exists, its frequency freq is incremented; otherwise, a new element $\mathsf{GrayPair}$, with freq equal to 1, is allocated and appended to the end of the list.

This simple but efficient encoding allows for removing all the zero elements inside the GLCM. In addition, when the GLCM symmetry is exploited, the length of the list is halved: indeed, the pairs $\langle i, j \rangle$ and $\langle j, i \rangle$ are considered as the same pair and the frequency of the pair $\langle i, j \rangle$ is doubled.

Considering that there are no dependencies between the sliding windows, we assigned each pixel of the input image to a GPU thread. In such a way, each thread computes all the features related to its pixel, which represents the center of the corresponding window. As a matter of fact, since a medical image could be often composed of more than 250 thousand pixels, involving the same number of sliding windows in the feature map computation, GPUs—thanks to their high number of threads that can be executed in parallel—are the most suitable co-processors to parallelize the required massive computational workload. In order to maximize the GPU performance and to fully exploit the GPU acceleration, we created a bi-dimensional structure for both the number of blocks and the number of threads. We fixed the number of threads to 16 for both the components of the

---

[1] https://uk.mathworks.com/matlabcentral/fileexchange/22187-glcm-texture-features.

bi-dimensional structure, while the number of blocks for each component of the corresponding bi-dimensional structure strictly depends on the number of the pixels (#pixels) composing the input image, and can be calculated as follows:

$$n_{\text{blocks}} = \begin{cases} \hat{n}, & \text{if } \hat{n}^2 \geq \lceil \frac{\#\text{pixels}}{256} \rceil \\ 1, & \text{otherwise} \end{cases}.$$

We used 16 threads in each component to take into consideration the CUDA warp size (i.e., 32 threads) as well as the limited number of registers.

Each thread processes a sliding window, that is, a subset of the pixels of the original image. Hence, all threads fetch from the GPU's global memory the pixels that are necessary for the calculations. However, some pixels may be shared by partially overlapping windows, a circumstance that introduces unnecessary latencies in the execution and might be mitigated by exploiting the shared memory. We will investigate this feature in a next release of HaraliCU.

5 Experimental Results

As described in the previous section, we validated HaraliCU by comparing the values of the features contrast, correlation, energy, and homogeneity with those extracted using the built-in functions `graycomatrix`.

5.1 Test Images

For the tests presented here, we considered two medical datasets characterized by different modalities and image size:

- axial T1-weighted Fast Field Echo contrast-enhanced MR sequences of brain metastases (matrix size: 256×256 pixels, pixel spacing: 1.0 mm, slice thickness: 1.5 mm), where the extracted features can be applied to segmentation and classification tasks [18,35];
- axial contrast-enhanced CT series of high-grade serous ovarian cancer (matrix size: 512×512 pixels, pixel spacing: ~0.65 mm, slice thickness: 5.0 mm), where texture features can evaluate intra- and inter-tumoral heterogeneity [36,37]. Pelvic lesions only were selected for this work.

In both cases, the intensity depth is 16 bits.

5.2 Computational Results

The existing versions of Haralick feature extraction tools are typically characterized by prohibitive running times, making them unfeasible in the clinical research. Moreover, these tools are not capable of taking into consideration the full dynamics of gray-scale levels; we therefore developed a memory-efficient CPU version of HaraliCU (coded in C++), which overcomes this limitation and was

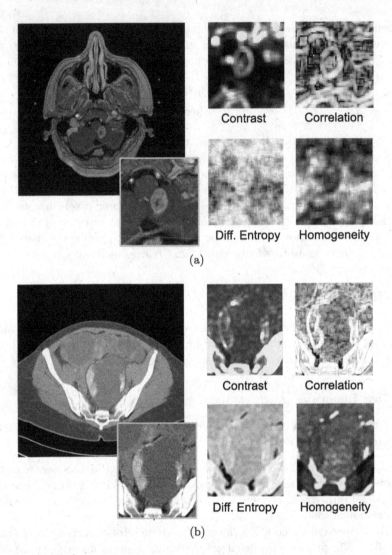

Fig. 1. Examples of feature maps obtained by HaraliCU by considering the full dynamics of gray-scale levels: (a) axial contrast-enhanced T1-weighted MR image of enhancing brain metastatic cancer; (b) axial venous phase contrast enhanced CT image of a patient with high-grade serous ovarian cancer showing the partly calcified and cystic ovarian tumor (red ROI) and omental disease (not outlined). The original images are shown in the leftmost panel. The ROIs (i.e., the tumor regions) are highlighted with a red contour and the corresponding cropped sub-images containing the ROIs are zoomed at the bottom right of each sub-figure. In the rightmost panel, we show four selected feature maps for the ROI-centered cropped images, namely: contrast, correlation, difference entropy and homogeneity. In both cases, the features were extracted by using $\delta = 1$ and averaging over $\theta \in \{0°, 45°, 90°, 135°\}$ to enrich the visual content. We selected $\omega = 5$ and $\omega = 9$ for the brain metastasis MR and the ovarian cancer CT images, respectively. (Color figure online)

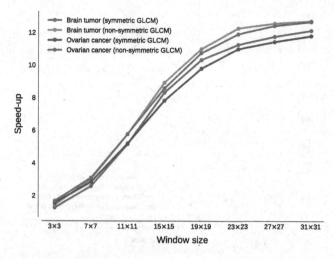

Fig. 2. Speed-up achieved by the GPU-powered version of HaraliCU, with respect to the C++ counterpart, on brain metastatic tumor MR and ovarian cancer CT images, by considering 2^8 intensity levels, enabling and disabling the GLCM symmetry, and considering $\omega \in \{3, 7, 11, 15, 19, 23, 27, 31\}$. Blue and green lines denote the speed-up trend considering brain metastasis MRI images, while red and violet lines are used for ovarian cancer CT images. (Color figure online)

also used as a benchmark to show the advantages of exploiting the GPUs to accelerate the calculations required by this computationally intensive task.

We first show in Figs. 1a and b two examples of input images along with the corresponding feature maps of four selected descriptors in the case of brain metastatic tumor MR and ovarian cancer CT images, respectively, to evaluate the correctness of our implementation. From the computational point of view, our C++ implementation resulted extremely efficient with respect to the Mat-Lab version, based on the `graycomatrix` and `graycoprops` functions, to extract Haralick features on a brain metastasis MR image. As a matter of fact, by varying the gray-scale range from 2^4 to 2^9 levels, we achieved speed-up values around $50\times$ and $200\times$, respectively.

As a second step, we compared the computational performance of our single core CPU version and GPU-powered versions of HaraliCU to extract all the provided features, to assess the capabilities of the parallel implementation. The GPU version of HaraliCU was run on an NVIDIA GeForce GTX Titan X (3072 cores, clock 1.075 GHz, 12 GB of RAM), CUDA toolkit version 8 (driver 387.26), running on a workstation with Ubuntu 16.04 LTS, equipped with a CPU Intel Core i7-2600 CPU (clock 3.4 GHz) and 8 GB of RAM. The CPU version of HaraliCU was run on the same workstation, relying on the computational power provided by the CPU Intel Core i7-2600 CPU. The CPU version was compiled by using the GNU C++ compiler (version 5.4.0) with optimization flag -O3,

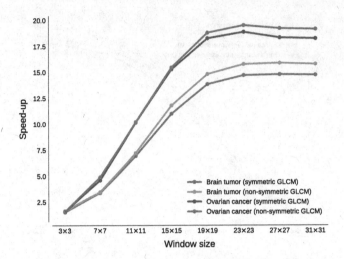

Fig. 3. Speed-up achieved by the GPU-powered version of HaraliCU, with respect to the C++ counterpart, on brain metastatic tumor MR and ovarian cancer CT images, by considering 2^{16} intensity levels, enabling and disabling the GLCM symmetry, and considering $\omega \in \{3, 7, 11, 15, 19, 23, 27, 31\}$. Blue and green lines denote the trend considering brain metastatic tumor MRI images, while red and violet lines are used for ovarian cancer CT images. (Color figure online)

while the GPU version was compiled with the CUDA Toolkit 8.0 by exploiting the optimization flag -O3 for both CPU and GPU code.

In order to collect statistically sound results and take into consideration the variability and the heterogeneity typically characterizing these images, we randomly selected 30 images from 3 different patients (10 per patient) affected by brain metastases and 30 images from 3 different patients affected by ovarian cancer. We tested both versions of HaraliCU by considering different window sizes, that is, $\omega \in \{3, 7, 11, 15, 19, 23, 27, 31\}$, as well as two different intensity levels (i.e., 2^8 and 2^{16}). For each combination of ω and intensity levels, we also enabled and disabled the GLCM symmetry to evaluate how the symmetry affects the running time of HaraliCU. It is worth noting that the measurements of the execution time of HaraliCU include the data transfer between the host memory and the device memory.

Figures 2 and 3 show the speed-up achieved by the GPU-powered version of HaraliCU. Considering only 2^8 intensity levels, the speed-up increases almost linearly; in addition, by disabling the GLCM symmetry and using $\omega = 31$ we obtained the highest speed-ups of 12.74× and 12.71× on brain metastasis (256×256 pixels) and ovarian cancer images (512×512 pixels), respectively. When the full dynamic of the gray-scale levels (i.e., 2^{16}) is considered, the GPU-powered version of HaraliCU outperforms the sequential counterpart, achieving speed-ups up to 15.80× with $\omega = 31$ and 19.50× with $\omega = 23$, on brain metastasis and ovarian cancer images, respectively. Taking into account ovarian cancer images, when ω is greater than 23 pixels the speed-up decreases for two reasons. Firstly, each thread, which is associated with a pixel, must consider more neighbor pixels that might have very different gray-level intensities, since their values are in the range $[0, 1, \ldots, 2^{16} - 1]$. This corresponds in increasing the required workload that each thread must perform; since the GPU cores have a lower clock frequency than CPU cores, the speed-up is reduced. Secondly, the GPU resources are saturated since the GLCM size associated with each thread may increase due to the high full-dynamic range. In this specific situation, the total GLCM size might overwhelm the dimension of the global memory and some threads handle different pixels, computing the corresponding Haralick features in a sequential way, decreasing the number of threads running in parallel.

The source code and the instructions for the compilation and execution of HaraliCU are available under the GNU GPL v3.0 license on GitHub at the following URL: https://github.com/andrea-tango/HaraliCU. HaraliCU requires an NVIDIA GPU, CUDA toolkit version 8 (or higher), OpenCV library version 3.4.1 (or higher).

6   Conclusion

Image texture extraction and analysis is playing a key role in quantitative biomedicine, leading to valuable applications in radiomic [5,6] and radiogenomic [38] research, by also combining heterogeneous sources of information. Therefore, advanced computerized medical image analysis methods, specifically designed to deal with the massive amount of extracted features, could be beneficial for the definition of imaging biomarkers, guiding towards personalized patient care. However, these large-scale studies need efficient techniques to drastically reduce the prohibitive running time that is typically required.

In this paper, we presented HaraliCU, a computationally efficient approach capable of effectively exploiting the power of the modern GPUs, which aims at accelerating the GLCM computation by keeping the full dynamic range in medical images. Our method was tested on a dataset composed of brain metastatic tumor MR images and ovarian cancer CT images. Our C++ coded sequential version showed to be $\sim 200\times$ faster than the corresponding MatLab implementation. In addition, the GPU-powered version was able to achieve speed-ups up to $15.80\times$ and $19.50\times$, with respect to the CPU version, on brain metastasis MR and ovarian cancer CT images, respectively. It is worth noting that neither the C++ version nor HaraliCU implementations have been optimized. Indeed, we expect to further increase their performance by exploiting vectorial instructions and multi-threading, in the case of the sequential version, and by carefully using the high-performance memories of the GPU (i.e., registers, shared memory), for what concerns HaraliCU.

Finally, thanks to this outstanding performance, the C++ version and even more so HaraliCU might enable multi-scale radiomic analyses by properly combining several values of distance offsets, orientations, and window sizes.

As a future development, we plan to develop an improved version of HaraliCU by exploiting the vectorization of the input image matrices for a better

GPU thread block managing. In order to improve the scalability of the proposed approach, the dynamic parallelism, supported by CUDA, could be exploited to further parallelize the computations when the workload increases (e.g., high window size). Moreover, even though the spatial and temporal locality are already exploited during the GLCM construction process, based on the sliding window, the usage of the GPU memory hierarchy might be optimized [39]. Finally, dealing with the clinical feasibility of radiogenomic studies, the integration of the imaging phenotype and genotype can provide valuable information about tumor heterogeneity as well as treatment response [40], by efficiently exploiting high-throughput techniques.

Acknowledgment. This work was partially supported by The Mark Foundation for Cancer Research and Cancer Research UK Cambridge Centre [C9685/A25177]. Additional support has been provided by the National Institute of Health Research (NIHR) Cambridge Biomedical Research Centre. The views expressed are those of the authors and not necessarily those of the NHS, the NIHR or the Department of Health and Social Care.

References

1. Trivedi, M.M., Harlow, C.A., Conners, R.W., Goh, S.: Object detection based on gray level cooccurrence. Comput. Vis. Graph. Image Process. **28**(2), 199–219 (1984)
2. Soh, L.K., Tsatsoulis, C.: Texture analysis of SAR sea ice imagery using gray level co-occurrence matrices. IEEE Trans. Geosci. Remote Sens. **37**(2), 780–795 (1999)
3. Torheim, T., et al.: Classification of dynamic contrast enhanced MR images of cervical cancers using texture analysis and support vector machines. IEEE Trans. Med. Imaging **33**(8), 1648–1656 (2014)
4. Yankeelov, T.E., et al.: Quantitative imaging in cancer clinical trials. Clin. Cancer Res. **22**(2), 284–290 (2016)
5. Lambin, P., et al.: Radiomics: extracting more information from medical images using advanced feature analysis. Eur. J. Cancer **48**(4), 441–446 (2012)
6. Lambin, P., et al.: Radiomics: the bridge between medical imaging and personalized medicine. Nat. Rev. Clin. Oncol. **14**(12), 749 (2017)
7. Yip, S.S., Aerts, H.J.: Applications and limitations of radiomics. Phys. Med. Biol. **61**(13), R150 (2016)
8. Stoyanova, R., et al.: Prostate cancer radiomics and the promise of radiogenomics. Transl. Cancer Res. **5**(4), 432 (2016)
9. Chen, C.C., DaPonte, J.S., Fox, M.D.: Fractal feature analysis and classification in medical imaging. IEEE Trans. Med. Imaging **8**(2), 133–142 (1989)
10. Galloway, M.M.: Texture analysis using gray level run lengths. Comput. Graph. Image Process. **4**(2), 172–179 (1975)
11. Thibault, G., et al.: Shape and texture indexes application to cell nuclei classification. Int. J. Pattern Recognit. Artif. Intell. **27**(01), 1357002 (2013)
12. Zhu, H., et al.: A new local multiscale Fourier analysis for medical imaging. Med. Phys. **30**(6), 1134–1141 (2003)
13. Arivazhagan, S., Ganesan, L.: Texture classification using wavelet transform. Pattern Recognit. Lett. **24**(9–10), 1513–1521 (2003)

14. Haralick, R.M., Shanmugam, K., Dinstein, I.: Textural features for image classification. IEEE Trans. Syst. Man Cybern. SMC **3**(6), 610–621 (1973)
15. Haralick, R.M.: Statistical and structural approaches to texture. Proc. IEEE **67**(5), 786–804 (1979)
16. Brynolfsson, P., et al.: Haralick texture features from apparent diffusion coefficient (ADC) MRI images depend on imaging and pre-processing parameters. Sci. Rep. **7**(1), 4041 (2017)
17. Gómez, W., Pereira, W., Infantosi, A.F.C.: Analysis of co-occurrence texture statistics as a function of gray-level quantization for classifying breast ultrasound. IEEE Trans. Med. Imaging **31**(10), 1889–1899 (2012)
18. Ortiz, A., Górriz, J., Ramírez, J., Salas-Gonzalez, D., Llamas-Elvira, J.M.: Two fully-unsupervised methods for MR brain image segmentation using SOM-based strategies. Appl. Soft Comput. **13**(5), 2668–2682 (2013)
19. Park, S., Kim, B., Lee, J., Goo, J.M., Shin, Y.G.: GGO nodule volume-preserving nonrigid lung registration using GLCM texture analysis. IEEE Trans. Biomed. Eng. **58**(10), 2885–2894 (2011)
20. Rundo, L., et al.: MedGA: a novel evolutionary method for image enhancement in medical imaging systems. Expert Syst. Appl. **119**, 387–399 (2019)
21. Dercle, L., et al.: Limits of radiomic-based entropy as a surrogate of tumor heterogeneity: ROI-area, acquisition protocol and tissue site exert substantial influence. Sci. Rep. **7**(1), 7952 (2017)
22. Gipp, M., et al.: Haralick's texture features computation accelerated by GPUs for biological applications. In: Bock, H., Hoang, X., Rannacher, R., Schlöder, J. (eds.) Modeling, Simulation and Optimization of Complex Processes, pp. 127–137. Springer, Heidelberg (2012). https://doi.org/10.1007/978-3-642-25707-0_11
23. Leijenaar, R.T., et al.: The effect of SUV discretization in quantitative FDG-PET radiomics: the need for standardized methodology in tumor texture analysis. Sci. Rep. **5**, 11075 (2015)
24. Orlhac, F., Soussan, M., Chouahnia, K., Martinod, E., Buvat, I.: 18F-FDG PET-derived textural indices reflect tissue-specific uptake pattern in non-small cell lung cancer. PLoS One **10**(12), e0145063 (2015)
25. Orlhac, F., Soussan, M., Maisonobe, J.A., Garcia, C.A., Vanderlinden, B., Buvat, I.: Tumor texture analysis in 18F-FDG PET: relationships between texture parameters, histogram indices, standardized uptake values, metabolic volumes, and total lesion glycolysis. J. Nucl. Med. **55**(3), 414–422 (2014)
26. Jen, C.C., Yu, S.S.: Automatic detection of abnormal mammograms in mammographic images. Expert Syst. Appl. **42**(6), 3048–3055 (2015)
27. Shafiq-ul Hassan, M., Latifi, K., Zhang, G., Ullah, G., Gillies, R., Moros, E.: Voxel size and gray level normalization of CT radiomic features in lung cancer. Sci. Rep. **8**(1), 10545 (2018)
28. Larue, R.T., et al.: Influence of gray level discretization on radiomic feature stability for different CT scanners, tube currents and slice thicknesses: a comprehensive phantom study. Acta Oncol. **56**(11), 1544–1553 (2017)
29. Luebke, D.: CUDA: scalable parallel programming for high-performance scientific computing. In: Proceedings 5th IEEE International Symposium on Biomedical Imaging: From Nano to Macro (ISBI), pp. 836–838. IEEE (2008)
30. Nobile, M.S., Cazzaniga, P., Tangherloni, A., Besozzi, D.: Graphics processing units in bioinformatics, computational biology and systems biology. Brief. Bioinform. **18**(5), 870–885 (2016)
31. Eklund, A., Dufort, P., Forsberg, D., LaConte, S.M.: Medical image processing on the GPU-past, present and future. Med. Image Anal. **17**(8), 1073–1094 (2013)

32. Smistad, E., Falch, T.L., Bozorgi, M., Elster, A.C., Lindseth, F.: Medical image segmentation on GPUs-a comprehensive review. Med. Image Anal. **20**(1), 1–18 (2015)

33. Shen, D., Wu, G., Suk, H.I.: Deep learning in medical image analysis. Annu. Rev. Biomed. Eng. **19**, 221–248 (2017)

34. Tsai, H.Y., Zhang, H., Hung, C.L., Min, G.: GPU-accelerated features extraction from magnetic resonance images. IEEE Access **5**, 22634–22646 (2017)

35. Militello, C., et al.: Gamma Knife treatment planning: MR brain tumor segmentation and volume measurement based on unsupervised Fuzzy C-Means clustering. Int. J. Imaging Syst. Technol. **25**(3), 213–225 (2015)

36. Vargas, H.A., et al.: A novel representation of inter-site tumour heterogeneity from pre-treatment computed tomography textures classifies ovarian cancers by clinical outcome. Eur. Radiol. **27**(9), 3991–4001 (2017)

37. Rizzo, S., et al.: Radiomics of high-grade serous ovarian cancer: association between quantitative CT features, residual tumour and disease progression within 12 months. Eur. Radiol. **28**, 4849–4859 (2018)

38. Pinker, K., et al.: Background, current role, and potential applications of radiogenomics. J. Magn. Reson. Imaging **47**(3), 604–620 (2018)

39. Gupta, S., Xiang, P., Zhou, H.: Analyzing locality of memory references in GPU architectures. In: Proceedings ACM SIGPLAN Workshop on Memory Systems Performance and Correctness. ACM (2013). 12

40. Sala, E., et al.: Unravelling tumour heterogeneity using next-generation imaging: radiomics, radiogenomics, and habitat imaging. Clin. Radiol. **72**(1), 3–10 (2017)

Cellular Automata

A Web-Based Platform for Interactive Parameter Study of Large-Scale Lattice Gas Automata

Maxim Gorodnichev[1,2,3]([✉]) [iD] and Yuri Medvedev[1] [iD]

[1] Institute of Computational Mathematics and Mathematical Geophysics SB RAS,
Novosibirsk, Russia
{maxim,medvedev}@ssd.sscc.ru
[2] Institute of Computational Technologies SB RAS, Novosibirsk, Russia
[3] Novosibirsk State Technical University, Novosibirsk, Russia

Abstract. A problem of development of user-friendly interfaces for high performance computing (HPC) applications is addressed. The HPC Community Cloud (HPC2C) service that provides a RESTful application programming interface for unified control of HPC jobs was used to develop a prototype of a web-based UI for cellular automata simulation package. The UI allows a user to easily run multiple simulations on remote HPC resources and, this way, study a parameter space of a cellular automaton. The interface was used to organize a series of numerical experiments resulting in reproduction of the Kármán vortex street.

Keywords: High performance computing · HPC cloud ·
User interfaces · Application programming interfaces ·
Cellular automata · Lattice Gas Automata · Turbulent flows ·
Kármán vortex street

1 Introduction

Development of new models such as a cellular automaton for simulation of turbulent flows requires conducting a lot of computational experiments in order to find model parameters that lead to reproduction of certain natural or artificial phenomena. Often, such experiments cannot be run and analysed automatically because researcher's intuition plays an important role in selection of meaningful subspaces of parameter values. This makes development of proper user interfaces a critical problem for productivity of a researcher's work. There are tools and projects that are focused on usability of model development tools (Matlab, Jupiter Notebooks, Wolfram Alpha, etc.). However, many models are developed as non-interactive programs that are controlled with command-line interface and

The Siberian Branch of the Russian Academy of Sciences (SB RAS) Siberian Supercomputer Center is gratefully acknowledged for providing supercomputer facilities.
Partially supported by the budget project of the ICMMG SB RAS No. 0315-2019-0007.

V. Malyshkin (Ed.): PaCT 2019, LNCS 11657, pp. 321–333, 2019.
https://doi.org/10.1007/978-3-030-25636-4_25

configuration files because the programs are targeted for remote runs on HPC hardware. The paper addresses the problem of creation of high-level user interfaces for such programs.

We study a particular cellular automata (CA) model implemented as a software package [1], and propose a web-interface that allows control of the package and running of simulations on the resources of the Siberian Supercomputing Center and other remote resources, access to which must be provided by a user. We implement a graphical user interface (GUI) and apply it to a systematic search of CA parameters that allows one to reproduce turbulent flow effects such as the Kármán vortex street [2].

Cellular automata began to be studied as a mathematical model of spatial processes at the end of the last century. At the present time, cellular automata are typically used to model least-studied phenomena. They are also used as an alternative to existing approaches. The demand for computer simulation led to an intensive development of CA-models, as evidenced by the emergence of works both on the theory of cellular automaton models and on their use [3–5].

The main advantages of CA-models are as follows.

1. Non-linear and discontinuous phenomena can be described using transitional rules for the cells comprising the cellular automata per cell basis. This allows for fine granularity of the model as contrasted to definition of the model with general laws expressed by differential equations [6]. The process of simulation becomes a simple application of transitional rules and a researcher is able to naturally describe complicated boundaries up to the formation of a porous structures [7].
2. Because of the natural parallelism of fine-grained models, they are well suited for scalable parallel implementations on supercomputers.
3. Since cellular automata are discrete, their software implementations are naturally processed by a discrete computer without rounding errors.
4. Using the available CA-models of simple processes, one can models a complex process as a combination of these simple processes in a natural way by composing simple automata.

The advantages listed above encourage researchers to study models based on cellular automata and create new ones. One of such models for gas flow simulation is a lattice gas automaton Frisch-Hasslacher-Pomeau—Multi-Particle (FHP-MP) [8]. A correspondence of the original FHP model to the Navier-Stokes equation is shown [9]. The FHP-MP cellular automaton is studied experimentally at the current stage.

Since research in CA-based simulation is a relatively new field, there are no established software packages allowing a researcher to efficiently implement new models. Researchers have to independently create software implementations of the models they study. Development of a user interface is a part of the complexity associated with implementation of the models, particularly, if HPC resources should be used for simulation.

Besides a need for high-level interfaces for particular remote running applications, research in the field of unified interfacing systems for development, deployment and use of simulation and data processing applications is motivated by multiple factors. These factors include concerns related to reproducibility, accessibility, and transparency of computational research [10–12], problems of application discovery [13], need for collaborative work, diversity of target high-performance computing systems that must be abstracted for a regular application user, and others.

An extensive survey of web-portals for HPC has been done recently in [14]. The problem of creation of such portals has a long history of discussion and a vast list of associated projects. During the last decade a number of projects emerged [15–21] building such portals around services that provide application programming interfaces (API) in the trending RESTful architectural style. The APIs allow developers to implement applications capable of data management and control of computing jobs on remote HPC resources. Having their specific traits, these projects seem to share a view on users' needs and associated problems, and, thus, are similar in approaches and solutions. The particular focus of the HPC Computing Cloud (HPC2C) project [16] used in the present work is on development of a platform for accumulation and reuse of user-developed applications within its web-application and collaboration of users over the development and use of the applications.

The development of the high-level HPC services should be based, among the other, on analysis of use cases as particularly underlined in [22]. In the present work, an attempt to develop a convenient environment for studying FHP-MP model was made as a use case for the HPC2C.

Section 2 introduces a FHP-MP model—a specific model in a class of Lattice Gas Automata [6]. Section 3 explains the process of parameter study and present the particular parameter set that can be used to reproduce a vortex street. Section 4 describes the HPC2C service and Sect. 5 explains how the service was used to implement a platform for interactive model parameters study. The summary of the work is given in the conclusion.

2   FHP-MP Cellular Automata

2.1   Basic Definitions

The cellular automaton FHP-MP is a triple of objects (W, A, N). Each element w of the set W is called a cell and is associated with a finite state machine A, called an elementary automaton. The state of a cell $w \in W$ is represented by the vector $s(w)$ with components $s_1(w), s_2(w), \ldots, s_6(w)$; their values are non-negative integers. The set of states $s(w)$ of all cells $w \in W$ at the discrete time t is called the global state of the cellular automaton.

Each cell $w \in W$ is associated with some coordinates $x(w)$ and $y(w)$ on the 2D Cartesian plane. Thus, one can imagine the set W as a 2D cellular array. Between any two cells $w_1 \in W$ and $w_2 \in W$ one can calculate the distance $d(w_1, w_2)$. Practical implementations of the cellular automaton contain finite number of cells in each direction.

For each cell $w \in W$, some ordered set $N(w) = \{N_j(w) : N_j(w) \in W \wedge d(w, N_j(w)) = 1, (j = 1, 2, \ldots, 6)\}$ is defined, whose elements are called its neighboring cells.

A term of a model particle is introduced. We will say that a cell contains particles, or particles are in a cell at a certain moment of time t. A particle will be said to have unit mass and unit velocity. An element $s_j(w)$ of the state $s(w)$ is interpreted as a number of model particles in the cell w that are directed toward its neighbour $N_j(w)$ and we denote the direction as a vector c_j. The mass of all particles in the cell w is $m(w) = \sum_{j=1}^{6} s_j(w)$. The model momentum of the particles in the cell w is $p(w) = \sum_{j=1}^{6} s_j(w) c_j(w)$.

2.2 Behavior of the Cellular Automaton

Each iteration of the cellular automaton consists of two steps: propagation and collision, i.e. the transition function δ of the elementary automaton A is the composition of the functions δ_1 (propagation) and δ_2 (collision): $\delta(s) = \delta_2(\delta_1(s))$.

At the propagation step, in each cell $w \in W$, each particle moves to the neighboring cell $N_j(w)$, corresponding to the vector of its velocity c_j. Thus, $\delta_1(s_j(w)) = s_j(N_{((j+2) \bmod 6)+1}(w)), j = 1, \ldots, 6$.

At the collision step, there is a change in the direction of movement of particles according to certain collision rules, which are independent of the neighboring cells' states, i.e δ_2 depends only on the value of δ_1. In the FHP-MP cellular automaton, the function δ_2 is probabilistic. Collision rules for cells of different types are described below: conventional cells $W_c \subset W$, walls $W_w \subset W$, inlet cells $W_{in} \subset W$, and outlet cells $W_{out} \subset W$. The pairwise intersections of these subsets are empty, and their union coincides with the set of all cells of the cellular automaton $W_c \cup W_w \cup W_{in} \cup W_{out} = W$. The behavior of the walls, the inlets, and the outlets determines the boundary conditions of the cellular automaton.

In the conventional cells $w \in W_c$, the function δ_2 is chosen so that the particles' mass $m(w)$ and the momentum $p(w)$ are preserved in the cell. The value of the function δ_2 equiprobably selected among all the possible values that satisfy these conditions.

In the wall cells $w \in W_w$, the particles change the direction of the velocity vector to the opposite, thus not preserving the momentum. Because the number of particles in the cell does not change, the mass is preserved.

At each iteration an inlet (outlet) cell generates $n_{in}(w)$ ($n_{out}(w)$) particles selecting their velocities from all possible directions equiprobably. One can create various objects from the inlet cells. For example, one can get a source of a uniform flow of particles of a given concentration by placing a line of inlet cells somewhere in the cellular array (typically, at the border of the cellular array). An isolated inlet cell will simulate a nozzle. One can do the same with the outlet cells. Obviously, neither mass nor momentum are preserved in the inlet and the outlet cells.

2.3   The Averaged Values

Gas flow simulation is performed to obtain velocity and pressure fields. The mass
and the velocity of a separate particle in a cell do not provide this information.
The flow velocity and gas pressure at a certain point (x, y) correspond to the
averaged vector of the particles' velocity and to particles' concentration computed in some vicinity $V(x, y)$ of the point (x, y). The vicinity of (x, y) includes
all cells $w \in W$ such that their distances to the point (x, y) do not exceed some
value r called the averaging radius.

The averaged velocity of particles in the vicinity $V(x, y)$ is calculated as

$$u(x, y) = \frac{1}{|V(x, y)|} \sum_{w \in V(x,y)} \sum_{j=1}^{6} s_j(w) c_j(w),$$

where $|V(x, y)|$ is the number of cells included in the vicinity, $c_j(w)$ is the unit
velocity vector corresponding to the jth component of the state vector $s(w)$, and
the $s_j(w)$ is the value of the jth component of the state vector $s(w)$ of the cell
$w \in V(x, y)$.

The particle concentration in the vicinity $V(x, y)$ is calculated as:

$$n(x, y) = \frac{1}{|V(x, y)|} \sum_{w \in V(x,y)} \sum_{j=1}^{6} s_j(w).$$

The averaged values of the model velocities and concentration correspond
to their physical counterparts only when the averaging vicinity $V(x, y)$ consists
solely of the conventional cells $w \in W_c$. Otherwise, their values are considered as
undefined. This condition does not allow one to calculate n and u at the distance
to walls, inlet, and outlet cells closer than the averaging radius r.

3   Simulation of a Vortex Street

"We model a gas flow in a pipe with an obstacle inside (Fig. 1)". The obstacle has
a shape of a straight line. The purpose of computer experiments is to determine
the parameters of a model that will lead to formation of a stable vortex street
behind the obstacle. The street is a sequence of vortices that detach from the
ends of the obstacle at approximately equal time intervals.

The simulation platform supports tree basic actions for a model researcher:

1. "construction of a cellular automaton global state (see Sect. 2.1) that takes a
   specification of the global state as an input parameter;"
2. simulation that takes a global state and apply transition rules iteratively to
   global states thus producing an array of consecutive global states;
3. post-processing that takes a global state, computes fields of averaged velocities
   and concentration, and visualize these fields.

These three actions related to computer experiments are described below.

3.1   The Cellular Automaton's Global State Construction

Construction of the initial global state (Fig. 1) results in a file in which states of all cells are stored separately. Construction is organized as follows. A researcher provides a specification of a global state where massively defines the states of the cells with basic declarations. In the case of the Fig. 1 the size of the 2D cellular array is set first, then all the cells are declared as conventional cells, then some of them are redefined as wall, inlet or outlet cells (see Sect. 2.2). Thus, the left-most cells are declared inlet cells, the right-most cells become outlet cells and the cells corresponding to the upper, lower boundaries and the straight line obstacle are declared as wall cells.

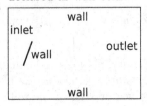

Fig. 1. Initial global state.

Among the parameters of the CA-model that need to be found during the experiments are the appropriate size of the pipe (in model units: the model unit of length is equal to the distance between the centers of neighboring cells). The criterion for choosing sizes is a compromise between the lack of space for the formation of vortices in the case of a small size of the pipe and a large computing time in the case of its large size.

As a result of numerous experiments, a compromise size of the simulated area was chosen as 3000×2000 in model units.

The boundary cells $y = 0$ and $y = 2000$ have the function of a wall. The $x = 0$ cells have the inlet function, the $x = 3000$ cells have the outlet function.

Another objective of the experiments is to choose the size, position and inclination angle of an obstacle. Too small size of the obstacle does not provide conditions for the stable vortices formation; too large obstacle deflects the flow to the lateral walls so that they significantly affect the vortices behaviour. Too small distance between the obstacle and the inlet adversely affects the stability of a vortex street. A distant position of the obstacle from the inlet leaves little room for a vortex street behind the obstacle. A "normal" vortex street is characterized by stable periodic formation of similar-sized vortices. Too high angle of attack of the obstacle slows down the moment of the beginning of a stable formation of vortices in the transition process (beginning of simulation). If the angle of attack is too low (this is similar to the case of a too small obstacle) then the stability of the vortices formation is reduced.

As a result of many experiments, an obstacle was chosen, having the shape of a straight line with ends at the points with coordinates $(350, 1250)$ and $(550, 750)$.

3.2   Running the Simulator

Running the simulator with the CA initial global state leads first to a transient process, during which vortices can form aperiodically. Then the motion of the simulated flow gets stabilized in the form of a vortex street.

One of the objectives of the experiments is to select the number of iterations required to achieve a stable vortex street. The results of these iterations should be excluded from consideration of the flow behaviour in a vortex street. Too few iterations devoted to the transition process will lead to the fact that in the first periods of the considered results there will be data distorted by the transition process. Too many iterations, rejected as a transition process, will lead to unnecessary computer time spent on calculations.

As a result of multiple experiments, 30, 000 iterations are chosen as an appropriate time for transition process. The period of separation of a pair of vortices from the ends of the obstacle was found to be approximately 11, 000 iterations. Thus, six complete periods were observed during simulation lasted from the iteration 30, 000 to the iteration 96, 000.

An appropriate number of particles generated inlet and outlet cells at each iteration must be selected. The concentration of particles is directly proportional to the pressure of the simulated gas. It is necessary to take into account that when the concentration of particles is too low, the artefacts caused by the discrete nature of the model, such as directional anisotropy and an increased level of automata noise, become more acute. Too large number of particles of particles will lead to an increased consumption of computer time, since the processing time of each cell at each iteration increases with the number of particles per cell.

The difference between the number of particles produced by the inlet border and the outlet border determines the magnitude of the pressure gradient, which affects the flow speed and, consequently, the conditions for the vortices formation. A too low gradient produces a flow with small velocity, unable to cause a vortex to come off. A too steep gradient causes the flow to move at an excessively high speed, at which the formation of vortices occurs not only on the obstacle, but also spontaneously, on flow fluctuations; and the effect of automata noise on the result also increases.

After performing a large number of experiments, the particle generation rate at the inlet was chosen to be 80 particles per cell per iteration, and at the outlet—40 particles per cell per iteration. To speed up the transition process, at the initial global state all the cells are filled with particles, 40 particles per cell.

3.3 Post-processing

The global state of a cellular automaton saved to a hard disk should be subjected to an averaging procedure in order to get pressure and velocity fields. The obtained averaged values are used for visualization (Fig. 2). The obstacle, the inlet and the outlet are shown in the picture. The lateral walls are cropped. The grayscale background depicts the distribution of gas pressure in the simulation space. Lighter areas correspond to higher pressure, darker—to lower. The maximum pressure is indicated by white, the minimum pressure in the simulation area is black. Arrows indicate flow direction. Their length is proportional to the local flow velocity.

Of many options for the averaging and visualization parameters, the following were chosen to better display the simulated process. The averaging region is a part of the simulation region located between the coordinates $y_1 = 350$ and $y_2 = 1850$. Areas excluded from visualization cannot be excluded from simulation

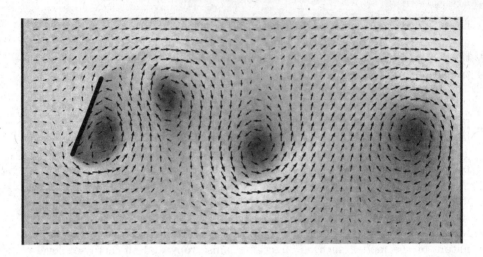

Fig. 2. Vortex street as a simulation result.

since the motion of the vortices will be affected by closely located walls. The selected area is displayed on a bitmap of 4000×2000 pixels. To calculate the averaged values of the concentration (pressure) of the gas, the averaging radius was chosen to be 15 cells, and when calculating the averaged values of the flow velocity—20 cells.

3.4 Interpretation of Simulation Results

As a result of the studies, the simulation parameters that provide a reproduction of the gas flow with the necessary properties were found. Behind an obstacle, a vortex street with vortices of equal sizes and with a stable period of vortex separation is formed.

The difficulty of the work lies mostly not in the large number of various parameters, but in the mutual influence of these parameters. For example, the obstacle of a certain size is too large for the selected size of the cellular array; or the necessary concentration gradient does not provide the stable vortices formation on the obstacle of the selected size. Therefore, the number of necessary experiments rapidly increases with an increase in the number of simulation parameters.

The process of finding suitable parameters requires interactive human participation. A brute-force search among all possible parameters is impossible due to the huge number of computer experiments that would require a very large amount of computer time. A researcher can significantly reduce this parameter space, based on the human experience and intuition. The researcher's work must be supported with an appropriate user interface system.

4 HPC Community Cloud

The HPC2C software consists of a management server that provides a RESTful application programming interface (API) to external software systems and a web application that provides a graphical interface to users. The HPC2C implements a platform for accumulation and reuse of content created by users and third-party developers: software development tools, data visualization tools, interactive training materials, numerical simulation and data analysis tools. The HPC2C web interface is developed to improve the productivity of users of research and educational computer centers and, particularly, reduce the threshold for users to enter the field of HPC. A prototype of the HPC2C service is available at http://hpccloud.ssd.sscc.ru.

4.1 HPC2C Management Server

The management server keeps track of users, registered computing resources, computing jobs. It organizes a storage for users' programs and data. HPC2C API is a basis for development of external software systems that can access the resources of computer centers for large-scale computations. The HPC2C hides specifics of interfaces of particular computing systems behind a single access point and a single management system. This is achieved by implementing modules that transmit user commands made with a unified interface to specific interfaces of the attached computing systems. Any user can register ("attach") a computing system with the HPC2C by providing its address, access credentials (as in [19]), and a type of the interface. Examples of interface types are: a TORQUE job management system on a Linux cluster head node and a regular Linux box with no job management system. Plain SSH protocol is used to control remote jobs as a sufficient solution for current use cases.

Users can be included in different user groups. The rights to perform various operations on various objects can be set at the group level and at the level of individual users. Users can provide access to the objects they have created to other users and groups.

A JavaScript library is implemented to support development of the JavaScript-based HPC2C API clients.

4.2 Usage Scenarios

One of the possible usage scenario for the HPC Community Cloud can be described as follows. The user uploads source files of a numerical simulation or data processing application to HPC Community Cloud, registers computing systems or selects computer systems from the list of systems provided by the service and/or other users, describes rules for building programs based on the Make automation tool, builds programs according to the rules and corresponding to an architecture of a target computing system, uploads input data files, submits the computing job for execution, tracks the status of the jobs, gets access to the output files produced by the job. An application, once uploaded, can be registered with the HPC2C system and reused in further job submissions. It is implied that such applications are controlled by input files and command-line parameters, they are called CL-applications in the rest of the paper.

A CL-application can be registered without uploading sources files and building them through the HPC2C systems. The only important point is that executable and configuration files are prepared according to the HPC2C rules for any computing system where users may want to employ the application.

In other scenarios, users can submit jobs based on CL-applications to which other users have granted access. All of these functions are available through the API, and end-users access these functions either through the HPC2C web interface or through external software systems.

4.3   GUI-Applications

A complex GUI application such as a proposed platform for CA-model parameter study should be able to support complex user flows and keep track of all the data belonging to the application and all the associated computing jobs. Such applications can be implemented externally, as a desktop, mobile or a separate web applications with calls to the HPC2C API. On the other hand, the HPC2C encourages developers to embed complex GUI applications into the HPC2C web-interface in order to build a collection of applications within the service. Thus, a concept of an (embedded) GUI-application is introduced. This is a numerical simulation or data processing application with an HPC2C-integrated graphical interface in which the user sets input parameters, submits jobs to chosen computing systems, etc. A GUI-application can provide tools for analysis and visualization of computing results. The user receives notifications on job state changes, returns to continue to interact with the application, analyses the results, prepares and launches further jobs.

The HPC2C system provides a model for embedding of GUI-applications. When a user account is registered with HPC2C a "home" directory is created in the file storage with the following subdirectories:

- apps: CL-applications are stored here,
- appstorage: GUI-applications store their data here,
- jobs: a directory for the files associated with computing jobs—management scripts, configuration files, input and output files,
- gui-apps: subdirectories contain files of GUI-applications.

In order to embed a GUI-application a developer should create a directory under gui-apps with the name of the GUI-application and place files implementing the GUI into this new directory. These are *.html, *.css, and *.js files. A file index.html should be included.

A list of all GUI-applications added in such a way becomes available in a dashboard of a user in the HPC2C web-application. A user can choose one of GUI-applications from a list, that makes elements defined in the corresponding index.html to be displayed as a part of a HPC2C interface. The index.html will include a JavaScript code that will manage user flows within the GUI-application and load other necessary resources. HPC2C API is called to manage files in the storage and jobs. The GUI-applications should remember all the associated files and jobs. Typically, an application will store such data under appstorage directory. Requirement to implementation of GUI-applications are provided in the HPC2C documentation.

5   Implementation of a Web GUI for FHP-MP in HPC2C

Section 3 describes basic actions a user takes to work with the FHP-MP model: construction of a CA global state, simulation, and post-processing. A first step to implement a the GUI-application is to support these basic actions.

The FHP-MP package consist of three command-line applications corresponding the above mentioned actions. They first should be registered with HPC2C as CL-applications: fhpmake, fhpsimulation and fhpvizualization.

All three CL-applications require configuration files to be provided as their inputs. That means each time a user wants to construct a new CA global state as an initial conditions for a simulation, or wants to run simulation, or wants to run visualization the user either selects an earlier prepared configuration files or create new ones, edit the files and store them to use for future job launches. GUI-forms for editing configuration files are generated automatically from their JSON-based specifications.

A simulation can start from a CA global state file prepared with fhpmake or continue from a certain global state obtained during previous simulations. This means that the GUI should provide a tool to select a global state to start with. In order to achieve this, each global state generated with fhpmake is placed in a new subdirectory under appstorage/fhp-mp directory. A user can select such a global state and run a simulation. The simulation will produce a series of files with global states that correspond to certain iterations of the simulation as requested by a user in a fhpsimulation configuration file. A user may wish to continue a previously fulfilled simulation with more iterations. This is done in the same directory with the same parameters. Then, a user may wish to select not the last state in a simulation directory and repeat computations with the same or modified parameters. Or, a user may wish to take the last state and continue with modified parameters. In these cases, the selected state and the parameters are placed in a new directory and a new computation experiment is conducted. This allows one to keep directories consistent such that all the global states in a directory are as produced in order from the initial state in this directory. The history of such forks is stored and can be researched later. A user can associate tags (key words) with global states. This allows a user to easily find global states by a list of tags.

Post-processing is called for a certain global state or for a series of global states and results in the files with numerical data for the averaged velocity and concentration fields for each global state and corresponding raster images.

6 Conclusion

A web-based platform for interactive parameter study of FHP-MP cellular automata has been developed as a GUI-application within the HPC Comunity Cloud service. Computer experiments have been carried out to find parameters of the FHP-MP model appropriate for reproduction of the Kármán vortex street. The conducted study is a typical use case for research in cellular automata models and particularly characterized by a large number of computer experiments needed to find the required parameters. These experiments cannot be run in a style of an automatic parameter sweep because of unreasonably large parameter spaces. A researcher's intuition remains an indispensable tool for optimization of the parameter study process. The provided high-level interactive user interface platform helps to boost the productivity of a researcher.

References

1. Medvedev, Y.G.: Lattice gas Cellular Automata for a flow simulation and their parallel implementation. In: Tarkov, M.S. (ed.) Parallel Programming: Practical Aspects, Models and Current Limitations. Series: Mathematics Research Developments, pp. 143–158. NSP Inc., New York (2014)
2. Kármán, T.: Aerodynamics, pp. 67–73. First McGraw-Hill Paperback Edition (1963). ISBN 07-067602-x
3. Bandman, O.L.: Relationships between cellular automata model parameters and their physical counterparts. Bull. Nov. Comp. Center, Series Comput. Sci. (42), 1–14 (2018). https://doi.org/10.31144/bncc.cs.2542-1972.2018.n42.p1-14
4. Vanag, V.K.: Study of spatially extended dynamical systems using probabilistic cellular automata. Phys. Usp. **42**(5), 413–434 (1999). https://doi.org/10.1070/PU1999v042n05ABEH000558
5. Chopard, B.: Cellular automata modeling of physical systems. In: Meyers, R. (ed.) Computational Complexity, pp. 407–433. Springer, New York (2012). https://doi.org/10.1007/978-1-4614-1800-9
6. Toffoli, T.: Cellular automata as an alternative to (rather than approximation of) differential equations in modeling physics. PhysicaD **10**, 117–127 (1984). https://doi.org/10.1016/0167-2789(84)90254-9
7. Bandman, O.L.: A discrete stochastic model of water permeation through a porous substance: parallel implementation peculiarities. Numer. Anal. Appl. **11**(1), 4–15 (2018). https://doi.org/10.1134/S1995423918010020
8. Medvedev, Y.: Cellular-automaton simulation of a cumulative jet formation. In: Malyshkin, V. (ed.) PaCT 2009. LNCS, vol. 5698, pp. 249–256. Springer, Heidelberg (2009). https://doi.org/10.1007/978-3-642-03275-2_25
9. Frisch, U., Hasslacher, B., Pomeau, Y.: Lattice-Gas Automata for the Navier-Stokes Equation. Phys. Rev. Lett. **56**(14), 1505–1508 (1984). https://doi.org/10.1103/PhysRevLett.56.1505
10. Goecks, J., et al.: Galaxy: a comprehensive approach for supporting accessible, reproducible, and transparent computational research in the life sciences. Genome Biol. **11**(8), R86 (2010). https://doi.org/10.1186/gb-2010-11-8-r86
11. Stodden, V., Seiler, J., Ma, Z.: An empirical analysis of journal policy effectiveness for computational reproducibility. Proc. Nat. Acad. Sci. USA **115**(11), 2584–2589 (2018). https://doi.org/10.1073/pnas.1708290115

12. Jiménez, R.C., Kuzak, M., Alhamdoosh, M., et al.: Four simple recommendations to encourage best practices in research software [version 1; peer review: 3 approved]. F1000Research **6**, 876 (2017) https://doi.org/10.12688/f1000research.11407.1

13. Hucka, M., Graham, M.J.: Software search is not a science, even among scientists: a survey of how scientists and engineers find software. J. Syst. Softw. **141**, 171–191 (2018). https://doi.org/10.1016/j.jss.2018.03.047. ISSN 0164–1212

14. Calegari, P., Levrier, M., Balczyński, P.: Web portals for high-performance computing: a survey. ACM Trans. Web **13**(1), 5:1–5:36 (2019). https://doi.org/10.1145/3197385

15. Afanasiev, A., Sukhoroslov, O., Voloshinov, V.: MathCloud: publication and reuse of scientific applications as RESTful web services. In: Malyshkin, V. (ed.) PaCT 2013. LNCS, vol. 7979, pp. 394–408. Springer, Heidelberg (2013). https://doi.org/10.1007/978-3-642-39958-9_36

16. Gorodnichev, M., Vaycel, S.: Organization of access to supercomputing resources in the HPC community cloud. Bull. South Ural State Univ. Ser. Comput. Math. Softw. Eng. **3**(4), 85–95 (2014). https://doi.org/10.14529/cmse140406

17. Sukhoroslov, O., Volkov, S., Afanasiev, A.: A web-based platform for publication and distributed execution of computing applications. In: 14th International Symposium on Parallel and Distributed Computing, Limassol, pp. 175–184 (2015). https://doi.org/10.1109/ISPDC.2015.27

18. Cholia, S., Sun, T.: The NEWT platform: an extensible plugin framework for creating ReSTful HPC APIs. Concurrency Computat. Pract. Exper. **27**, 4304–4317 (2015). https://doi.org/10.1002/cpe.3517

19. OLeary, P., Christon, M., Jourdain, S., Harris, C., Berndt, M., Bauer, A.: HPC-Cloud: a cloud/web-based simulation environment. In: IEEE 7th International Conference on Cloud Computing Technology and Science (CloudCom), Vancouver, BC, pp. 25–33 (2015). https://doi.org/10.1109/CloudCom.2015.33

20. Cao, R., Xiao, H., Lu, S., Zhao, Y., Wang, X., Chi, X.: SCEAPI: a unified restful web API for high-performance computing. J. Phys. Conf. Ser. **898**(9), 092022 (2017). https://doi.org/10.1088/1742-6596/898/9/092022

21. Bychkov, I.V., Oparin, G.A., Bogdanova, V.G., Pashinin, A.A., Gorsky, S.A.: Automation development framework of scalable scientific web applications based on subject domain knowledge. In: Malyshkin, V. (ed.) PaCT 2017. LNCS, vol. 10421, pp. 278–288. Springer, Cham (2017). https://doi.org/10.1007/978-3-319-62932-2_27

22. Struckmann, N., et al.: MIKELANGELO: MIcro KErneL virtualizAtioN for hiGh pErfOrmance cLOud and HPC systems. In: Mann, Z.Á., Stolz, V. (eds.) ESOCC 2017. CCIS, vol. 824, pp. 175–180. Springer, Cham (2018). https://doi.org/10.1007/978-3-319-79090-9_15

A Probabilistic Cellular Automata Rule Forming Domino Patterns

Rolf Hoffmann[1]([✉]), Dominique Désérable[2], and Franciszek Seredyński[3]

[1] Technische Universität Darmstadt, Darmstadt, Germany
hoffmann@informatik.tu-darmstadt.de
[2] Institut National des Sciences Appliquées, Rennes, France
domidese@gmail.com
[3] Department of Mathematics and Natural Sciences,
Cardinal Stefan Wyszynski University, Warsaw, Poland
fseredynski@gmail.com

Abstract. The objective in this study is to form a domino pattern by Cellular Automata (CA). In a previous work such patterns were formed by CA agents, which were trained with high effort by the aid of Genetic Algorithm. Now two probabilistic CA rules are designed in a methodical way that can perform this task very reliably even for rectangular fields. The first rule evolves stable sub–optimal pattern. The second rule maximizes the overlap between dominoes thereby maximizing the number of dominoes.

Keywords: Pattern formation · Probabilistic cellular automata · Matching templates · Asynchronous updating · Parallel Substitution Algorithm

1 Introduction

Pattern formation is an area of active research in various domains as in physics, chemistry, biology, computer science or natural and artificial life. Cellular Automata (CA) are suitable and powerful tools for catching the influence of the microscopic scale onto the macroscopic behavior of such complex systems [1–3]. At the least, the 1–dimensional Wolfram's "Elementary" CA can be viewed as generating a large diversity of 2–dimensional patterns whenever the time evolution axis is considered as the vertical spatial axis, with patterns depending or not on the random initial configuration [4]. Regarding the agent–based Yamins–Nagpal "1D spatial computer" [5,6] the authors emphasize therein how the local-to-global CA paradigm can turn into the inverse global-to-local question, namely "given a pattern, which CA rules will robustly produce it?" Such CA rules can be found by (i) proper design, (ii) by exhaustive search, or (iii) by heuristics like Genetic Algorithm (GA) or Simulated Annealing, methods which were applied to solve the Density Classification Problem [7], for instance.

© Springer Nature Switzerland AG 2019
V. Malyshkin (Ed.): PaCT 2019, LNCS 11657, pp. 334–344, 2019.
https://doi.org/10.1007/978-3-030-25636-4_26

The arrangement of dominoes in a grid of cells is a special case of pattern formation. Possible applications are: parcel packing encountered in different logistics settings, such as loading boxes on pallets, arrangements of pallets in trucks, or cargo stowage [8]; the design of a sieve for rectangular particles with a maximum flow rate; or an optimal arrangement of nanoparticles; and so forth.

Previous and Related Work. In further previous work [9–11], different patterns were generated by agents with embedded finite state control which was evolved by GA. Matching pattern templates were also applied during the training period, but are not part of the CA rule as in our current proposal. They were also defined in a different simple way in order to count the number of dominoes for the fitness function during the evolutionary process.

In [12] domino patterns were formed by moving agents. Agents' behavior was controlled by a finite state machine, evolved by GA. The effort to find such agents was quite high, especially to find agents that work on any field size. In order to avoid such a computational effort, a novel approach to construct directly the required CA rule was proposed that will be presented thereafter. It has also the potential to be applied to further pattern formations.

Parallel Substitution Algorithm (PSA) [13] is a powerful generalization of CA, which was also inspiring this work. PSA allows to substitute small locally defined patterns P by other patterns Q in a non–conflicting way. Thereby very complex computations and transformations can be performed in a decentralized and parallel way.

The problem of optimal domino layout is presented in Sect. 2 and the probabilistic CA rules derived from the Parallel Substitution method are discussed in Sect. 3. Results of simulation, performance evaluation and robustness are discussed in Sect. 4 before Conclusion.

2   Optimal Arrangements of Dominoes

Given is a square array of $N = n \times n$ cells with values $\in \{0, 1\}$. It is enclosed by a border of constant value 0. So the whole array is of size $(n + 2) \times (n + 2)$. In our representation, state 0 is colored white or green, and state 1 is colored black or blue.

2.1   The Problem

The objective is to find a CA rule that can form a *Domino Pattern* with a maximum number of dominoes. A domino consists of two black cells (the *kernel*) and 10 surrounding white cells (the *hull*). Two types of dominoes are distinguished, the horizontal oriented domino (D_H) and the vertical oriented (D_V) (Fig. 1(a)). It is allowed –and even necessary for a good solution– that white cells from the hull of a domino can overlap with other white cells (border cells or hull cells of other dominoes). The possible levels of overlapping, from 2 to 4, are displayed in Fig. 1(b–c–d).

The number of dominoes is denoted as $d = d_H + d_V$, where d_H is the number of horizontal dominoes and d_V is the number of vertical dominoes. A further requirement can be that the number of domino types is equal (or almost equal) (*balanced pattern*): $d_H = d_V$ if d_{max} is even, and $d_H = d_V \pm 1$ if d_{max} is odd, where $d_{max}(n)$ is the maximal possible number of dominoes that can be placed into the field with overlapping. Some optimal solutions are shown in Fig. 2.

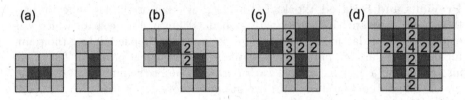

Fig. 1. (a) Horizontal and vertical domino tile, (b) two cells of two domino hulls are overlapping, marked by *2*, (c) the cell marked by *3* is the overlap of three domino hulls, (d) a case with *4* overlapping hull cells. (Color figure online)

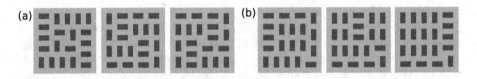

Fig. 2. $N = 10 \times 10$. (a) Three optimal balanced domino patterns with 20 dominoes, (b) three optimal unbalanced domino patterns with $d_H = 7$ and $d_V = 13$. (Color figure online)

2.2 Domino Enumeration

The maximal domino number is derived from an inductive formula in [12]. Let ν_n be the void index in a $n \times n$ field, with n even. Setting $m = n/2$ and $p = \lfloor m/3 \rfloor$ we got

$$\nu_n = \begin{cases} 4p\,(3p - 2) & (m \equiv 0) \\ 4p\,(3p) & (m \equiv 1) \quad (\mathrm{mod}\ 3) \\ 4p\,(3p + 2) & (m \equiv 2) \end{cases} \tag{1}$$

whence the maximal domino number

$$\xi_n = \frac{n^2 - \nu_n}{4} \tag{2}$$

as denoted therein, which will be compared, for $2 \leq n \leq 16$, with the d_{max} column of Table 1 in Sect. 4.

3   The Designed CA Rules

The first approach was to design a *deterministic* rule with *synchronous* updating. After some experiments and experience from previous work it showed to be very difficult if not even impossible to design such a rule that can converge always or with a high probability to the optimal or near-optimal aimed pattern.

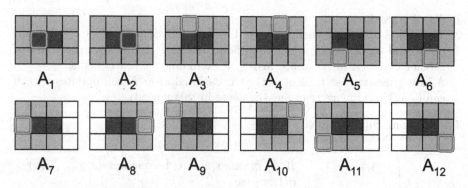

Fig. 3. The 12 templates of the horizontal domino. The value of the template center (marked) is used for cell updating if all remaining template cells match with the corresponding cells of the current configuration. (Color figure online)

The second approach was the construction of a *probabilistic* rule with *synchronous* updating. Indeed, such a rule was found for a field of size 6×6 by GA, where each cell is modeled as an agent that can turn in any direction. But the effort to find such rules is high and the good behavior can not be guaranteed in general.

The third and successful approach used here is the design of a *probabilistic* rule with *asynchronous* updating in a methodical way.

3.1   The First Rule

The basic idea is to modify the current configuration in a systematic way such that increasingly more dominoes appear and at last the CA evolves to a stable pattern. To do this, correct domino parts have to be detected and completed, and some random *noise* has to be injected where no correct domino parts are detected.

The domino parts are called *templates* A_i. They are derived from the two domino tiles. For the horizontal domino the 12 templates are shown in Fig. 3. For each cell value (marked in Fig. 3) of a domino a template is defined. A marked value is also called the *correct value* $v(A_i)$. It is placed at the relative position (used for addressing inside a template) $(\Delta x, \Delta y) = (0, 0)$, which is the *focus* of the template. The remaining space-ordered values define the correct neighbors A_i^* of a template. Note that many of these templates are similar under mirroring, which can facilitate an implementation. For the vertical domino a similar set of 12 templates is defined by $90°$ rotation, altogether we need 24 templates. The templates $A_7 - A_{12}$ show white cells that are not used because the maximal Manhattan distance $(\Delta x + \Delta y)$ from the marked cell to the neighbors was limited to $MH = 2 + 2$. The implementation with these incomplete templates worked very well, but further investigations are necessary to prove to which extent templates can be incomplete.

The rule updating scheme is:

1. A cell is randomly selected.
2. The rule is applied asynchronously. The new cell state $s' = f(s, B^*)$ is computed and immediately updated without buffering. B^* denotes the states of the neighbors within distance $MH \leq 4$, excluding the center cell s.
   A new generation at time–step $t + 1$ is declared after N cell updates (A cell can be updated more than once or never in this period).

The following rule is applied:

$$
s'(x, y) = \begin{cases} v(A_i) & \text{if } A_i^* \text{ matches with } CA\text{–}Neighbors(x, y) \quad (a) \\ & \textbf{otherwise} \\ random(0, 1) & \text{with probability } \pi_0 \qquad\qquad\qquad (b1) \\ s(x, y) & \text{with probability } 1 - \pi_0 \qquad\quad (b2) \end{cases}
$$

The neighbors A_i^* of every template with the marked center $(\Delta x, \Delta y) = (0, 0)$ are tested on each site (x, y) of the current CA cell field configuration, i.e. A_i^* is tested against the corresponding neighbors $B^*(x, y)$. If all values match then the value of cell at (x, y) is set to the correct value $v(A_i^*)$, and we have then a correct tile in the configuration. Otherwise with probability π_0, the cell is set randomly to either 0 or 1, or remains unchanged with probability $1 - \pi_0$.

It is important to note that the rule obeys the criterion of stability, which means that a field filled with dominoes without gaps (uncovered cells) is stable because we have matching hits at every site. Otherwise some random noise is injected in order to drive the evolution to the aimed pattern.

The rule was tested on 1,000 10×10 fields with a random initial configuration up to 2,000 time–steps (generations), with $\pi_0 = 0.2$. The CA system converges relatively fast to a sub–optimal domino pattern after 28.5 ± 13.6 time–steps on average. These patterns contain 13–19 dominoes (16 on average).

All patterns reached a stable domino arrangement. But some of them contained gaps; the lower the number of dominoes, the larger the number of gaps, up to 5. With ongoing time, the gaps change their color to black randomly, but the number of dominoes does not increase. Some of these patterns evolved are shown in Fig. 4.

3.2 The Second, Improved Rule

The purpose of this enhancement was to improve the rule in such a way that the number of dominoes reaches a maximum or be close to it.

(a)   **(b)**   **(c)**   **(d)**

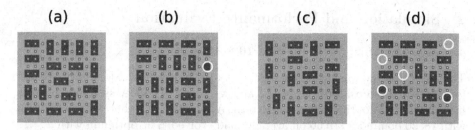

Fig. 4. Field of size 10×10. Some patterns evolved with the first rule. (a) RUN284: $t = 40$, $d = 19$ dominoes, no gap. (b) RUN699: $t = 18$, $d = 19$, one gap $g = 1$. (c) RUN875: $t = 40$, $d = 15$, $g = 0$. (d) RUN165: $t = 40$, $d = 15$, $g = 5$. Template hits are marked by dots, and gaps by white circles. (Color figure online)

A hit matrix was introduced. It stores the number of template hits for every site (x, y) that was selected for computation and updating. The number of hits on a site $h(x, y)$ is:

- 0, if no template matches or there is a gap.
- 1, if it results from one template match where the template focus is white.
- 2–4, if it results from the overlap at the same site (x, y) of 2–4 template matches with white focus, that means that 2–4 tiles are overlapping.
- 100, if it results from one template match where the template focus is black. Note that black cells are not allowed to overlap. The number 100 was chosen in order to differentiate such hits from the other.

The idea is to maximize the overlap between tiles by destroying non–overlapping situations ($h = 1$) through noise, allowing a reordering with high hit rates. First the new state s' is computed, and then the hit matrix. Then the new state is modified to s'':

$$s''(x, y) = \begin{cases} random(0, 1) & \text{with probability } \pi_1 \text{ if } hit(x, y) = 1 \\ s'(x, y) & \textbf{otherwise} \end{cases}$$

Now with the second rule it is not clear whether the stability criterion is still fulfilled because of the additional noise. In fact, it turned out that stability can only be reached it there exists a tiling where every tile overlaps with at least another or the border (called *totally overlapping tiling*). E.g. a totally overlapping tiling exists for 10×10 fields but not for 8×8 fields. Therefore the number of dominoes will reach the maximum and remain stable in a 10×10 field whereas the number of dominoes in a 8×8 field is reaching a maximum, and then it is decreasing and fluctuating and is driving again towards another maximum, and so forth. A deeper analysis is subject to further research.

4   Simulation and Performance Evaluation

4.1   Performance for Field Size 10 × 10

The improved rule was tested on 10,000 10×10 fields with random initial colors, for 2,000 time–steps, with $\pi_0 = 0.2$ and $\pi_1 = 0.1$. The CA system converges most often to an optimal domino pattern. After 708 ± 528 time–steps a pattern with 18–20 dominoes (19.66 on average) had evolved. The optimum with $d = 20$ dominoes was reached in 6,630 cases; 675 of them were balanced $(d_H, d_V) = (10, 10)$, 2,983 were unbalanced $(13, 7)$ and 2,972 were unbalanced $(7, 13)$. 3,369 contained 19 dominoes, and only one 18. This means that the probability to find

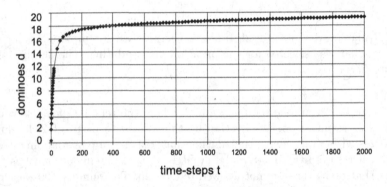

Fig. 5. Number of dominoes d vs. number of time–steps t. Average over 10,000 simulations. The CA system converges safely to the optimum ($d = 20$). (Color figure online)

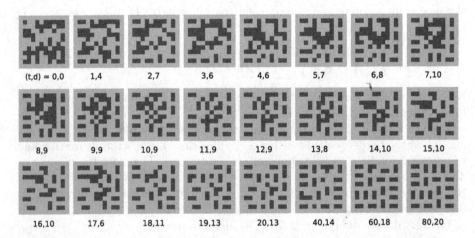

Fig. 6. Starting from a random initial configuration a balanced optimal pattern evolves. Example: 10×10 field. One fastest evolution taken from 10,000 recorded simulations. (Color figure online)

Table 1. 1,000 simulation runs were performed with time limit of 2,000 generations (time–steps) for different field sizes. The number of optimal patterns with d_{max} dominoes was evaluated, and the needed number of time–steps (on average, min – max) was computed. Optimal patterns are reliably generated (e.g. with a high probability of 66% for a 10×10 field).

Field size $n \times n$	Max. number of dominoes d_{max}	$b =$ optimal patterns found in 1000 runs	Balanced optimal p. found	Average time to find an optimal pattern
2×2	1	1000	1000	1.74 (0−19)
4×4	4	1000	1000	105 (1−800)
6×6	8	831	831	732 (8−2000)
8×8	13	1000	593	331 (10−2000)
10×10	20	663	68	916 (40−2000)
12×12	28	77	12	1233 (120−2000)
14×14	37	275	99	1091 (120−2000)
16×16	48	28	17	1283 (540−1940)

an optimal (balanced or unbalanced) pattern is 66 %, and to find an optimal balanced pattern is 6.75 %, for fields of size 10×10.

Figure 5 shows that the CA system converges quickly to the optimum, on the average of 10,000 simulation runs for 10×10 fields. A simulation sequence is depicted in Fig. 6 where an optimal balanced pattern appeared after at $t = 80$. Note that the number of dominoes is not increasing steadily but reaches finally an optimum after reorganization through the injected noise.

4.2   Performance for Different Field Sizes

The system with the second rule was simulated on different fields of size $n \times n$, for a limit of 2,000 time-steps, 1,000 runs for each size. Table 1 shows the results for even n. In all the cases, optimal patterns were found, and also balanced ones. For $n = 2, 4, 8$ only optimal patterns were generated. For $n = 16$, the success rate (percentage of optimal patterns found) was only 2.8%, but this rate can be enhanced by increasing the simulation time limit beyond 2,000. The average time to find an optimal pattern increases roughly with n, but not strictly. One explanation could be that the percentage of optimal patterns of all possible patterns is not strictly increasing with n. Note that the variance between the recorded *min* and *max* time (in brackets) is high. It can also be observed that the fluctuation of the domino number can be quite high during the evolution.

Table 2. The success rate (for finding optimal patterns) for $n \times n$ fields, where n is odd. Average over 1,000 runs, time limit = 2,000 generations.

n	3	5	7	9	11	13	15
d_{max}	2	6	10	16	24	32	42
Success rate [%]	100	100	100	100	42.6	59.7	20.1

4.3 Robustness

Further experiments have shown that the CA second rule also works well if (i) the initial configuration is totally black or white, (ii) the field size n is odd, or (iii) the field is rectangular.

Initial Configuration White *or* Black. One thousand runs were performed on a 10×10 field where initially all cells were *exclusively* colored (WH) white *or* (BL) black.

- (WH) After 723 time–steps on average (min 40 – max 2,000) a pattern with 19–20 dominoes ($d = 19.65$ on average) had evolved. The optimum with $d_{max} = 20$ dominoes was reached for 64.6% of the cases, where 8.4% of them were balanced.
- (BL) After 690 time–steps on average (min 40 – max 2,000) a pattern with 19–20 dominoes ($d = 19.65$ on average) had evolved. The optimum with $d_{max} = 20$ dominoes was reached in 64.6% of the cases, where 5.9% of them were balanced.

These results are statistically very close to the ones gained from random initial configurations (as evaluated before). Thus the evolution of domino patterns does not depend significantly on the initial configuration.

Odd Field Size. By one thousand simulation runs, the maximal number of dominoes and the percentage of optimal generated pattern were found (Table 2). The results show that the (second) CA rule works very well for $odd(n)$, too.

Rectangular Fields. The second rule was also tested successfully on several rectangular fields with different sizes. The different simulations confirmed that two cases have to be distinguished, (i) field sizes that do not allow a total overlapping of the tiles, and (ii) field sizes that allow it. In the (i) case, different optimal patterns appear in sequence, with changing non–optimal patterns in between. The search for an alternate pattern is never ending, because the noise injection at the non–overlapping sites does not vanish (for example, marked in yellow in Fig. 7a). In the (ii) case, an optimal pattern remains stable (for example, Fig. 7b), because no more noise is injected.

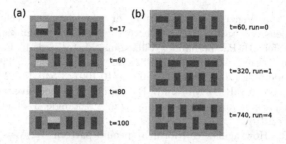

Fig. 7. (a) 2×10 field: no solution with total overlapping exists; non-overlapping cells are marked in yellow. Optimal patterns appear from time to time. (b) 4×10 field: there are solutions with total overlapping. Some stable optimal patterns are shown. (Color figure online)

5 Conclusion

Two probabilistic CA rules were designed that can form high quality domino patterns. 24 matching templates were derived from the two 3×4 domino tiles. Each selected cell was tested against the templates and was adjusted in the case that a template matches in the neighborhood. The first rule is sub–optimal with respect to the number of placed dominoes because non–overlapping tiles are allowed, however the reached patterns are always stable. The second rule introduces additional noise where domino tiles do not overlap. Thereby the overlap between domino tiles and the number of dominoes is maximized, leading to a global optimum with high probability. The reached optimal pattern remains stable, if there exists a totally overlapping tiling.

References

1. Chopard, B., Droz, M.: Cellular Automata Modeling of Physical Systems. Cambridge University Press, Cambridge (1998)
2. Deutsch, A., Dormann, S.: Cellular Automaton Modeling of Biological Pattern Formation. Birkäuser (2005)
3. Désérable, D., Dupont, P., Hellou, M., Kamali-Bernard, S.: Cellular automata in complex matter. Complex Syst. **20**(1), 67–91 (2011)
4. Wolfram, S.: Statistical mechanics of cellular automata. Rev. Mod. Phys. **55**(3), 601–644 (1983)
5. Nagpal, R.: Programmable pattern-formation and scale-independence. In: Minai, A.A., Bar-Yam, Y. (eds.) Unifying Themes in Complex Sytems IV, pp. 275–282. Springer, Heidelberg (2008). https://doi.org/10.1007/978-3-540-73849-7_31
6. Yamins, D., Nagpal, R.: Automated Global-to-Local programming in 1-D spatial multi-agent systems. In: Proceedings 7th International Conference on AAMAS, pp. 615–622 (2008)

7. Tomassini, M., Venzi, M.: Evolution of asynchronous cellular automata for the density task. In: Guervós, J.J.M., Adamidis, P., Beyer, H.-G., Schwefel, H.-P., Fernández-Villacañas, J.-L., (eds.): Parallel Problem Solving from Nature – PPSN VIIPPSN 2002. LNCS, vol. 2439, pp. 934–943. Springer, Heidelberg (2002). https://doi.org/10.1007/3-540-45712-7_90
8. Birgin, E.G., Lobato, R.D., Morabito, R.: An effective recursive partitioning approach for the packing of identical rectangles in a rectangle. J. Oper. Research Soc. **61**, 303–320 (2010)
9. Hoffmann, R.: How agents can form a specific pattern. In: Wąs, J., Sirakoulis, G.C., Bandini, S. (eds.) ACRI 2014. LNCS, vol. 8751, pp. 660–669. Springer, Cham (2014). https://doi.org/10.1007/978-3-319-11520-7_70
10. Hoffmann, R.: Cellular automata agents form path patterns effectively. Acta Phys. Pol. B Proc. Suppl. **9**(1), 63–75 (2016)
11. Hoffmann, R., Désérable, D.: Line patterns formed by cellular automata agents. In: El Yacoubi, S., Wąs, J., Bandini, S. (eds.) ACRI 2016. LNCS, vol. 9863, pp. 424–434. Springer, Cham (2016). https://doi.org/10.1007/978-3-319-44365-2_42
12. Hoffmann, R., Désérable, D.: Generating maximal domino patterns by cellular automata agents. In: Malyshkin, V. (ed.) PaCT 2017. LNCS, vol. 10421, pp. 18–31. Springer, Cham (2017). https://doi.org/10.1007/978-3-319-62932-2_2
13. Achasova, S., Bandman, O., Markova, V., Piskunov, S.: Parallel Substitution Algorithm, Theory and Application. World Scientific, Singapore (1994)

Synchronous Multi-particle Cellular Automaton Model of Diffusion with Self-annihilation

Anastasiya Kireeva[1]($\boxtimes$) (iD), Karl K. Sabelfeld[1,2](iD), and Sergey Kireev[1,2](iD)

[1] Institute of Computational Mathematics and Mathematical Geophysics,
6, Prospekt Lavrentjeva, Novosibirsk 630090, Russia
{kireeva,kireev}@ssd.sscc.ru, karl@osmf.sscc.ru
[2] Novosibirsk State University, Pirogova str., 2, Novosibirsk, Russia

Abstract. In this paper a synchronous multi-particle cellular automaton model of diffusion with self-annihilation is developed based on the multi-particle cellular automata suggested previously by other authors. The models of pure diffusion and diffusion with self-annihilation are described and investigated. The correctness of the models is tested separately against the exact solutions of the diffusion equation for different 3D domains. The accuracy of the cellular automata simulation results is investigated depending on the number of cells per a single physical unit. The calculation time of cellular automaton simulation of diffusion with self-annihilation is compared with the calculation time of the Monte Carlo random walk on parallelepipeds method for different domain sizes. The parallel implementation of the cellular automaton model is developed and efficiency of the parallel code is analyzed.

Keywords: Multi-particle cellular automaton · Synchronous mode ·
Diffusion · Self-annihilation · Monte Carlo

1 Introduction

Many complex phenomena include diffusion phase. There are different approaches for simulation of diffusion. One of the conventional approaches is the use of the finite-difference and finite-element methods [1]. In [2], a random walk in a bounded domain is considered and related with a diffusion boundary value problem. There exist other random walk based Monte Carlo algorithms [3] where instead of a detailed trajectory tracking one simulates jumps to a surface of certain subdomains like spheres, cubes, cylinders. We mention the Random Walk on Spheres method which is developed for isotropic stationary and transient drift-diffusion equations [4,5]. In [6], the Random Walk on Parallelepipeds (RWP) method is suggested for solving the transient anisotropic diffusion equation.

Supported by the Russian Science Foundation under Grant 19-11-00019.

V. Malyshkin (Ed.): PaCT 2019, LNCS 11657, pp. 345–359, 2019.
https://doi.org/10.1007/978-3-030-25636-4_27

Another approach to diffusion simulation is the use of cellular automata modeling. Cellular automaton (CA) is a discrete dynamical system consisting of a set of cells which evolve according to local rules in discrete time steps [7]. There are different CA models of diffusion. Let us mention some classes of these CA. A class of CA is constructed to simulate processes governed by partial differential equations. Rules of these CA are based on simple finite difference schemes and discretization operators [8–10]. The next class of CA is based on the lattice-gas CA models. The lattice-gas model's rules consist of two stages: collision and propagation [11,12]. The collision describes how particles entering the same cell change their moving directions. According to the propagation rule, the particles shift to the neighboring cell in the direction of their moving. Simultaneously only one particle can move in the same cell in the same direction. The lattice-gas CA diffusion models usually use the propagation rule as it is, and modified collision rules which imitate random rotation of particles [13,14]. These CA, as a rule, save the restriction of the lattice-gas CA: a single cell cannot contain two particles moving in the same direction, so, the number of particle in the cell is up to the number of neighboring cells. Another class of CA simulates the diffusion as a substance transport [15]. Unlike the lattice-gas CA, here particles have not any velocities and at each time step move to one of the neighboring cells selected at random. There are different modes of particle moving: synchronous, when all particles are shifted simultaneously, block-synchronous, when particles are shifted only inside their block, and asynchronous, when all particles are shifted at random and independently [7,16]. A single cell can contain only one particle, these models are Boolean CA [17]. Also, a single cell can contain several particles, theses models are integer multi-particle CA [18,19]. The multi-particle CA allow to extremely decrease the automaton noise with respect to the Boolean ones [20].

CA approach is a useful method for simulation of nonlinear spatially distributed phenomena. CA rules are much simpler than mean-field methods used for solving nonlinear partial differential equations. They usually mimic elementary acts of modeled process on the micro-level, for example, moving molecule at unit distance or an interaction of two molecules in a collision. CA rules are free from restrictions on the temporal and spatial steps of the difference schemes. In addition, CA are appropriate for high efficient implementation on supercomputers. The CA approach however has also many restrictions and disadvantages, for instance, often there is a need to use very small time and space step. Combining the CA approach with the developed random walk methods may result in new efficient simulation algorithms.

The main problems of CA approach are to determine the physical characteristics that correspond to the CA model, and to prove the coincidence of the CA simulation results with the modeled phenomenon. To solve these problems one matches the results of CA simulation with the results obtained by the method, assumed to be correct. In [16,21] the diffusion coefficients are determined for different CA models by comparison of the CA characteristics with the diffusion equation solution obtained by the finite-difference method. In [22,23] the results of the reaction-diffusion simulation by the Monte Carlo method and CA models are compared for proving CA models correctness. Also, analytical solution of equations, that describe the modeled phenomenon, is used for analyzing CA simulation results [13, 14].

The purpose of this paper is to construct a CA model of diffusion of non-interacting particles. Monte Carlo models [5, 6] of the diffusion simulate many independent trajectories and calculate such characteristics as the flux to the boundary or adsorbed particle concentration. The multi-particle CA allow to simulate a large number of particles at once, therefore, one single computing may be enough to obtain characteristics with a good accuracy. The question under study is: what is the comparative efficiency of the multi-particle CA and Monte Carlo RWP algorithm. Therefore, based on [18, 19] the synchronous multi-particle CA model of diffusion with self-annihilation is developed. To check the CA simulation, the results are tested against the exact expression for the probability density of some features characterizing the diffusion process. The calculation time of CA simulation of diffusion with self-annihilation is compared with the calculation time of Monte Carlo RWP algorithm. Obtained results show that the CA simulation of a diffusion process in comparatively large domains requires more computer time than that of the Monte Carlo RWP algorithm. Therefore, the parallel implementation of the CA model of diffusion with self-annihilation is performed and analyzed.

This paper is organized as follows. In Sect. 2 the synchronous multi-particle CA model of diffusion is described and its correctness is proved. In the third section the CA model of diffusion is extended to take into account the self-annihilation, and the accuracy of this model is studied. Section 4 presents the results of parallel implementation of the CA model of diffusion with self-annihilation.

2 Cellular Automaton Model of Diffusion of Non-interacting Particles (CAM-DNIP)

2.1 Description of CAM-DNIP

Based on the notation [16], a CA is a set of cells denoted by the pairs (a, x), where a is the state of the cell, and x is the coordinate of the cell in a finite d-dimensional discrete space. All admissible in a model states are named an alphabet A. All admissible cell coordinates are named a set of names X. The set of cells $\Omega = \{(a, x) : a \in A, x \in X\}$, that do not contain cells with the same coordinates, are formed a cellular array. The cell states are updated by local rules Θ that depend on states of cell (a, x) and its neighbors replace an old state a to a new state a'. The rules Θ can be applied at different order named mode. The application of the rule Θ to all cells of the cellular array is called an iteration. The CA diffusion model is described by the following notations:

$$\aleph = \langle A, X, \Theta_{dif}, \mu \rangle. \tag{1}$$

Here, the alphabet $A \in Z^+$ is a set of non-negative integer numbers that denote the number of particles in a cell. The set of names

$$X = \{x = (i, j, k) : i = 0, ..., Size_X, \ j = 0, .:., Size_Y, \ k = 0, ..., Size_Z\} \quad (2)$$

is a three-dimensional Cartesian lattice. The rules Θ_{dif} of particle moving are based on CA rules given in [18]. On each time step each particle in a cell can jump to one of the neighboring cells with probability P_{move} or remain in the cell. For the cell (a, x) the neighboring cells are the cells having coordinates different from x by 1 only in one of the directions. That is, the set of neighboring cells is defined by the template $T(i, j, k) = \{(i \pm 1, j, k), (i, j \pm 1, k), (i, j, k \pm 1)\}$. As well as in [18], the cell state a is divided into two parts: moving particles a_m and remaining particles a_l:

$$a_m = \lfloor a \cdot P_{move} \rfloor + b, \quad a_l = a - a_m,$$

$$\text{where } b = \begin{cases} 1, \text{ if } rand < (a \cdot P_{move} - \lfloor a \cdot P_{move} \rfloor), \\ 0, \text{ otherwise}. \end{cases} \quad (3)$$

Here, $\lfloor a \rfloor$ denotes the integer part of a. The variable b simulates a rounding residue of a. With probability $a \cdot P_{move} - \lfloor a \cdot P_{move} \rfloor$ the $a \cdot P_{move}$ is rounded up, otherwise it is rounded down. The variable $rand$ is a random number uniformly distributed on $[0, 1]$.

Thus, an application of the rule Θ_{dif} to a cell (a, x) is as follows. One calculates the number of moving a_m and remaining in the cell a_l particles by formula (3). One of the neighboring cells (c, x_k) is selected with probability $1/6$. The moving particles jump to the selected neighboring cell, i.e. a state of cell (a, x) is changed to $a' = a - a_m$ and a state of cell (c, x_k) is changed to $c' = c + a_m$.

Due to independence of the moving particles, the rule Θ_{dif} simulates only jumping of moving particles to the neighboring cell, unlike [18], where the rule simulates exchange of moving particles between cell (a, x) and one of the neighboring cells. This difference is due to the mode of application of the rule. We use the synchronous mode as well as in [19], but in [18] the asynchronous mode is used. According to synchronous mode, the rule Θ_{dif} is applied to all cells of the cellular array and changed their cells simultaneously. This mode is an analog of the explicit scheme of the finite-difference methods [1]. It is implemented in a code by the two arrays: for the *current* cell states and for the *new* states. For each cell (a, x) the values of a_m and a_l are computed based on the current state a and added to the state of cell x in the new array. After each iteration we copy the cell values from the new array at the current array and then set the new array to zero.

2.2 Verification of CAM-DNIP

To prove the correctness of CAM-DNIP we need to compare results of the CA simulations with the results calculated by some exact method of solving the diffusion equation:

$$\frac{\partial u(\mathbf{r}, t)}{\partial t} = D\Delta u(\mathbf{r}, t), \tag{4}$$

where $\mathbf{r}$ is a space coordinate, D is a constant diffusion coefficient.

In [19,21], the correspondence between a physical diffusion coefficient, for which Eq. (4) is solved, and a diffusion coefficient, which is obtained by the multi-particle CA diffusion models, is given in the following form:

$$D = \frac{P_{move}}{2d} \frac{l_c^2}{t_{itr}}, \tag{5}$$

where P_{move} is the moving probability, d is the space dimension, l_c is a cell length, t_{itr} is a time of a single iteration. For the three-dimensional space $d = 3$.

In the multi-particle CA models the variable parameters are the cell length l_c, the iteration time t_{itr} and the moving probability P_{move}. Choosing the values of any two parameters, one can calculate a value of the third parameter by formula (5). Thus, we fix the values of l_c and P_{move} and calculate the value of t_{itr}.

In [6], a Random Walk on Cubes Monte Carlo method is suggested to solve anisotropic transient drift-diffusion-reaction problems. It is a meshfree method which is based on simulation of the particle trajectories exactly according to their temporal and spatial distributions. Different concentration-related functions can be calculated by this method without calculating the whole concentration field, for example, the fluxes to the domain boundaries among them. The fluxes are calculated as follows. A cube (or, generally, a parallelepiped) is constructed with its center at the point of a particle source r_0 and length side equal to the distance to the closest boundary. The first passage time τ and the random position r_1 on the sides of the cube are sampled according to the densities explicitly derived in [6]. The living time of the particle is increased by τ. If the random position r_1 hits a small layer along the boundary, the particle is adsorbed on it and the score f_i for the flux to this boundary is added. Otherwise, the next cube is constructed with its center at the point r_1 and length side equal to the distance to the closest boundary and the simulation steps are repeated as in the previous cube. After simulation of N_{tr} particle trajectories, the fluxes to the domain boundaries are calculated as the total scores f_i divided by N_{tr}.

The probability density of the first exit time of the particle to some domain's boundary uniquely characterizes the diffusion process and can be taken as an exact solution used to verify the CA diffusion model.

The probability density of the first passage time τ of a particle starting from the center of a cube to the cube sides is derived in [6], and in the case of isotropic diffusion it has the following form:

$$p(\tau) = \frac{192D}{\pi l^2} \cdot F_1(D, \tau) \cdot (F_2(D, \tau))^2 , \tag{6}$$

where l is a length of the cube side and functions $F_1(D, \tau)$ and $F_2(D, \tau)$ are defined by the following formulae:

$$F_1(D, \tau) = \sum_{m=1}^{\infty} (-1)^{m+1}(2m-1) \exp\left[-\frac{(2m-1)^2\pi^2 D}{l^2}\tau\right],$$

$$F_2(D, \tau) = \sum_{n=1}^{\infty} (-1)^{n+1}\frac{1}{2n-1} \exp\left[-\frac{(2n-1)^2\pi^2 D}{l^2}\tau\right]. \tag{7}$$

In addition, the probability density of the first passage time of a particle starting from the height z is derived in the case of diffusion in an infinite layer of height l whose plane boundaries are both absorbing:

$$p_l(\tau) = \frac{1}{2t\sqrt{\pi Dt}} \sum_{n=-\infty}^{\infty} \left\{p_l^1 - \frac{1}{2}\left(p_l^2 + p_l^3\right)\right\},$$

$$p_l^1 = (2nl + z)\exp\left[-\frac{(2nl+z)^2}{4Dt}\right],$$

$$p_l^2 = [(2n+1)l + z]\exp\left[-\frac{[(2n+1)l+z]^2}{4Dt}\right], \tag{8}$$

$$p_l^3 = [(2n-1)l + z]\exp\left[-\frac{[(2n-1)l+z]^2}{4Dt}\right].$$

Also, the probability density of the first passage time of a particle starting from the height z is derived in the case of diffusion in an infinite layer of height l with an absorption on the plane $z = 0$ and reflection on the plane $z = l$:

$$p_{ref}(\tau) = \frac{1}{2t\sqrt{\pi Dt}} \sum_{n=-\infty}^{\infty} (-1)^n(2nl + z)\exp\left[-\frac{(2nl+z)^2}{4Dt}\right] \tag{9}$$

A histogram is an approximation of the probability distribution of a continuous variable [24]. Therefore, we compare the histograms of the first passage time obtained by the CA simulation of the diffusion inside the cube and layer with the exact probability densities $p(\tau)$, $p_l(\tau)$ or $p_{ref}(\tau)$ calculated by formulae (6), (8), (9), respectively. To compute the histogram of the first passage time to the cube sides the following computational experiment is performed. The diffusion of particles starting from the center is simulated in the cube with length side l. The length of a single cell l_c and the value of the moving probability P_{move} is selected. For chosen l_c and P_{move} the number of iterations n_{itr} needed to reach the cube boundaries is computed and multiplied by the time of a single iteration t_{itr}. For the obtained values of the exit time t_{exit} the histogram is computed as follows. A sufficiently large interval $[0, T_{max}]$, where T_{max} is a maximum time needed to reach the cube boundaries, is taken. This interval is divided into subintervals, or bins. The number of time values $n_{t_{exit}}$ falling in each bin is calculated and divided by the bin width. The obtained values of $n_{t_{exit}}$ for each bin is the histogram of first passage time $h(t_{exit})$. Analogously, for the layer, the number of iterations n_{itr} needed to reach the layer boundaries is computed and multiplied by the time of a single iteration t_{itr}. For the obtained values of the exit time t_{exit} the histogram is computed as well as it is described above.

2.3   Results of Simulation by CAM-DNIP

The CA diffusion model $\aleph$ is tested on the diffusion task inside the cube with side length $l = 10\,\text{nm}$, with the diffusion coefficient $D = 1\,\text{nm}^2/\text{ns}$. The moving probability P_{move} is taken equal to 0.7. The cell length l_c is determined on the basis of how many cells accounted per a physical unit of length. The value of l_c is calculated as $1/n_c$, where n_c is the number of cells in a single physical unit, here in $1\,\text{nm}$. The value of n_c is varied from 1 to 10 cells. The cellular array size $Size_x \cdot Size_y \cdot Size_z = Size^3$, where $Size$ is the number of cells in a single side of the cube. The value of $Size$ is equal to the ratio l/l_c. The absorbing boundary conditions are used. It means that the particles hitting the boundary cell disappear.

At the initial time, the state of the central cell is $(a, x) = (10^{14}, x)$, $x = (Size/2, Size/2, Size/2)$, the other cell states equal to zero. The value $a = 10^{14}$ has been chosen in order to decrease the automaton noise [20] and obtain the histograms of the first exit time with a sufficient accuracy. It is worth noting that small values of n_c do not provide a sufficient accuracy and require the averaging the results over several computational experiments with different sequences of random numbers. For example, when $n_c < 4$ the histograms $h(t)$ are obtained by the averaging over $N = 1000$ computational experiments, for $n_c = 4$ $N = 100$, and the values of $n_c > 4$ provide a sufficient accuracy of CA simulation for a single computational experiment.

Figure 1 presents the exact probability density $p(t)$ and the histograms of the first passage time $h(t)$ obtained by the CA simulation for the different number of cells accounted for $1\,\text{nm}$. The plots are shown in a logarithmic scale of both axis. The histograms $h(t)$ coincide with the exact probability density $p(t)$ for large time values, but they differ at small time values. This is because the accuracy of CA simulation depends on the number n_c of cells per a single physical unit. The small values of $n_c < 4$ do not allow to simulate a small value of the first exit time. For example, $n_c = 1$ allows to obtain the first exit time of order 10^{-3}. However, increasing n_c leads to the possibility of simulating the smaller time values. For instance, the CA with $n_c = 4$ is able to simulate the first exit time of order 10^{-7}.

The next test is a simulation of diffusion in an infinite layer of height $l = 10\,\text{nm}$, with the diffusion coefficient $D = 1\,\text{nm}^2/\text{ns}$. A particle starts from the height $l/2$. As well as in the previous test, the moving probability is $P_{move} = 0.7$, the cellular array size is $Size_x \cdot Size_y \cdot Size_z = Size^3$, at the initial time the state of the central cell is $(a, x) = (10^{14}, x)$. The absorbing boundary conditions for top and down cube sides is used. The infinity along axes X and Y is simulated by the periodic boundary conditions.

The plots of the exact probability density $p_l(t)$ and the histograms of the first exit time computed by the CA $\aleph$ in the case of a layer are given in Fig. 2. As well as in the case of the cube, the CA $\aleph$ with $n_c = 5$ simulates the diffusion with a good accuracy (10^{-9}). However, the CA with the smaller number of cells $n_c \leq 4$ provides a smaller accuracy: 10^{-5} at $n_c = 3$ and 10^{-7} at $n_c = 4$.

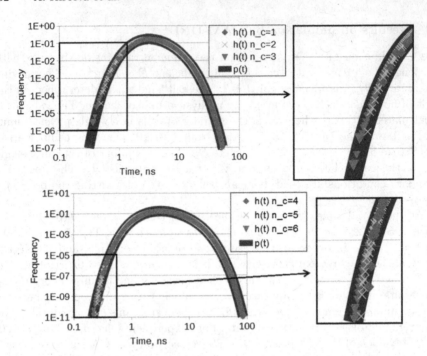

Fig. 1. Comparison of the probability density $p(t)$ and the histograms $h(t)$ computed by the CA simulation for different number n_c of cells per a single physical unit

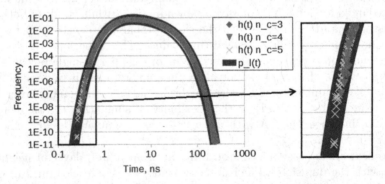

Fig. 2. Comparison of the probability density $p_l(t)$ and the histograms $h(t)$ computed by the CA model $\aleph$ in an infinite layer with absorbing boundary conditions

Another test was a simulation of diffusion in an infinite layer with an absorption on the plane $z = 0$ and with the reflection on the plane $z = l$. As well as in the previous test, the layer height $l = 10\,\text{nm}$, a particle starts from the height $l/2$, the diffusion coefficient $D = 1\,\text{nm}^2/\text{ns}$, the moving probability is $P_{move} = 0.7$, the cellular array size is $Size^3$. The periodic boundary conditions along the axes X and Y and the absorbing boundary condition for the top cube face are used. The reflecting boundary condition is simulated as follows. When a particle jumps from a cell with coordinate $z = Size - 2$ to a cell with coordinate $z = Size - 1$, it is reflected back to the cell with coordinate $z = Size - 2$.

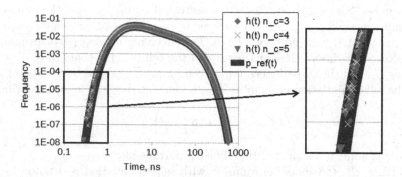

Fig. 3. Comparison of the probability density $p_{ref}(t)$ and the histograms $h(t)$ obtained by the CA simulation of diffusion in an infinite layer with absorption on the plane $z = 0$ and reflection on the plane $z = l$

Figure 3 shows the histograms of the first exit time obtained by the CA $\aleph$ and the exact probability density $p_{ref}(t)$. Like the previous cases, the CA simulation results with $n_c = 5$ agree well with the exact probability density.

Thus, we can conclude that the CA model $\aleph$ is able to simulate the diffusion in a domain with different boundary conditions (absorbing, periodic, reflecting) with a good accuracy.

3   Cellular Automaton Model of Diffusion with Self-annihilation (CAM-DSA)

3.1   Description of CAM-DSA

The multi-particle CA model of diffusion can be extended to simulation of reaction-diffusion processes. Here, we consider a particular case of reaction, the self-annihilation of particles in a domain volume at some annihilation rate.

The diffusion rule Θ_{dif} remains the same as described in Sect. 2.1. The self-annihilation is simulated based on the diffusion rule as follows. When we apply the rule Θ_{dif} to a cell (a, x), after calculation of the number of moving a_m and remaining in the cell a_l particles, it is computed what parts of a_m and a_l are annihilated. For that, each number a_m and a_l is divided into two parts according to the self-annihilation probability P_{ads}:

$$a_{m\_ads} = \lfloor a_m \cdot P_{ads} \rfloor + b_m, \quad a_{m\_live} = a_m - a_{m\_ads},$$

$$\text{where } b_m = \begin{cases} 1, \text{ if } rand < (a_m \cdot P_{ads} - \lfloor a_m \cdot P_{ads} \rfloor), \\ 0, \text{ otherwise}. \end{cases} \quad (10)$$

and $\quad a_{l\_ads} = \lfloor a_l \cdot P_{ads} \rfloor + b_l, \quad a_{l\_live} = a_l - a_{l\_ads},$

$$\text{where } b_l = \begin{cases} 1, \text{ if } rand < (a_l \cdot P_{ads} - \lfloor a_l \cdot P_{ads} \rfloor), \\ 0, \text{ otherwise}. \end{cases} \quad (11)$$

The number of moving particles a_m is decreased by a_{m_ads}, and the number of remaining in a cell particles a_l is decreased by a_{l_ads}. One of the neighboring cells (c, x_k) is selected with probability $1/6$. The surviving moving particles jump to the selected neighbor cell. Thus, a state of the cell (a, x) is changed to $a' = a - a_m - a_{l\_ads}$, and the state of neighboring cell (c, x_k) is changed to $c' = c + a_{m\_live}$. The self-annihilation probability P_{ads} is calculated by the following formula:

$$P_{ads} = 1 - \exp\left\{-\frac{\tau}{\bar{\tau}}\right\}, \quad (12)$$

where $\bar{\tau}$ is a mean life time of the diffusing particle [3].

Further, the CA model of diffusion with self-annihilation is denoted by the $\aleph_{ads}$.

3.2   Verification of CAM-DSA

In [25], in the case of diffusion with self-annihilation, the probability density of the first passage time of the particle starting from the center of a cube to the cube faces has the form:

$$p_{ads}(\tau) = \frac{192D}{\pi l^2} \cdot \exp\left\{-\frac{\tau}{\bar{\tau}}\right\} \cdot F_1(D, \tau)(F_2(D, \tau))^2. \quad (13)$$

Analogously to the pure diffusion case, we compute the histograms of the first exit time by the CA simulation of the diffusion with self-annihilation inside the cube. To prove a correctness of the CA model $\aleph_{ads}$ we compare these histograms with the exact probability density $p_{ads}(\tau)$.

3.3   Results of Simulation by CAM-DSA

In this section, for the CA simulation mainly the same model parameter values are used as in Sect. 2.3. The diffusion with self-annihilation is simulated inside the cube with side length $l = 10$ nm, with the diffusion coefficient $D = 1$ nm^2/ns and with the mean life time of the diffusing particle $\bar{\tau} = 5$ ns. The moving probability is $P_{move} = 0.7$. The absorbing boundary conditions are used for all cube sides. At the initial time, the state of the central cell is $(a, x) = (10^{14}, x)$, where $x = (Size/2, Size/2, Size/2)$, the other cell states equal to zero.

As well as in the case of the CA model $\aleph$, the number of cells in 1 nm affects the accuracy of the CA model $\aleph_{ads}$. The histograms of the first passage time $h(t)$ obtained by the CA simulation of diffusion with self-annihilation for the different values of n_c are given in Fig. 4. A comparison of the histograms $h(t)$ with the exact probability density $p_{ads}(t)$ shows that the model $\aleph_{ads}$ provides the accuracy of order 10^{-7} for using of $n_c \geq 5$ cells per 1 nm.

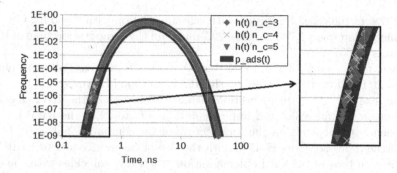

Fig. 4. Comparison of the probability density $p_{ads}(t)$ and the histograms $h(t)$ obtained by the CA simulation of diffusion with annihilation in the cube

4   Performance of CAM-DSA Implementation

4.1   Comparison of CAM-DSA and RWP Implementations

In this section we compare the calculation time of the implementations of the synchronous multi-particle CA and the Monte Carlo Random Walk on Parallelepipeds (RWP) method [6], both simulating diffusion with self-annihilation.

The same model parameter values as in Sect. 3.3 are taken in calculations. We consider the both cases: a pure diffusion with $\bar{\tau} = 0$ ns, and diffusion with self-annihilation with the mean life time of particle $\bar{\tau} = 5$ ns. The number of cells in 1 nm is $n_c = 4$, that provides the accuracy of the simulation results of order 10^{-6}. The same accuracy is obtained by the RWP method with the number of trajectories equal to $N_{tr} = 10^8$.

Computational tests are executed on the processor Intel(R) Core(TM) i7-4770 CPU 3.40 GHz.

In the case of pure diffusion, the calculation time of the CA $\aleph$ implementation $T_{\aleph}$ is 6.76 s, and the time of RWP method implementation T_{RWP} is 52.316 s. In the case of diffusion with self-annihilation, the CA $\aleph_{ads}$ implementation takes 6.34 s, and RWP method takes 36.175 s.

Now, we investigate how the model parameters affect the computation time of the both models. In the case of pure diffusion, when varying the diffusion coefficient, the calculation time of both models is practically not changing. For example, for $D = 10$ nm^2/ns the computation time of the CA $\aleph$ is 6.75 s and $T_{RWP} = 49.977$ s. However, in the case of diffusion with self-annihilation, the execution time is increased for the both models: $T_{\aleph\_ads} = 9.858$ s and $T_{RWP} = 54.741$ s. When increasing the mean life time of particle $\bar{\tau}$, the calculation time of both method is increased as well. For example, for $D = 1$ nm^2/ns and $\bar{\tau} = 10$ ns the time of CA $\aleph_{ads}$ execution is 8.298 s, and the time RWP method is 43.728 s. Increasing the domain size has a crucial impact on the computation time of CA implementation and almost no effect on the time of RWP method. For instance, in case of the pure diffusion with $D = 1$ nm^2/ns in the cube with $l = 50$ nm, the execution time of CA $\aleph$ is 22,902,983 s, i.e. 6.36 h, and the time of RWP method is 63.507 s. In the case of diffusion with self-annihilation with $\bar{\tau} = 5$ ns, CA $\aleph_{ads}$ implementation takes 1,783.25 s, i.e. 29.7 min, and the computation time of RWP method is 28.657 s.

Thus, for the small domain size the CA models $\aleph$ and $\aleph_{ads}$ allow to obtain the result 6–8 faster compared to the RWP method. However, for the large domain size the CA implementation is much slower than the RWP method. The domain size for which CA implementation is faster than the RWP method implementation depends on the architecture of the computational node. In the case of the processor Core i7-4770 with the size of cache L2 equal to 8 Mb, the computation time of the CA implementation practically coincides with the time of the RWP method execution for the domain size $l = 15$ nm and $D = 1$ nm^2/ns. A more detailed analysis of the cellular array size for which CA takes less time than RWP method is an issue of a separate study.

4.2 Parallel Implementation of CAM-DSA

To simulate the diffusion with self-annihilation for large domains by the CA model its parallel implementation is needed. We employ the MPI standard for the parallel execution of the CA model $\aleph_{ads}$. The general approach to CA parallel implementation is a domain decomposition method. The cellular array is divided into subdomains which is distributed between available MPI processes. Each MPI process computes states of cells of its subdomain. After each iteration MPI processes exchange new states of their boundary cells by non-blocking communication functions MPI_Isend and MPI_Irecv. The boundary exchange is executed during the time of computation of cell states of an internal part of a subdomain. Also, on each iteration each MPI process calculates the number of particles adsorbed on the domain boundaries and annihilated in the volume. The sum of these values over all MPI processes is obtained by the non-blocking collective function MPI_Iallreduce.

To estimate the performance of the parallel code its speedup $S(n_{mpi})$ and efficiency $E(n_{mpi})$ are computed as follows: $S(n_{mpi}) = T(1)/T(n_{mpi})$, $E(n_{mpi}) = S(n_{mpi})/n_{mpi}$, where n_{mpi} is the number of MPI processes used, $T(n_{mpi})$ is a calculation time obtained when a task is executed on n_{mpi} processes. The partition "Broadwell" of the cluster "MVS-10P" of the Joint Supercomputer Center of RAS [26] is employed for calculations. The "Broadwell" node consists of two processors Intel Xeon CPU E5-2697A v4 2.60 GHz, each contains 16 cores with 2 threads and 40 MB SmartCache.

The performance of the parallel code is tested for the following model parameter values. The simulating domain is a cube of side length $l = 50$ nm, the diffusion coefficient is $D = 1$ nm^2/ns, the mean life time of particle $\bar{\tau} = 5$ ns. The moving probability is $P_{move} = 0.7$. The absorbing boundary conditions are used for all cube sides. At the initial time, the state of the central cell is $(a, x) = (10^{14}, x)$, where $x = (Size/2, Size/2, Size/2)$, and the other cell states equal to zero. The number of cells in 1 nm is $n_c = 5$. The cellular array size is $Size^3 = (l \cdot n_c)^3 = 250^3 = 15,625,000$ cells.

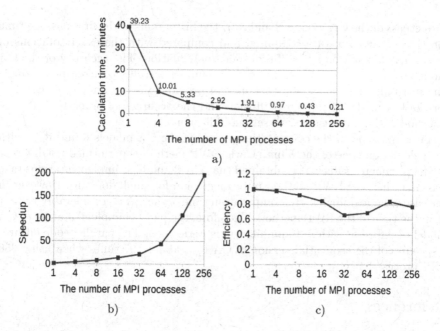

Fig. 5. The values of characteristics of the parallel implementation of the CA model $\aleph_{ads}$: (a) computation time, (b) speedup and (c) efficiency

Figure 5 shows the plots of characteristics of the parallel implementation of the CA model $\aleph_{ads}$: the computation time, speedup and efficiency, depending on the number of MPI processes. The parallel code speedup is 197 times when using 256 MPI processes comparing with the sequential code. The speedup strictly increases with the increase of MPI processes till 256. However, the parallel code efficiency is strongly decreasing when using 32 nodes. It is explained by the memory bandwidth limiting. Each processor E5-2697A v4 has 16 cores and 40 MB SmartCache. The computational node has two such processors. Thus when we execute the task using 32 MPI processes, each 16 of these processes uses the same 40 MB SmartCache. The data needed to calculation of each processes does not fit in the cache until using 128 MPI processes. When increasing the number of processes the intensity of memory access increases and the memory bandwidth constraints the code performance. Next increasing the number of MPI processes causes the decreasing of the subdomain size. Therefore, the efficiency of using 64 and 128 processes increases. However, further increasing of the number of MPI processes causes the slowdown of the code execution since the data exchange becomes the limiting factor.

5   Conclusion

Synchronous multi-particle CA models of the pure diffusion ($\aleph$) and diffusion with self-annihilation ($\aleph_{ads}$) are developed based on [18,19]. To prove the correctness of the CA models $\aleph$ and $\aleph_{ads}$ the histograms of the first passage time for different domains are computed and compared with the exact solutions of the diffusion equation and diffusion-reaction equation. The accuracy of the CA models is investigated depending on the number of cells n_c taken per a single physical unit. It is concluded that $n_c = 4$ provides the accuracy of order 10^{-7} for the CA model of the pure diffusion and the accuracy of order 10^{-6} for the CA model of the diffusion with self-annihilation.

A comparison of the computational time of the CA models $\aleph$ and $\aleph_{ads}$ with the calculation time of the Monte Carlo RWP method is performed for different model parameter values. It is obtained that CA $\aleph$ and $\aleph_{ads}$ implementations take less time than the RWP method implementation for small domains. However in the case of large domains, the computation time of the CA is considerably greater than the time of the RWP method. Therefore, the parallel implementation of CA model of diffusion with self-annihilation is performed. The parallel code ensures a speedup of the sequential version of CA $\aleph_{ads}$ about 197 times when using 256 MPI processes.

References

1. Smith, G.D.: Numerical Solution of Partial Differential Equations (Finite Difference Methods). Oxford University Press, Oxford (1990)
2. Courant, R., Friedrichsund, K., Lewy, H.: Über die partiellen Differentialgleichungen der mathematischen Physik. Math. Annalen **100**, 32–74 (1928)
3. Sabelfeld, K.K.: Monte Carlo Methods in Boundary Value Problems. Springer, Heidelberg (1991)
4. Sabelfeld, K.K.: Random walk on spheres method for solving drift-diffusion problems. Monte Carlo Methods Appl. **22**(4), 265–275 (2016)
5. Sabelfeld, K.K.: Random walk on spheres algorithm for solving transient drift-diffusion-reaction problems. Monte Carlo Methods Appl. **23**(3), 189–212 (2017)
6. Sabelfeld, K.: Stochastic simulation methods for solving systems of isotropic and anisotropic drift-diffusion-reaction equations and applications in cathodoluminescence imaging. Submitted to Probabilistic Engineering Mechanics (2018)
7. Toffoli, T., Margolus, N.: Cellular Automata Machines: A New Environment for Modeling. MIT Press, USA (1987)
8. Weimar, J.R.: Cellular automata for reaction-diffusion systems. Parallel Comput. **23**, 1699–1715 (1997)
9. Weimar, J.R.: Three-dimensional cellular automata for reaction-diffusion systems. Fundamenta Informaticae **52**(1–3), 277–284 (2002)
10. Weimar, J.R., Tyson, J.J., Watson, L.T.: Diffusion and wave propagation in cellular automaton models of excitable media. Physica D **55**(3–4), 309–327 (1992)
11. Chopard, B.: Cellular automata modeling of physical systems. In: Meyers, R. (ed.) Computational Complexity, pp. 407–433. Springer, New York (2012). https://doi.org/10.1007/978-1-4614-1800-9_27
12. Frenkel, D., Ernst, M.H.: Simulation of diffusion in a two-dimensional lattice-gas cellular automaton: a test of mode-coupling theory. Phys. Rev. Lett. **63**(20), 2165–2168 (1989)
13. Chopard, B., Droz, M.: Cellular automata model for the diffusion equation. J. Stat. Phys. **64**(3–4), 859–892 (1991)

14. Dab, D., Boon, J.-P.: Cellular automata approach to reaction-diffusion systems. In: Manneville, P., Boccara, N., Vichniac, G.Y., Bidaux, R. (eds.) Cellular Automata and Modeling of Complex Physical Systems, pp. 257–273. Springer, Heidelberg (1989). https://doi.org/10.1007/978-3-642-75259-9_23

15. Karapiperis, T., Blankleider, B.: Cellular automaton model of reaction-transport processes. Physica D **78**, 30–64 (1994)

16. Bandman, O.L.: Comparative study of cellular-automata diffusion models. In: Malyshkin, V. (ed.) PaCT 1999. LNCS, vol. 1662, pp. 395–409. Springer, Heidelberg (1999). https://doi.org/10.1007/3-540-48387-X_41

17. Bandman, O.: Cellular automata diffusion models for multicomputer implementation. Bull. Nov. Comp. Center Comp. Sci. **36**, 21–31 (2014)

18. Medvedev, Y.: Multi-particle Cellular-automata models for diffusion simulation. In: Hsu, C.-H., Malyshkin, V. (eds.) MTPP 2010. LNCS, vol. 6083, pp. 204–211. Springer, Heidelberg (2010). https://doi.org/10.1007/978-3-642-14822-4_23

19. Chopard, B., Frachebourg, L., Droz, M.: Multiparticle lattice gas automata for reaction diffusion systems. Int. J. Mod. Phys. C **05**(01), 47–63 (1994)

20. Medvedev, Yu.: Automata noise in diffusion cellular-automata models. Bull. Nov. Comp. Center Comp. Sci. **30**, 43–52 (2010)

21. Bandman, O.: The concept of invariants in reaction-diffusion cellular-automata. Bull. Nov. Comp. Center Comp. Sci. **33**, 23–34 (2012)

22. Kortlüke, O.: A general cellular automaton model for surface reactions. J. Phys. A Math. Gen. **31**(46), 9185–9197 (1998)

23. Mai, J., von Niessen, W.: Diffusion and reaction in multicomponent systems via cellular-automaton modeling: $A + B_2$. J. Chem. Phys. **98**(3), 2032–2037 (1993)

24. Rice, J.A.: Mathematical Statistics and Data Analysis, 3rd edn. Thomson Brooks/Cole, USA (2006)

25. Sabelfeld, K.K., Kireeva, A.E.: A meshless random walk on parallelepipeds algorithm for solving transient anisotropic diffusion-recombination equations and applications to cathodoluminescence imaging. Submitted to Numerische Mathematik (2018)

26. MVS-10P cluster, JSCC RAS. http://www.jscc.ru. Accessed 22 May 2019

Pseudorandom Number Generator Based on Totalistic Cellular Automaton

Miroslaw Szaban$^{(\boxtimes)}$

Institute of Computer Science, Siedlce University of Natural Sciences
and Humanities, Siedlce, Poland
mszaban@uph.edu.pl

Abstract. In this paper, is considered a problem of selection rules for
one-dimensional (1D) totalistic cellular automaton (TCA), which is used
for generation of pseudorandom sequences which could be useful in cryptography. The quality of pseudorandom bit sequences generated by TCA-
based pseudorandom number generator (PRNG) depends on appropriately selected totalistic rules assigned to CA cells. There is presented a
methodology of selecting TCA rules, starting from initial selection based
on application Entropy of bit streams generated by the TCA. Next, the
selected rules were examined with the use of the NIST SP 800-22rev1a
tests and the Diehard set of Marsaglia tests. In the paper was analyzed,
the uniform TCA with totalistic rules with neighborhood radius equal
to 1, 2, 3, and 4. During the studies, selected sets of TCA are presented
as a new set of CA rules, which can be used as quite cryptographically
strong TCA-based PRNG, supplying a new huge space of keys.

Keywords: Cellular automaton · Pseudorandom number generator ·
Totalistic rules · Cryptography

1 Introduction

Development of digital techniques and their expansion can be observed today in
almost every area of human activity. Large quantities of digital data are created
every minute, so the need for securing safety and privacy of digital information
stored or transmitted over global networks is growing. Cryptographic techniques
are among others used to provide information security, being essential components of any secure communication tools. Nowadays, two core cryptography systems are used: secret and public-key systems. An extensive overview of currently
known or emerging cryptography techniques used in both type of systems can
be found, e.g., in [12]. One of such a promising for cryptography technique is the
application of CA.

Intended for public-key cryptosystems, CA was proposed by Guan [2] and
Kari [6]. Such systems require two types of keys: one key for encryption and
the other one for decryption. One is held in private, the other rendered public.
The main concern of this paper are cryptosystems with a secret key, also called

© Springer Nature Switzerland AG 2019
V. Malyshkin (Ed.): PaCT 2019, LNCS 11657, pp. 360–370, 2019.
https://doi.org/10.1007/978-3-030-25636-4_28

the symmetric key cryptography systems. In such systems, the encryption and the decryption key are the same. The encryption process is based, in particular, on the generation of high-quality pseudorandom bit sequences, and CA can be effectively used for this purpose. CA for symmetric cryptography was first studied by Wolfram [20], who proposed 1D CA-based PRNG with rule 30, and later by Habutsu et al. [3], Hortensius et al. [4] and Nandi et al. [10], who proposed rules 90 and 150. After that, this subject was studied by Tomassini et al. [16], [18], where the set of rules was enlarged to rules: 90, 105, 150, 165. Afterward, in the paper [13] authors presented a new larger set of rules {86, 90, 101, 105, 150, 153, 165, 1436194405}, discovered with use of evolutionary technique called cellular programming (CP) [14]. This set of rules consists of rules with the neighbourhood of radius $r = 1$ and $r = 2$ (last rule in the set). Correspondingly, this set gives similar results in the sense of passed tests like entropy test and also FIPS 140-2 (standard tests for basic analysis of the PRNG's quality), but offered larger space of keys (different bit sequences) than previous proposals. Lately, in the paper [15] were presented techniques of appropriate selection of rules for one-dimensional (1D) cellular automata (CA), which is used for generation PRNG, based on a cryptographic criterion known as a balance. The paper [15] present a new set of CA rules with neighborhood radius $r = 2$. For a selected set of CA rules, the statistical testing approach was applied. As a result, the whole general set and each subset of these rules can be used in CA-based PRNG and provide cryptographically strong bit sequences. Similar statistical testing approach for analysing CA usefulness in cryptography and as PRNG used author in papers: [1,5,7], etc.

Cryptographic techniques and ciphers require secure keys, being high pseudo-random or almost random sequences. Many generators of such keys are known, but they did not supply demanded today quantities of applied keys. CA are known to be a powerful tool for supplying creation of such amounts of number sequences. In literature, for this purpose are applied an elementary CA [21], but also totalistic CA [19] seems to be promising and able to enlarge set of existing tools for generating PRNS.

The paper is organized as follows. The next section presents an idea of an encryption process based on Vernam cipher. Section 3 outlines the concept of 1D Totalistic CA and its relation with CA-based symmetric cryptography. In Sect. 4, a construction of TCA-based PRNG is described. Section 5 describes the sets of quality tests for examining obtained sequences of bits. Section 6 presents the results of selecting proper totalistic rules, testing these rules, and analysis of their cryptographical quality. The last section concludes the paper.

2   Symmetric Key Cryptography and Vernam Cipher

The main idea of cryptography using a symmetric key is that both sides of the cryptographic process apply the same key to encrypt and decrypt the message. The key is secret and most secure because only two persons can use it while other people can only know the encrypted message, which is too challenging to encrypt without knowing the key. In this study, we continue Vernam's approach to cryptography with the secret key.

Let P be a plain-text message consisting of m bits $(p_1p_2...p_m)$ and $(k_1k_2...k_m)$ a bit stream of a key k. Let c_i be the $i-th$ bit of a cipher-text obtained by applying XOR (exclusive-or) enciphering operation:

$$c_i = p_i XOR k_i. \tag{1}$$

The original bit p_i of a message can be recovered by applying the same operation XOR on c_i (bit of a cipher-text) using the same bit stream key k:

$$p_i = c_i XOR k_i. \tag{2}$$

The enciphering algorithm called Vernam Cipher is known (see, [8,12]) as perfectly safe if the keystream is genuinely unpredictable and used only once. In this paper, we answer the questions: how to select an appropriate set of CA rules providing near pure randomness of key bitstreams, and how to obtain such a key with length large enough to encrypt real-world amounts of data.

3   Totalistic Cellular Automata and Symmetric Cryptography

A cellular automaton (CA) is a discrete, dynamical system consisted of identical cells arranged in a regular grid, in one or more dimensions [21]. In this paper, one-dimensional CA is considered. 1D CA is in the simplest case a collection of two-state elementary cells arranged in a lattice of the length N, and locally interacting in a discrete time t. For each cell i, called a central cell, a neighborhood of a radius r is defined. The neighborhood consists of $n = 2r + 1$ cells, including the cell i. A cyclic boundary condition is applied to a finite size CA, which results is in a circle grid. Initial states of all cells (an initial configuration of a CA) and states of cells are updated synchronously at discrete time steps, according to a local rule defined on a neighborhood. In this paper, finite CA with the totalistic type of CA rule (TCA) [19] is considered. It is assumed that a state q_i^{t+1} of a cell i at the time $t+1$ depends only on states of its neighborhood at the time t, i.e.:

$$q_i^{t+1} = TF_t(\sum_{j=-r}^{r} q_{i+j}^t) = TF_t(q_{i-r}^t + ... + q_{i-1}^t + q_i^t + q_{i+1}^t + ... + q_{i+r}^t), \tag{3}$$

where TF_t is a totalistic transition function called also a totalistic rule, defining the way of updating the cell i. The length L of a totalistic rule and the number of neighboring states for a binary CA is $L = 2^n = 2(r+1)$, where n is a number of cells of a given neighborhood. The number of such rules can be expressed as 2^L. For CA with e.g., $r = 1(r = 3)$ the length of the rule is equal to $L = 4(L = 8)$, while number of such rules is $2^4(2^8)$ and grows very fast with L.

CA can change its state in time with the use of a single rule assigned to all CA cells, and it is called a uniform CA. If two or more different rules are assigned to update cells, CA is called nonuniform CA. Wolfram system [20] was uniform; the other mentioned above systems were non-uniform. In this paper will analyse uniform TCA.

4 A Concept of 1D TCA-Based PRNG

Similarly like in the case of elementary CA [4,10,13,15,16,18,20], let us consider a PRNG based on 1D TCA with a lattice of the length N consisting of cells locally interacting in a discrete time t. A rule (rules) of the TCA controlling cells are described as in equation (3). A corresponding seed of the generator will consist of few elements. The first element is an initial configuration of CA, second is a set of TCA rules, third is an index of a cells (i), which generate bit sequence used for encryption, and last is a number of time steps (T), which correspond to the length of a bit sequence. During CA work, fixed cell i changes its states. The next states of the cell i create the bit sequence. Such proposed construction is TCA-based PRNG. The TCA-based PRNG should generate cryptographically strong bit sequences independently to an initial configuration of CA, a selected cell i and time step T. The set of totalistic rules for managing the CA should be carefully chosen, and a particular assignment of rules to cells should not be conflicting. To satisfy these requirements, key streams generated by selected rules will be put under dedicated for these purpose cryptographic tests, like the Entropy test, the NIST SP 800-22 tests and also a Diehard set of Marsaglia tests.

5 Quality Tests for Number Generators

5.1 The Entropy Test

The entropy E_h is used to specify the statistical quality of each PNS. We used Shannon's equation of even distribution as an entropy function. To calculate a value of the entropy each PNS is divided into subsequences of size h ($h = 4$). Let k be the number of values, which can construct a single element of a sequence (for binary values $k = 2$) and k^h a number of possible states of each sequence of length h (if $h = 4$ than $k^h = 16$). E_h can be calculated in the following way:

$$E_h = -\sum_{j=1}^{k^h} p_{h_j} \log_2 p_{h_j},$$ (4)

where p_{h_j} is a probability of occurrence of a sequence h_j in a PNS.

The entropy achieves its maximum $E_h = h$ when the probabilities of the h_j (possible sequences of the length h) are equal to $\frac{1}{k^h}$.

5.2   NIST SP 800-22

NIST SP 800-22rev1a (dated April 2010) is a Statistical Test Suite for the Validation of Random Number Generators and Pseudo Random Number Generators
for Cryptographic Applications [11]. These tests may be useful as a first step
in determining whether or not a generator is suitable for a particular cryptographic application. However, none set of statistical tests can certify a generator
as appropriate for usage in a particular application, i.e., statistical testing cannot serve as a substitute for cryptanalysis. The NIST SP 800-22 contains 15
hard tests. Additionally, it is recommended that the Spectral Test should only
be used for sequences of lengths 10^6 bits, so each bit sequence verified with NIST
SP 800-22 test was conducted with length 10^6 bits.

The NIST SP 800-22 test utilizes statistic to calculate a P-value that summarizes the strength of the evidence against the null hypothesis. For these tests,
each P-value is the probability that a perfect random number generator would
have produced a sequence less random than tested one, given the kind of non-
randomness assessed by the test. If a P-value for a test is determined to be equal
to 1, then the sequence appears to have perfect randomness. A P-value of zero
indicates that the sequence appears to be entirely non-random. The P-value is
interpreted concerning a specified confidence interval, and then the proportion
of tests passing is calculated.

5.3   Diehard - Marsaglia Battery of Tests

The diehard tests are a battery of statistical tests for measuring the quality of
a random number generator [9]. They were developed by George Marsaglia over
several years and first published in 1995. The Diehard contains 18 tests. For
proper examining of the number sequence, it is recommended that the sequence
lengths should be not shorter than 80,000,000 bits. Most of the tests in Diehard
set return a p-value, which should be uniform on [0,1) if the input file contains
genuinely independent random bits. Those p-values are obtained by $p = F(X)$,
where F is the assumed distribution of the sample random variable X - often
normal. However, that assumed F is just an asymptotic approximation, for which
the fit will be worst in the tails. When a bitstream really "fails big," obtained
p-values will be equal or near to 0 or 1 in many places.

6   Experimental Results

6.1   Selection of Totalistic Rules for Application in TCA-Based
PRNG

The starting point is an analysis of a uniform CA with all totalistic rules
with neighborhood radius $r = 1$ and examination of all these 16 rules (i.e.,
$\{t0, ..., t15\}$), and also all of TCA with $r = 2, 3$, and 4 [17]. These set of TCA rules
consist of $64, 256$ and 1024, respectively. In all experiments, CA size was equal to
100, and CA was working in 4096-time steps (a value suitable for Entropy test),

which examine the distribution of subsequences consisted of 4 elements, in the whole bit sequence. From one CA run was selected one-bit sequence obtained from states of randomly selected CA cell. For each rule test was repeated 100 times with random initial configuration of CA state, and average Entropy of each rule was calculated.

Table 1. Values of entropy for each TCA rule with $r = 1$, obtained from 100 tests with the random initial configuration of CA state.

Rule	Binary rule	Entropy min.	Entropy ave.	Entropy max.
t9	1001	3,992149182	3,994392436	3,99615006
t5	0101	3,989524766	3,993392964	3,995395635
t10	1010	3,990261905	3,992207791	3,995395635
t6	0110	3,893282765	3,950631179	3,98135886
t2	0010	3,530754338	3,547633488	3,579125378
...	...	...	...	...
t0	0000	0	0,002011287	0,011173819

In the Table 1 are presented values of minimal, average and maximal entropy for each rule with $r = 1$ for TCA. We can see that only four of all set of TCA rules are good quality, and obtained entropy values near to maximal equal to 4. The best rules are contained in a sequence $\{t9, t5, t10$ and $t6\}$, while the best rule is t9 with average entropy equal to 3,994392436.

Results for the entropy test performed on the whole set of TCA rules with $r = 2$ presents Table 2. We can see that only 13 rules from the whole set of 64 TCA rules with $r = 2$ are of good quality (ave. entropy $\geq 3, 9$) and reached entropy values near or equal to 4, i.e., maximal. The best rules enclose in set $\{t21, t42, t25, t38, t30, t51, t33, t10,\ t43, t14,\ t35, t41$ and $t26\}$. The best rule in the set is t21 with average entropy equal to 3,9999817 and t42 with average entropy equal to 3,999980847. These rules attain the maximal value of entropy equal to 4.

Examining each TCA rules with $r = 3$ gave similar results. The best observed values of minimal, average and maximal entropy were obtained for rules enclosed in a set $\{t60, t204, t51, t42, t171, t213, t43, t84, t154, t212, t166, t203,$ $t102, t153, t44$ and $t28\}$, this set of 16 rules was selected from 256 existing rules. Selected rules were characterized by average entropy not lower than 3,97. The best rule is t60 has average entropy equal to 3,994149919.

Testing each TCA rules with $r = 4$, resulted with the best observed values of minimal, average and maximal entropy obtained for rules from the set $\{t614, t409, t340, t903, t852, t683, t542, t120, t171, t481, t853, t170, t682, t819, t715,$ $t342, t308, t229, t211, t597, t106, t341, t343, t85, t204, t854, t84, t665, t178, t596,$ $t679, t105$ and $t820\}$, this set of 33 rules was selected from 1024 existing rules. Selected rules were characterized by average entropy not lower than 3,99. The best rule is t614 with average entropy equal to 3,999959717.

Table 2. Values of entropy for the best TCA rules with $r = 2$, obtained from 100 tests with random initial configuration of CA state.

Rule	Binary rule	Entropy min.	Entropy ave.	Entropy max.
t21	010101	3,999977985	3,999981728	4
t42	101010	3,999977985	3,999980847	4
t25	011001	3,992381858	3,993456931	3,994802592
t38	100110	3,987154196	3,990798927	3,995433061
t30	011110	3,985916981	3,989140583	3,991474277
t51	110011	3,980464284	3,989084668	3,995395256
t33	100001	3,958708982	3,984050161	3,994774355
t10	001010	3,962158839	3,970533038	3,978475795
t43	101011	3,941343981	3,969315173	3,980360415
t14	001110	3,922575717	3,93316915	3,953488068
t35	100011	3,900546949	3,924284356	3,951504197
t41	101001	3,897518965	3,913300287	3,929707915
t26	011010	3,896005621	3,911179089	3,921030796
t22	010110	3,864817845	3,881104557	3,90548812
...	...	...	...	...
t24	011000	0	0,001340858	0,011173819

6.2 Testing of Selected Totalistic Rules and Analysis of Their Cryptographical Quality

Entropy test was used for examining all of TCA with $r = 1, 2, 3$, and 4. During these test large set of TCA rules was reduced. The selected sets of TCA rules contain $\{t9, t5, t10 \text{ and } t6\}$, $\{t21, t42, t25, t38, t30, t51, t33, t10, t43, t14, t35, t41 \text{ and } t26\}$, $\{t60, t204, t51, t42, t171, t213, t43, t84, t154, t212, t166, t203, t102, t153, t44 \text{ and } t28\}$, and also $\{t614, t409, t340, t903, t852, t683, t542, t120, t171, t481, t853, t170, t682, t819, t715, t342, t308, t229, t211, t597, t106, t341, t343, t85, t204, t854, t84, t665, t178, t596, t679, t105 \text{ and } t820\}$, for radius of neighbourhood $r = 1, 2, 3$ and 4 respectively. Selected rules are characterized by the best entropy values from all rules in examined set. These selected rules were examined in further stages of analysis of its the usefulness for application as the TCA based PRNG.

Next stage of examining totalistic rules was testing selected rules with the use of NIST SP 800-22rev1a - Statistical Test Suite for the Validation of Random Number Generators and Pseudo Random Number Generators for Cryptographic Applications [11]. Each TCA rule from selected sets was examined by each of 15 tests from NIST SP 800-22rev1a. These tests require the length of single sequence equal to 10^6 bits. Thus a number of examined such bit sequences generated by CA with selected totalistic rules was equal to 100. The parameters of some tests were defaulted, i.e., blocks length for the test: 2. Frequency Test within a Block

Table 3. Comparison of NIST SP 800-22 tests results for the best selected totalistic CA rules. Values are given as percentage.

Test	Rule t170	Rule t171	Rule t340	Rule t683	Rule t852	Rule t853
1. Frequency (Monobit) Test	5%	99%	2%	8%	100%	1%
2. Frequency Test within a Block	100%	100%	97%	99%	79%	90%
3. Runs Test	4%	0%	0%	4%	0%	0%
4. Test for the Longest Run of Ones in a Block	99%	98%	98%	100%	96%	96%
5. Binary Matrix Rank Test	99%	98%	100%	96%	99%	98%
6. Discrete Fourier Transform (Spectral) Test	100%	99%	98%	98%	99%	100%
7. Non-overlapping Template Matching Test	98%	96%	97%	99%	94%	97%
8. Overlapping Template Matching Test	97%	97%	97%	99%	72%	69%
9. Maurer's "Universal Statistical" Test	99%	100%	100%	98%	99%	98%
10. Linear Complexity Test	99%	98%	98%	99%	98%	100%
11. Serial Test	98%	98%	100%	99%	99%	99%
12. Approximate Entropy Test	93%	58%	92%	98%	60%	86%
13. Cumulative Sums (Cusum) Test	6%	99%	4%	10%	100%	2%
14. Random Excursions Test	100%	99%	94%	100%	98%	95%
15. Random Excursions Variant Test	98%	99%	100%	99%	99%	97%

was equal to 128, 7. Non-overlapping Template Matching Test was equal to 9, 8. Overlapping Template Matching Test was equal 9, 10. Linear Complexity Test was equal to 500, 11. Serial Test was equal to 16 and 12. Approximate Entropy Test was equal to 10.

Conducted experiments lead to the strict selection of TCA rules. From quite a broad set of rules selected by the entropy test, after analysis with the use of NIST SP 800-22 tests, only six 1D TCA rules with $r = 4$ passed the tests quite good and much better than other selected rules (see, Table 3). In Table 3 we can see detailed information about the pass rate of these rules for each performed test. The problem with passing occurred only in Test number 3 and Runs Test. Other tests were passed almost in 100% (mostly in 98–99%). The average tests pass rate for rule t170 is equal to 80%, for rule t171 89%, for rule t340 78%, for rule t683 80%, for rule t852 86% and for rule t853 75%. So, we can see that each of these six rules is characterized by similar average pass rate, but for rule t171 it is higher than for others (equal to 89%). Some other rules are characterized by much lower pass rate (lower than 70%), while the most extensive collection of examined rules has pass rate lower than 50% or near to 0%.

After examining the selected rules with the use of NIST SP 800-22rev1a, each TCA rule from sets selected by Entropy tests was examined by each of 19 Diehard tests and Marsaglia statistical tests for measuring the quality of a random number generator [9]. The one assumptions, recommended for these tests was the sequence's lengths being not shorter than $8 * 10^7$ bits. Each parameter of the tests had the default value.

From quite a broad set of rules selected by the entropy test, after analysis with the use of Diehard tests and selection with the use of NIST SP 800-22rev1a tests, only 6 1D TCA rules with $r = 4$ were selected. These selected rules passed the tests quite good, and what is more, definitely better than other rules (see, Table 4). Table 4 presents in details capacity for passing Diehard tests in particular K-S tests for these rules for each test. All presented rules are characterized by similar capacity for passing the tests, however the calculated p-values do accurately determine its capacity.

Table 4. Comparison of Diehard tests for the best selected totalistic CA rules. A number of p-values calculated for each test and capacity for passing K-S tests for presented rules, where: n-d means 'no data' (Diehard did not calculate K-S p-value) and $\sqrt{}$w mean 'weak passing.'

Test	Number of p-values	K-S Test					
		Rule t170	Rule t171	Rule t340	Rule t683	Rule t852	Rule t853
1. Birthday Spacings Test	9	√	√	√	√	√	√
2. Overlapping 5-permutation	2	n-d	n-d	n-d	n-d	n-d	n-d
3. Binary Rank Test for 31 × 31 Matrices	1	n-d	n-d	n-d	n-d	n-d	n-d
4. Binary Rank Test for 32 × 32 Matrices	1	n-d	n-d	n-d	n-d	n-d	n-d
5. Binary Rank Test for 6 × 8 Matrices	25	√	√	√w	√	√w	√
6. Monkey Tests 20-bit (Bitstream Test)	20	n-d	n-d	n-d	n-d	n-d	n-d
7. Monkey Test OPSO	23	n-d	n-d	n-d	n-d	n-d	n-d
8. Test OQSO	28	n-d	n-d	n-d	n-d	n-d	n-d
9. Test DNA	31	n-d	n-d	n-d	n-d	n-d	n-d
10. Count 1's in Stream of Bytes Test	2	n-d	n-d	n-d	n-d	n-d	n-d
11. Count 1's in Specific Bytes Test	25	n-d	n-d	n-d	n-d	n-d	n-d
12. Parking Lot Test	10	√	√	√	√	√	√
13. Minimum Distance Test	20	√	√	√	√	√	√
14. 3D spheres Test	20	√	√	√	√	√	√
15. Sqeeze Test	1	√	√w	√	√	√	√
16. Overlapping Sums Test	1	√	√	√	√	√	√
17. Runs Test	4	n-d	n-d	n-d	n-d	n-d	n-d
18. Craps Test	5	n-d	n-d	n-d	n-d	n-d	n-d

Conducted studies show that selected set of totalistic rules $\{t170, t171, t340,$ $t683, t852$ and $t853\}$ could be applied in cellular automaton for constructing PRNG based on TCA. CA with these rules generates bit sequences, which passed NIST SP 800-22rev1a - Statistical Test Suite for the Validation of Random Number Generators and Pseudo Random Number Generators for Cryptographic Applications and also Diehard Marsaglia set of tests for measuring the quality of a random number generator. TCA with these rules could be indeed named Pseudo Random Number Generator characterized by good cryptographic quality.

7   Conclusions an Future Works

In this paper was presented a problem of generation of a high cryptographic quality pseudorandom bit sequences, useful in cryptography as cryptographic keys. The pseudorandom number generator based on one-dimensional totalistic cellular automaton was proposed to fulfil this requirement. The quality of pseudorandom bit sequences generated by TCA-based pseudorandom number generator depends on applied totalistic rules assigned to CA cells. In this paper the one-dimensional totalistic rules with neighborhood radius equal to 1, 2, 3 and 4 were analyzed.

To select appropriate TCA rules, different statistical tests were performed. The first selection was conducted with the use of Entropy test, which reduced the broad set of TCA rules to a smaller subset. This subset was further examined with the use of two sets of tests: NIST SP 800-22rev1a and Diehard set of Marsaglia tests. As a result was obtained set of six totalistic rules $\{t170, t171, t340, t683, t852$ and $t853\}$ with neighborhood radius equal to 4, other considered rules failed the tests. Applied sets of tests determined TCA with selected rules as a generator of pseudorandom numbers (PRNG based on TCA) and confirmed the quite good quality of such generators.

In the future work is planned analysis of one-dimensional nonuniform TCA with neighborhood radius equal to 1, 2, 3 and 4. The n-element sets of totalistic rules, which applied in CA probably could be a good quality PRNG based on TCA (better than in the case of uniform TCA), will be examined. Due to the huge space of such subsets of rules, probably will be necessary to use for selection some Nature inspired algorithm.

References

1. Formenti, E., Imai, K., Martin, B., Yunès, J.-B.: Advances on Random sequence generation by uniform cellular automata. Computing with New Resources - Essays Dedicated to J. Gruska on the Occasion of His 80th Birthday, pp. 56–70 (2014)
2. Guan, P.: Cellular automaton public-key cryptosystem. Complex Syst. 1, 51–56 (1987)
3. Habutsu, T., Nishio, Y., Sasase, I., Mori, S.: A secret key cryptosystem by iterating a chaotic map. In: Davies, D.W. (ed.) EUROCRYPT 1991. LNCS, vol. 547, pp. 127–140. Springer, Heidelberg (1991). https://doi.org/10.1007/3-540-46416-6_11

4. Hortensius, R.D., McLeod, R.D., Card, H.C.: Parallel random number generation for VLSI systems using cellular automata. IEEE Trans. Comput. **38**, 1466–1473 (1989)
5. Hosseini, S.M., Karimi, H., Jahan, M.V.: Generating pseudo-random numbers by combining two systems with complex behaviors. J. Inform. Secur. Appl. **19**(2), 149–162 (2014)
6. Kari, J.: Cryptosystems based on reversible cellular automata (1992)
7. Leporati, A., Mariot, L.: Cryptographic properties of bipermutive cellular automata rules. J. Cell. Automata **9**(5–6), 437–475 (2014)
8. Menezes, A., van Oorschot, P., Vanstone, S.: Handbook of Applied Cryptography. CRC Press, Boca Raton (1996)
9. Marsaglia, G.: The Marsaglia Random Number CDROM including the Diehard Battery of Tests of Randomness, Florida State University (1995)
10. Nandi, S., Kar, B.K., Chaudhuri, P.P.: Theory and applications of cellular automata in cryptography. IEEE Trans. Comput. **43**, 1346–1357 (1994)
11. National Institute of Standards and Technology (NIST), Special Publication 800–22 (2010), A Statistical Test Suite for Random and Pseudorandom Number Generators for Cryptographic Applications. http://csrc.nist.gov/publications/nistpubs/800-22-rev1a/SP800-22rev1a.pdf
12. Schneier, B.: Applied Cryptography. Wiley, New York (1996)
13. Seredynski, F., Bouvry, P., Zomaya, A.: Cellular automata computation and secret key cryptography. Parallel Comput. **30**, 753–766 (2004)
14. Sipper, M., Tomassini, M.: Generating parallel random number generators by cellular programming. Int. J. Mod. Phys. C **7**(2), 181–190 (1996)
15. Szaban, M., Seredynski, F.: Designing conflict free cellular automata-based PRNG. J. Cell. Automata **13**(3), 229–246 (2018)
16. Tomassini, M., Perrenoud, M.: Stream cyphers with one- and two-dimensional cellular automata. In: Schoenauer, M., Deb, K., Rudolph, G., Yao, X., Lutton, E., Merelo, J.J., Schwefel, H.-P. (eds.) PPSN 2000. LNCS, vol. 1917, pp. 722–731. Springer, Heidelberg (2000). https://doi.org/10.1007/3-540-45356-3_71
17. Sienkiewicz, M.: Project, implementation and analysis of pseudorandom number generator based on one dimensional totalistic cellular automata. Master thesis (2017). (in Polish)
18. Tomassini, M., Sipper, M.: On the generation of high-quality random numbers by two-dimensional cellular automata. IEEE Trans. Comput. **49**(10), 1140–1151 (2000)
19. Wolfram, S.: Statistical mechanics of cellular automata. Rev. Mod. Phys. **55**, 601–644 (1983)
20. Wolfram, S.: Cryptography with cellular automata. In: Williams, H.C. (ed.) CRYPTO 1985. LNCS, vol. 218, pp. 429–432. Springer, Heidelberg (1986). https://doi.org/10.1007/3-540-39799-X_32
21. Wolfram, S.: A New Kind of Science. Wolfram Media (2002)

Distributed Algorithms

An Adaptive Bully Algorithm for Leader Elections in Distributed Systems

Monir Abdullah[1,2(✉)], Ibrahim Al-Kohali[2], and Mohamed Othman[3,4(✉)]

[1] Computer Science Department, University of Bisha, Bisha, Saudi Arabia
mkaid@ub.edu.sa
[2] Information Technology Department, Thamar University, Dhamar, Yemen
legend22013@hotmail.com
[3] Laboratory of Computational Science and Mathematical Physics,
Institute for Mathematical Research, Universiti Putra Malaysia,
UPM, 43400 Serdang, Malaysia
[4] Department of Communication Technology and Network,
Universiti Putra Malaysia, UPM, 43400 Serdang, Malaysia
mothman@upm.edu.my, mothman@ieee.org

Abstract. Leader election is a classical problem in distributed system applications. There are many leader election algorithms, but we focus here on Bully Algorithm (BA). The main drawback of BA algorithm is the high number of messages passing. In BA algorithm, the message passing has order O (n^2) that increases heavy traffic on the network. In this paper, an Adaptive BA (ABA) is proposed to reduce the number of messages and make the leader election operation more flexible and safer. The proposed algorithm is based on the Highest Process Identification (HPI) and the Next HPI (NHPI) to facilitate the leader election operation. Moreover, the repetition of the leader election is stopped when the candidate coordinator fails. Our analytical equations show that the ABA algorithm is more efficient rather than BA algorithm, in both, the number of message passing and the latency, and the message passing complexity decreased to O(n).

Keywords: Bully algorithm · Election system · Message passing

1 Introduction

Leader election is considered as an important problem, classical and fundamental problem which happens in distributed systems [1]. Leader election is to select one process or node in the system to become the new coordinator after the previous coordinator fail. The purpose of the leader election is to complete the same job as the ex-coordinator and to avoid any delay in tasks execution. Failures happen because of the occurrence of failures in the software, or hardware or maybe maintenance. Leader election operation occurred when there was no response from the coordinator, thus we were encouraged to start leader election. There are several algorithms had been introduced for electing coordinator process that based

© Springer Nature Switzerland AG 2019
V. Malyshkin (Ed.): PaCT 2019, LNCS 11657, pp. 373–384, 2019.
https://doi.org/10.1007/978-3-030-25636-4_29

on two basic algorithms, i.e. BA algorithm [2] and Token Ring algorithm [3]. In the coordinator election, our objective is to select a coordinator process among various processes that reside in a distributed environment. In this research, we are specifically focusing on BA algorithm. BA algorithm is an important algorithm used in leader election operation which is considered more popular [4]. Not only this, but it is recently used and implemented in Big Data and NoSQL [5] and IoT [6]. There is a plethora of research on BA algorithm and that helped in renewing related studies in this study [2,4,7,9–14]. The main drawback of BA algorithm is the high number of message passing. In this method, the message passing has order O (n^2) that increases heavy traffic on the network. Our proposed adaptive algorithm successfully reduced the number of message passing to $O(n)$. The rest of the paper is organized as follows. Section 2 reviews the related works. In Sect. 3, the original BA algorithm is presented. Section 4 presents the ABA algorithm. The experimental results and discussion will be presented in Sect. 5. Finally, Sect. 6 concludes the paper.

2   Related Works

Several coordinator election algorithms have been proposed over the years some of the main election algorithms are BA algorithm, Ring algorithm. Garcia-Molina [2] proposed a BA algorithm in which they introduce an election mechanism for the selection of the coordinator. While undertaking this procedure the number of messages increased, i.e. the identification of the failed node, then starting an election procedure and the process that having the highest identification process number will be selected as a coordinator. After selecting a coordinator we make an announcement of the selection of new coordinator among various processes in the network. This whole procedure requires a number of messages is to be exchanged which increases the traffic in the network. The researchers discuss the shortcoming of synchronous BA algorithm and propose a modified version. They maintain that their modified algorithm is more efficient than the traditional BA algorithm because it decreases the number of passing messages, and it has fewer stages [9]. Some researchers added an additional feature to the original algorithm [10]. This method uses an assistant as a leader when ex-leader fails. Therefore, there is no need to stop the execution of tasks when a leader crashes. The performance increases when the numbers of node increase. The modified bully election proposes a linear time algorithm for leader election using heap structure that deals with the leader election algorithm for a set of connected processes like a tree network [11]. The researchers discuss the shortcomings of three algorithms of the original and modified BA algorithms. They propose the same traditional BA algorithm but using a new concept called election commission, with the addition of Failure Detector (FD) and a Helper processes (H) to have a unique election with the Election Commission (EC). This method is more efficient and decreases the number of passing messages [12]. A new method is based on electing a leader and an alternative is proposed [16]. In this method, if the leader fails, the alternative takes care of the leader's responsibilities. This way is more effective, messages will be less complexity in the fewer stages. The researchers proposed a new method that uses fault tolerant mechanisms to improve the BA and Ring algorithms [13]. They present a new algorithm called a heap tree algorithm, based on the max-heap data structure. Their results show a fewer number of passing messages. Furthermore, a new algorithm is proposed in which a new leader is elected immediately after the leader fails. It depends on a process status table which contains the number of each process and its status in the current system [14]. The researchers present a safety strengthened leader election protocol with an unreliable failure detector. By analysis, it appears as more efficient in safety and liveness properties in asynchronous distributed systems [15]. A new method which uses a flag that works to reduce the number of passing messages when a failure discovered by more than one process is presented [8]. The results show a relative success in decreasing the number of passing messages and the number of steps. In [7], the researchers proposed a new method reduced passing messages between the coordinator and processes. This mean, when a process starts sending a request to the coordinator, it stores them in a list. Every period the coordinator sends messages to other processes that it has the higher id number. But when the coordinator failed, we will compare the processes between id number of process and id number which sent by the coordinator [7]. The researchers in [17] proposed a new method that uses a proxy server for leader election by performing an analytical simulation. Their results show a decreasing in the number of passing messages and waiting time. A comparative study discussed the concept of four election algorithms, BA [2], Modified Bully Election [9], Improved Bully Election [20], Ring Election [18]. In [19], a slight modification in the classic BA algorithm is proposed which reduces the number of messages that are needed to elect the leader and also proposes new methods of how to react when the dead leader recovers again. The result of the modified BA algorithm is more efficient than the existing leader election algorithms. The researchers in [4] put forward a new method which depends on the distance. They assumed that there exist a node is called centroid. If the distance between a centroid and a node is short, the node has the highest priority and if the distance between the centroid and the node is long, the node has the lowest priority. Recently, BA algorithm is implemented on a specific and low-performance Internet of Thing (IoT) devices [6]. The implementation of the BA algorithm for leader election is achieved in a two-stage process.

3   Bully Algorithm

Based on message generation in the system, a comparative analysis of [2] and our proposed algorithm would be appropriate to determine which algorithm performs better than the others. BA algorithm requires $n - 1$ messages to elect a leader node in the best case, where n is the number of nodes. The best case happens when the node having the next highest id number detects the failure of the leader node and hence announces an election [4].

In the worst case, it requires $O(n^2)$ messages to elect a leader node. The worst case happens when the lowest id node of the system detects the failure

of the leader node. It will send election messages to $n - 1$ nodes having higher
id than itself. Each of the nodes eventually initiates a separate election one
by one. In this algorithm, a previously failed node which was not a leader node
initiates an election after recovery. But if it was a former leader, it just broadcasts
coordinator messages to other nodes to announce itself as the new leader. Hence,
it requires O(n^2) messages to elect a leader node in the worst case and $n - 1$
messages in the best case. The BA algorithm steps are as follows:

1. The process (Pd) that discovers a failure sends a message to all processes in
   the system. The message contains the id of a process (Pd).
2. When the process (P_i) receives the message, it starts comparing the received
   id with its id.
3. If the id of process (Pd) is lower than the id of process (P_i), Then process
   (P_i) returns a message: "Ok" to process (Pd).
4. the process (Pd) continues steps 1, 2, 3 even coordinator selected.
5. If process (Pd) does not receive a message: "Ok" from the other processes,
   and then it will be chosen as a coordinator (Fig. 1).

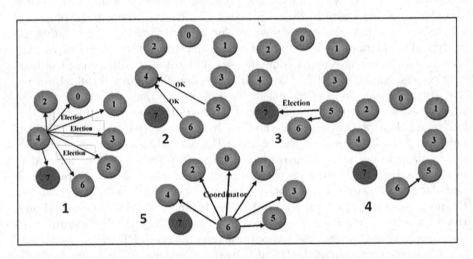

Fig. 1. Bully leader election algorithm.

The drawback of BA algorithm is that if the process that discovers the failure
has a lower Id, this leads to the increase of the number of messages in the election
operation. In this method, the message passing has order O (n^2) that increases
heavy traffic on the network.

4   Adaptive Bully Algorithm

In this section, our proposed ABA algorithm is presented. Firstly, we will explain
the four important variables:

1. The Election Variable (EV): is a variable that stores the node id of the coordinator.
2. Node ID: is a variable that stores the id number of the process itself. It cannot be modified.
3. The Highest Process Identification (HPI) and the Next HPI $(NHPI)$: are variables which store the highest two numbers during election operation.

To implement our algorithm, we adapt a new structure for every node in the system which contains the above four variables as shown in Fig. 2:

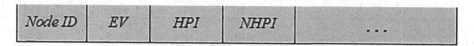

Fig. 2. ABA algorithm node structure.

4.1 Adaptive Bully Election Algorithm

When a process (P_i) requests any task from the coordinator and it does not receive any response within time $(T1)$, this signifies the coordinator fails. This action is called: failure check. Failure Check "is a procedure that is immediately executed whenever any process makes a request to the coordinator. This procedure will detect a failure if it occurs". The failure check is the first step in any election operation. Afterwards, the election operation starts. Now, process Pd sends "Start Election" message to all the processes in the system: The message contains the id of the process that discovered the failure. Time $T2$ starts when this message is sent. During this time, the process Pd receives messages from the other processes. We have two cases:

1. If a process P_i does not receive a response within the specified time, it sends a message to all the processes in the system: "I'm Coordinator".
2. If a process P_i receives a response within the specified time, then the main operation, which stores the HPI and $NHPI$ starts.

When time $(T2)$ finishes, process P_i sends a message to the winning process containing the highest NID: (Highest Value) and: "Tell everyone you are the coordinator". Time $(T3)$ begins when process (P) receives the message. The winning process returns a message: "Ok" to process (P). If process (P_i) does not receive the message: "Ok" within time $(T3)$, this means the process fails. Hence, process (P) sends to the second winning process, which has the second highest ID, a message contains $NHPI$ and: "Tell everyone you are the coordinator". Time $(T4)$ begins when process (P_i) receives the message: "Ok". If process (P_i) does not receive the message "Ok" within time $(T4)$, this means the process fails. The process (P_i) sends a message to all the processes in the system: "I'm Coordinator" as shown in Fig. 3.

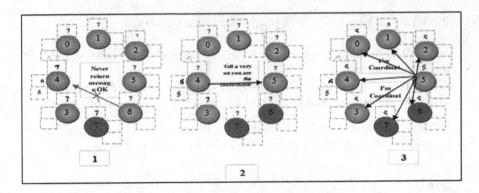

Fig. 3. Leader election operation in case of a failure.

When a process receives the message: "I'm the Coordinator", this signifies the end of the leader election operation, and the receiving process updates the value of *EV* which is attached to the message received. The ABA algorithm is shown in Fig. 4.

Adaptive Bully Election Algorithm

Begin

  1. IF P_i does not receive a response within *T1* from the coordinator then P_i starts the leader election.

  1.1 P_i sends to all nodes (processes) to compare its *ID* with their *IDs*.

    if *[P(d)> P(q)]*, Do not send *NID*.

  1.2 If *P (d) < P (q)*, return message of *NID* to this process *p (d)*.

  1.3 When Process *P(d)* receives messages:

    • Compare received *NIDs*.

    • Store the highest two *NIDs* in two variable (*HPI, NHPI*).

  1.4 *p (d)* sends to the winning process *p (q)*: "Tell everyone you are the coordinator".

  1.5 The winning process *p (q)* returns message to process *p (d)*: "Ok".

  1.6 The winning process *p (q)* tells every node "I'm the Coordinator".

  2. IF *p (d)* does not receive the "Ok" message from *p (q)* within *T2*, this signifies a new failure and for that we must follow these steps:

  2.1 *P (d)* sends a message to the next process which has the second highest *NID* in the table. *p (q)* sends a message: " Tell everyone you are the coordinator ".

  2.2 *p (q)* returns message to process *p (d)*: "Ok".

  2.3 The new winning process *p (q)* sends to every node: "I'm the Coordinator"

  3. If more than one process discovers the failure, every process will repeat steps 1.1 to 1.6.

End

Fig. 4. Adaptive bully election algorithm.

Before ending the election, there are important points that should be tackled. These points relate to what happens to the other processes when they receive the messages: "Start Election" and "I'm the Coordinator". When a process receives the message: "Start Election", it starts comparing the EV and the received ID (NID):

1. If Node ID is 0 or less than the EV, then do not return a message.
2. If Node ID is higher than the EV, then update the value of the EV and return a message to the sender which contains the value of (NID).

4.2 Notations and Definitions

Before discussing the cost model and its related equations, it is necessary to clarify the notations and the definitions used throughout this paper as shown in Table 1.

Table 1. Notations and definitions

Notation	Definition
n	number of processes
P_d	process that discover the failure
P_w	wining process
id	process identification
EV	election variable
HPI	highest identification
$NHPI$	next highest identification
N_{MP}	number of message passing
P_{HPI}	the process that has the highest priority identification
P_{NHPI}	process that has the next highest priority identification
l	constant latency
L	latency cost

4.3 Cost Model

HPI and NHPI Variables. For the best case, the number of messages passing that we need to complete the election operation in our proposed algorithm is calculated by:

$$N_{MP} = (n - 1) * 2 \tag{1}$$

where n is the number of processes that discovers the failure. Where the process that discovers the failure has a higher (id) number.

For the worse case, when the process that discovers failure has not the highest (id) number and there is more than one process discover the failure. Here, we will have two equations as follows:

When a process Pd discovers a failure, then the leader election starts:

1. Process Pd sends its id to all processes to compare it with their ids. If $Pd > Pq$, do not send your id. It needs $n - 1$ operations.
2. If $Pd < Pq$, then return a message of your id. It needs $n - Pd$.
3. When the process Pd receives the messages, the following steps take place:
   - Compare the received ids.
   - Store the highest two ids in two variables ($HPI, NHPI$).
4. Process Pd sends a message to the winning process $P - w$, which has the highest id, telling it that it is the coordinator.
5. Process Pw sends a message: "Ok" back to process Pq. It needs only 2 operations.
6. The winning process P_w sends to everyone: "I'm Coordinator".

Based on steps (1–6), N_{MP} will be calculated by Eq. (2):

$$N_{MP} = (n - 1) + [n - Pd] + 2 + (n - 1) \tag{2}$$

Equation (2) used when the election starts and there is no problem in the candidate coordinator.

However, when there is no response from Pd within ($T2$):

1. Process P_{NHPI} sends a message to process Pd that has the next highest priority id ($NHPI$) telling it that it is the coordinator now.
2. Process Pq sends a message: "Ok" back to process P_{NHPI}.
3. The winning process P_{NHPI} sends to everyone: "I'm the Coordinator". It needs $n - 1$ operations.

Based on (1–3), Eq. (3) will be used:

$$N_{MP} = 2 + (n - 1) \tag{3}$$

Latency. Another parameter used to compare our method is the latency (L). Latency is the time of sending a message from a source to the destination. However, the latency calculation in distributed system is difficult because of the different distances between devices. For this we assume the latency as stated in [21]. Equation (4) will be used to calculate the latency when using our algorithm:

$$L = [N_{MP} * l) \tag{4}$$

where N_{MP} is the number of message passing that calculated by Eqs. (1), (2) and (3) and l is a constant number (200 µs [4]).

The adaptive BA algorithm decreases the number of massages passing and latency. Four variables (VE, NID, HPI, NHPI) successfully decreased message passing complexity from O(n^2) to O(n). We can say when two processes discover failure, the election process is more flexible and safer.

5 Experimental Results and Discussions

In order to compare the performance of our algorithm with the other algorithms, we execute them in five test cases where the systems comprised 5, 10, 15, 20, and 25 nodes, respectively. We simulate our proposed algorithm using Java language on NetBeans editor. We used mesh topology to evaluate the cost model. Firstly, we will use Eqs. (1) and (2) mentioned above. We use Eq. (1) when the number of processes is equal to n. We assumed that the process $n - 1$ discovered the failure, which means that there is no process higher than it. Secondly, we use Eq. (2) when there are processes higher than the process that discovered the failure. The number of messages and latency is presented in Table 2.

Table 2. Number of passing messages and latency of the ABA algorithm.

No. of processes	Eqs. (1), (2)		Eq. (3)	
	Latency (μs)	Number of messages	Latency	Number of messages
5	1600	8	1200	6
10	5200	26	2200	11
15	8200	41	3200	16
20	11200	56	4200	21
25	14200	71	5200	26

As shown in Table 2, we observed that Eq. (3) produces better results compared with the results of Eqs. (1) and (2). In addition, when we compare our ABA with the BA algorithm [2] and Modified BA algorithm [8], it produces better results. The three algorithms are compared based on Messages passing and the results are shown in Table 3.

Table 3. Number of messages of the three algorithms.

No. of processes	BA	MBA	ABA	
			Eqs. (1), (2)	Eq. (3)
5	8	13	8	6
10	69	28	26	11
15	209	43	41	16
20	424	58	56	21
25	804	73	71	26

As shown in Table 3, it can be said that our method is better than Bully algorithm [2] and modified Bully algorithm [8] when there is no failure during the algorithm execution. That is because the number of passing messages in our method is less as clearly shown in Fig. 5.

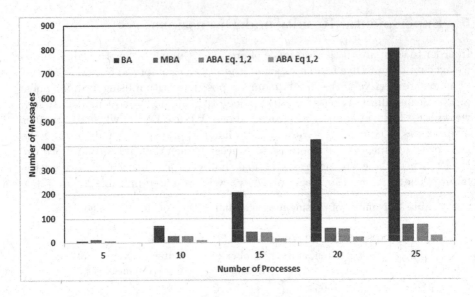

Fig. 5. Comparison between three algorithms.

As clearly shown in Table 3 and Fig. 5, it can be observed that our method is better than original Bully algorithm [2] and modified Bully algorithm [8] when repeating the leader election operation which occurs when the candidate coordinator fails too.

Latency. Another parameter compared in our work is latency. As shown in Tables 2 and 3, we created Table 4 and Fig. 6. Which contains the latency of the three algorithms.

Table 4. Latency (μs) of the three algorithms.

No. of processes	BA	MBA	ABA	
			Eqs. (1), (2)	Eq. (3)
5	1600	2600	1600	1200
10	13800	5800	5200	2200
15	41800	8600	8200	3200
20	84800	11600	11200	4200
25	160800	14600	14200	5200

As shown in Table 4, it can be observed that our method has a higher speed than the original Bully algorithm and modified Bully algorithm. When there is no failure during the algorithm execution it is safer. Overall, our experimental result shows that in the proposed algorithm, the number of messages and latency are very less as compared to the previous algorithms.

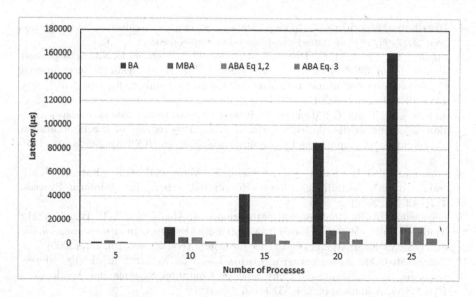

Fig. 6. Latency (μs) of the three algorithms.

6   Conclusion

In this paper, we successfully proposed ABA algorithm. Our ABA is better and more effective than BA algorithm and modified BA algorithm. It decreased the numbers of passing messages. Moreover, our ABA algorithm is safe (reliable) if failure for candidate coordinator happened. During the implementation of the algorithm, if errors occur for candidate coordinator, our method leads to stopping the repetition of algorithm implementation when failed in starting. In addition, four variables (VE, NID, HPI, NHPI) successfully decreased message passing complexity from $O(n^2)$ to $O(n)$.

Acknowledgment. The authors would like to thank everyone who provided valuable suggestions and support to improve the content of the paper. This research work is partial financially supported by the Malaysian Ministry of Education under the Fundamental Research Grant Scheme (FRGS/1/2018/STG06/UPM/01/2).

References

1. Coulouris, G., Dollimore, J., Kindberg, T., Blair, G.: Distributed System Concept and Design, 5th edn. Addison Wesley, USA (2011)
2. Garcia-Molina, H.: Elections in a Distributed Computing System. IEEE Trans. Comput. **100**(1), 48–59 (1982)
3. van Steen, M., Tanenbaum, A.S.: Distributed Systems. 3rd edn. CreateSpace Independent Publishing Platform (2017)
4. Murshed, Md.G., Allen, A.R.: Enhanced bully algorithm for leader node election in synchronous distributed systems. J. Comput. **1**(1), 3–23 (2012)

5. Distributed Algorithms in NOSQL Databases. https://highlyscalable.wordpress. com/2012/09/18/distributed-algorithms-in-nosql-databases/
6. Méndez, M., Tinetti, F.G., Duran, A.M., Obon, D.A., Bartolome, N.G.: Distributed algorithms on IoT devices: bully leader election. In: Proceeding of the International Conference on Computational Science and Computational Intelligence (CSCI), pp. 1351–1355, December 2017
7. Chhabra, S., Tyagi, G., Mundra, A., Rakesh, N.: Location based coordinator election algorithm in distributed environment. In: Proceedings of the International Conference on Computer and Computational Sciences (ICCCS), Noida, pp. 183– 188 (2015)
8. Soundarabai, P.B., Sahai, R., Thriveni, J., Venugopal, K.R., Patnaik, L.M.: Improved bully election algorithm for distributed systems. Int. J. Inform. Process. 7(4), 43–54 (2013)
9. Kordafshari, M.S., Gholipour, M., Mosakhani, M., Haghighat, A.T., Dehghan, M.: Modified bully election algorithm in distributed systems. In: Proceedings of the 9th WSEAS International Conference on Computers, Greece, pp. 1–6 (2005)
10. Zargarnataj, M.: New election algorithm based on assistant in distributed systems. In: ACS International Conference on Computer Systems and Applications (AICCSA), Amman, pp. 324–331 (2007)
11. Sepehri, M., Goodarzi, M.: Leader election algorithm using heap structure. In: 12th WSEAS International Conference on Computers, Heraklion, pp. 668–672 (2008)
12. Rahman, M.M., Nahar, A.: Modified bully algorithm using election commission. MASAUM J. Comput. (MJC) 1(3), 439–446 (2009)
13. EffatParvar, M.R., Yazdani, N., EffatParvar, M., Dadlani, A., Khonsari, A.: Improved algorithms for leader election in distributed systems. In: Proceedings of the 2nd International Conference on Computer Engineering and Technology, Chengdu, China (2010)
14. Basu, S.: An efficient approach of election algorithm in distributed systems. Indian J. Comput. Sci. Eng. (IJCSE) 2(1), 16–21 (2011)
15. Park, S.-H.: A stable election protocol based on an unreliable failure detector in distributed systems. In: Proceedings of the 8th International Conference on Information Technology: New Generations, pp. 979–984. IEEE Computer Society (2011)
16. Kordafshari, M M.S., Gholipour, M., Rahmani, A.M., Jahanshahi, M.: A New Approach for Election Algorithm in Distributed System, pp. 70–74 (2009)
17. Mishra, B., Singh, N., Singh, R.: Master-slave group based model for co-ordinator selection, an improvement of bully algorithm. In: Proceedings of the International Conference on Parallel, Distributed and Grid Computing, Solan, India, pp. 457–460 (2014)
18. Garg, D., Suman, N.: Study of assorted election algorithms in distributed operating system. In: Proceedings of the National Conference on Innovative Trends in Computer Science Engineering, pp. 132–134 (2015)
19. Sathesh, B.M.: Optimized bully algorithm. Int. J. Comput. Appl. 121(18), 24–27 (2015)
20. Arghavani, A., Ahmadi, A.E., Haghighat, A.T.: Improved bully election algorithm in distributed systems. In: Proceedings of the 5th International Conference on Information Technology & Multimedia, pp. 14–16 (2011)
21. Fredrickson, G.N., Lynch, N.A.: Electing a leader in asynchronous ring. J. ACM 34(1), 98–115 (1987)

Affinity Replica Selection
in Distributed Systems

W. S. W. Awang[1]([⊠]), M. M. Deris[2], O. F. Rana[3], M. Zarina[1],
and A. N. M. Rose[1]

[1] Faculty Informatics and Computing, University Sultan Zainal Abidin,
Besut Campus, Besut, Terengganu, Malaysia
{suryani, zarina, anm}@unisza.edu.my
[2] Faculty of Computer Science and Information Technology,
Universiti Tun Hussein Onn Malaysia, Batu Pahat, Johor, Malaysia
mmustafa@uthm.edu.my
[3] University of Cardiff, Cardiff, UK
ranaof@cardiff.ac.uk

Abstract. Replication is one of the key techniques used in distributed systems to improve high data availability, data access performance and data reliability. To optimize the maximum benefits from file replication, a systems that includes replicas need a strategy for selecting and accessing suitable replicas. A replica selection strategy determines the available replicas and chooses the most access files. In most of these access frequency based solutions or popularity of files are assuming that files are independent of each other. In contrast, distributed systems such as peer-to-peer file sharing, and mobile database, files may be dependent or correlated to one another. Thus, this paper focused on the combination of popularity and affinity files as the most important parameters in selecting replicas in distributed environments. Herein, a replica selection is proposed focusing on popular files and affinity files. The idea is to improve data availability in distributed data replica selection strategy. A P2P simulator, *PeerSim*, is used to evaluate the performance of the dynamic replica selection strategy. The simulation results provided a proof that the proposed affinity replica selection has contributed towards a new dimension of replica selection strategy that incorporates the affinity and popularity of file replicas in distributed systems.

Keywords: Replica selection · Affinity files · Popularity files · Data availability · Distributed systems · Replication strategy

1 Introduction

Data replication strategies have been widely employed in large-scale data intensive application such as high energy particle physics, climate simulation, genomics, molecular docking, and bioinformatics. The identical copies of data are generated and stored at various distributed sites to improve data access performance and data availability. As the demand for data increases, the centralized replication strategies are liable to a single point of failure and become a bottleneck when dealing with huge amount of data trying to access the same data simultaneously [11, 14–16]. Moreover, if a single

© Springer Nature Switzerland AG 2019
V. Malyshkin (Ed.): PaCT 2019, LNCS 11657, pp. 385–399, 2019.
https://doi.org/10.1007/978-3-030-25636-4_30

data file is only placed at a single server, in case of server crashes or does not respond, this data file becomes unavailable. In contrast, if a replica of the data file is stored on multiple servers, this additional server can provide the data file in case of a server or network failure. Thus, the availability of data can be improved even in the event of natural disasters like Tsunami or earthquakes. Since the similar data can be found at multiple servers, availability of data is assured in case of servers' failure. Additionally, data replication can provide increased fault tolerance, improved scalability, reduced bandwidth consumption and improved response time [12, 13].

When designing replication strategies one of the important parameters taken into consideration is the popularity of a file or popular group of files [15–17]. A file is determined by the most accessing files. Some files may be popular than others and data access pattern may change over time. Most of the popularity files are assuming that files are independent of each other. However, in distributed systems such as peer-to-peer, files may be dependent or correlated to one another. Correlated or affine files refer to the files that are accessed by the same transaction or more than one transaction accessing the same files. For example, a client or a query accessing multiple queries accesses the same data or a set of files accessed by one user is also likely to be accessed together by other users. This set of files has common features that bind or stick them together. Therefore, this paper focused on the notion of affinity as a binding feature in selecting and accessing the best replicas to improve data availability in distributed systems. An Affinity Replica Selection Mechanism (ARSM) is proposed to highlight the importance of affinity relationship to improve file access performance and assist replica selection decisions.

In this paper, two query scenarios were considered. The first query scenario refers to Single Query-Single file case. Whilst the second scenario refers to Single Query-Multi files case. The files in the distributed system were randomly broadcasted. The objective of the proposed model is to minimize access latency and optimize availability by allowing files to be replicated based on their high popularity and strong affinity degree. The rest of this paper is organized as follows: Sect. 2 discusses previous works on replication strategies. Section 3, the proposed ARSM model is presented. Next, we presented the simulation results in Sect. 4. Finally, Sect. 5 concludes our work.

2   Related Work

One of the practical techniques to enhance the efficiency of data sharing in distributed systems is data replication. In addition, load balancing, fault tolerance, reliability, and the quality of service can be improved with the help of data replication strategy [5–7]. When the data are placed at a single data server, that server can be a bottleneck if too many requests need to be served at the same time. Consequently, the whole system slows down. The major features of replication algorithms for distributed systems are the criteria for the selection of suitable objects for replication and selection of suitable sites for hosting new replica. These two important aspects have a direct impact on the performance of the system. If a node decides to replicate all the objects present in its shared directory to other nodes, it will increase the overhead in the network. The replica should be maintained in sites which are close to the source nodes to increase the search performance. The site selection policy of a replication technique decides where the replica should be stored. The number of sites may vary based on the replication scheme being employed. For example, if popular files are not replicated appropriately, overwhelming requests from peers can cause network congestions and slow download speed [14–16].

In addition, a system that includes replicas also requires a mechanism for selecting the right files based on the data access patterns. Choosing and accessing appropriate replicas are very important to optimize the use of distributed resources. Replica selection criteria might include access time as well as the source node that initiate the request, and the number of accesses. Slow network access hinders the efficiency of data transfer regardless of client and server implementation. In the real world, some files may be popular than others and data access pattern may change over time. The popularity of a file is determined by its recent access rate. Therefore, any dynamic replication strategies must keep track of file access histories to decide on when, what and where to replicate. The dynamic replication algorithm proposed by [8–12], [20] determines the popularity of a file by analyzing data access history.

Most of the related works [1–4] have concentrated on replication of a popular file or popular groups of files. However, not enough attention was paid to affinity or dependency among the files. An Affinity Replica Selection Mechanism (ARSM) is proposed in this paper as a new replica selection strategy that combines the popularity and affinity files. ARSM incorporates the popularity and the affinity among files; popularity and affinity are used to replicate a group of files that shows high access frequency and a strong affinity degree.

The notion of affinity in general refers to the close similarity, likeness, relationship or correspondence. However, in this paper, we defined an affinity as the correlated files, and dependency between two or more files. Inspired by the ancient social systems and human behavior, Larbani and Chen [19] explore the concept of affinity further in fuzzy and rough set framework, data mining and other applications. An affinity also means a meeting between friends with the same hobbies, various relationships with people such as friend to friend, parent to offspring, employee to boss and so on. These are some examples in relationship and social behavior of an affinity [19].

Depending on how affinity is defined, it can be used to examine, describe and predict the behavior of access pattern or data similarity in placing replica in distributed organizations. Different measurement systems lead to various affinity degrees and more importantly may lead to the dynamic decision or strategy in replica selection. The affinity replica location policy algorithm proposed by [18] replicates data near the user nodes where the file is accessed most. A file is copied and placed near to the user that generates access traffic the most. The algorithm is similar to the cascading replica placement algorithm discussed in [16].

3 Affinity Replica Selection Mechanism

This section presents a model for replica selection called Affinity Replica Selection Mechanism (ARSM). The ARSM selects popular files and affinity files for replication and calculates sufficient number of copies on the source node. The objective of ARSM is to improve data access performance through minimizing the access time and to ensure data availability in distributed systems.

In this paper, the access time is minimized by replicating the popular and affinity files to the requesting node(s). Likewise, to ensure data availability in the distributed systems, sufficient number of replicas is maintained in the system. The popular and affinity files were the two dominant factors proposed in ARSM. The access frequency determines the popularity of the access files whilst the affinity degree determines the binding feature between two nodes.

3.1   The Affinity

Data affinity in this paper is defined as the similarity between two or more correlated data. The affinity set is a set of any data that creates an affinity between files. Thus, the affinity between sets A and B is the set consisting of the intersection of elements between A and B plus the requested file in the destination node, and is not a null set. The requested file in the destination node is defined as $f_{qid}(B)$ where f is a file and qid refers to the identity of a queried or requested file.

Definition 1: Let $A = \{f_{a1}, f_{a2}, \ldots f_{an}\}$ and $B = \{f_{b1}, f_{b2}, \ldots f_{bn}\}$, f_{jk} is a requested file from the source node j to destination node k. The sets A and B are said to have affinity denoted by aff_{AB}:

$$aff_{AB}^A = \{x \mid x \in (A \cap B + \{f_{qid}(B)\}) \neq \phi\} \tag{3.1}$$

where $f_{qid}(B)$ is the requested file in B.

Definition 2: The affinity degree between A and B with respect to A, aff_{AB}^A, is defined as

$$aff_{AB}^A = \frac{|aff_{AB}| + |f_{qid}(B)|}{|A| + |f_{qid}(B)|} \tag{3.2}$$

where the symbol $|aff_{AB}|$ is the cardinality of affinity set A and B over A including $f_{qid}(B)$ which refers to the requested file in node B.

The value of aff_{AB}^A as shown in Eq. 3.2, expressing the degree of affinity between the dataset A and the affinity sets AB with respect to A.

The affinity function is defined as the cardinality of the affinity dataset between A and B over the cardinality A. Likewise the degree of affinity between B and A with respect to B is defined as the cardinality of the affinity set A and B over B.

Example 1 below shows how the proposed affinity degree is calculated.

Example 1: Let $A = \{f_{11}, f_{12}, f_{13}, f_{14}, f_{15}\}$ and $B = \{f_{21}, f_{22}, f_{23}, f_{13}, f_{14}, f_{15}, f_{26}, f_{27}, f_{28}\}$ and the requested fileId is f_{28}. Therefore the affinity degree over A

$$= \frac{|\{f_{13}, f_{14}, f_{15}\}| + |f_{28}|}{|\{f_{11}, f_{12}, f_{13}, f_{14}, f_{15}\}| + |f_{28}|}$$
$$= 4/6$$
$$= 0.67 \text{ (moderate)}$$

Table 1 shows a categorization of affinity correlation adapted from Dancey and Reidy [20]. The correlation of an affinity degree indicates that not every correlation deserves to investigate and some filtering mechanisms can be adopted to remove those files with weak correlation. In general, the higher the absolute value of affinity correlation coefficient, the stronger the relationship between the two nodes in the P2P network. For example, in Table 1, if the value of the aff_{AB}^A is equal to 0.49 or below, it indicates that the degree of the affinity files is weak and thus can be ignored. In this case, the files has weak affinity and will not be replicated.

Table 1. The affinity degree indicator (Adapted from Dancey and Reidy [20])

Value of the aff_{AB}^A	The degree of the affinity files
$0.9 \leq x < 1.0$	Very strong
$0.7 \leq x < 0.9$	Strong
$0.5 \leq x < 0.7$	Moderate
$0.1 \leq x < 0.5$	Weak
<0.1	Zero

Likewise, if the value of the affinity degree is either moderate, strong or very strong, then the file will be replicated. The explanation is detailed in the next paragraph. The representations of the affinity files are as follows Table 2:

Table 2. Example of affinity degree

A	B	f_{qid}	$(A \cap B) + f_{qid}$	$aff_{AB}^A = \frac{\|aff_{AB}\| + f_{qid}(B)}{\|A\| + f_{qid}(B)}$	Affinity indicator
$\{1, 2, 3, 4\}$	$\{1, 2, 3, 4, 5, 6\}$	6	5	$5/5 = 1.0$	Very strong
$\{1, 2, 3, 9\}$	$\{1, 2, 3, 4, 5, 6, 7, 8\}$	5	4	$4/5 = 0.8$	Strong
$\{1, 2, 3, 4, 7, 9, 10\}$	$\{1, 2, 3, 4, 5\}$	5	5	$5/8 = 0.61$	Moderate
$\{1, 2, 3, 4, 5, 6, 7, 8, 9, 10, 11, 12\}$	$\{1, 13\}$	13	2	$2/13 = 0.15$	Weak
#500	#1000	#20	300	$300/520 = 0.58$	Moderate
#1000	#5000	#50	300	$300/1050 = 0.29$	Weak

\# is the number of files

If the value of is near to 1, we can say that the affinity set between files is very strong whilst if the value of aff_{AB}^A is near to zero, we can say that the degree of affinity set between files is very weak or zero affinity. Through the affinity indicators, we can predict on how strong or high and how weak or low the affinity set between files in the nodes. This means that if the strength of similarity files is high, and if the average frequency of the access number of the file requested is also high, ARSM will choose the file to be replicated. This answers the issue of which file to replicate in replica selection problems. Despite this, if the degree of the affinity set is weak or zero, ARSM will NOT consider the file to be replicated regardless of how high the value of the file access frequency. The decision of replica selection depends on the affinity degree and the average number of access frequency. In the next section, the access frequency as another criteria for replica selection is discussed.

3.2   Access Frequency

ARPM only consider affinity and popular files to replicate (deciding which file to replicate). An access frequency, AF is calculated to represent the importance of access histories in different cycle number. Assume N_t is the cycle number passed, F is the set of files that have been requested and a_f^t indicates the number of accessed files in each cycle. Then AF is adapted from the calculation of AF in (Chang and Chang [17]):

$$Access\,Frequency = AF_{Nt}(f) = \sum \left(a^t f \times 2^{-(N_t - t)} \right), \quad \forall f \in F_t = 1 \qquad (3.3)$$

For example, if an affinity file has been accessed 7 times and 10 times in the first cycle and second cycle, respectively, then $AF\,(f)$ is $(7 \times 2^{-1}) + (10 \times 2^0)$. AF assigns different weights to access files for a different cycle number. The highest or largest AF is chosen as the popular files. Next we compare the average AF per cycle number of the popular files. The average AF is calculated as:

$$Average\,Access\,Frequency = AF_{N_i}^{average}(f) = \sum AF_{N_i}(f)/N_c, \quad \forall f \in F \qquad (3.4)$$

$N_F = |F|$ is the number of different files that have been requested by any nodes. The threshold value of access frequency is considered as the average of access frequencies in the systems. If the access frequency is above or equal to the average access frequency, then we categorize it as "high" or "popular". Likewise, if the access frequency is below than the average frequency, then we categorize it as "low" or "unpopular". Table 3 shows which file to replicate based on the two dominant factors proposed in this paper.

The primary goal of the algorithm is to increase data access performance from the perspective of the clients by dynamically creating replicas for "popular" files. In the real world, some files will be more popular than others and data access patterns may change over time, so any dynamic replication strategy must keep track of file access histories to decide on when, what and where to replicate. The "popularity" of a file is determined by its recent access rate by the clients. Identifying popular files is thus one of the dominant factors of ARSM. In ARSM, popular data files are identified by analyzing the file access histories.

Table 3. Dominant factors which file to replicate

Affinity indicator	#Average access frequency	Replicate	Not replicate
Very strong	High	1	
	Low		0
Strong	High	1	
	Low		0
Moderate	High	1	
	Low		0
Weak	High		0
	Low		0
Zero	High		0
	Low		0

Note: 1 = Yes 0 = No

3.3 Replica Selection Decisions

This section focuses on the decisions in replica selection phase. In this section, the affinity properties from Table 3 has been transformed into Table 4 in *Boolean-valued* data. In *Boolean-valued* data, the dominant factor is holding either a value 0 or 1. In this *Boolean* representation, the aim is to qualify the different importance of linguistic terms of vague terms of affinity factors which include very strong, strong, moderate, weak and zero.

Table 4. Dominant factors which file to replicate in *Boolean* representation

Affinity indicator	#Average access frequency	Replicate	Not replicate
1	1	1	
	0		0
1	1	1	
	0		0
1	1	1	
	0		0
0	1		0
	0		0
0	1		0
	0		0

Definition 3: Let affinity and average access frequency be two dominant factors for replica placement. The replica placement occurs when both dominant factors are equal to 1 respectively.

The *Boolean* representation in Table 4 are used as indicators to decide whether to replicate or not. The replica placement occurs when both dominant factors are equal to 1. Indeed, if the affinity degree is high and the access frequency exceeds the threshold value of the average number of accesses, or if both values are equal to 1, then the decision to replicate is made.

3.4   Access Frequency as Dominant Factor

This section describes two cases considered in this paper in selecting popular data files and calculating the files affinity degree. Case-1: *Single-Query to Single-File*, Case-2: *Single-Query to Multiple*. Based on these two queries, both dominant factors play an important role in influencing the decision of replica placement. Table 5 shows the two cases scenarios between the requestor/source node(s) and the query file(s). During experimentation, the number of cycles and files are increased whilst the number of nodes simulated is up to 10000 nodes.

Table 5. The single query to single file and single query to multiple files scenarios

Cycle Number	Requestor NodeId	FileId	
0	3	28	
1	39	23	} Case-1
2	92	31	
3	67	25	
4	97	15	
5	63	6	
6	42	19	
7	69	25	Case- 2
8	31	3	
9	1	29	
10	97	17	
11	50	21	
12	54	8	
13	32	12	
14	46	12	
15	71	3	
16	25	22	
17	31	6	
18	14, 15, 37	9, 27,33	
19	91	30	
20	28	19	

3.4.1   Case 1: Single-Query to Single-File

In Table 5 during *cycle1*, a *NodeId 39* requests for a *FileId23*. This is a case of a *Single-Query to Single-File* request whereby only one client node is requesting for one file in the systems during a period of time. This refers to the cycle number between *cycle0* to *cycle20*. This is the case of no replication.

3.4.2   Case 2: Single-Query to Multiple-Files

In *cycle4* and *cycle10*, the same *NodeId 97* was requesting two different files, *FileId15* and *FileId17*. This is the case of the same client node requesting two files in the systems during a period of cycles. Tables 6 and 7 show an example of historical records of the *NodeId97* during the first and the second time interval respectively. Assume N_T is the number of time interval passed, F is the set of files that have been requested and a_f^t indicates the number of accesses for file f at time interval t. In the first time interval, $t = 1$, *FileId15* have been requested by *NodeId4* times and 10 times during the second time interval, $t = 2$. Then The *Access Frequency (AF)* for each file can be calculated as:

$$Access\,Frequency = AF_{Nt}(f) = \sum \left(a_f^t \times 2^{-(N_t - t)} \right), \quad \forall f \in F_t = 1$$

Thus for *FileId15*, *Access Frequency*

$$= AF_{N_t}(f) = \left(4 \times 2^{-(1-1)} \right) + \left(10 \times 2^{-(2-1)} \right) = 12$$

Based on Eq. 3.3, number of access frequency for file 15, 17, and 21 were 5, 2.5, and 1, 5 respectively. Therefore, the threshold of the average access frequency in the period of cycle can be calculated as in 3.4. The average threshold is 4.17. Therefore two files with *fileId* 15 and 17 are above the threshold value that are considered as popular files. These files will be selected to be replicated if the affinity degree for these files are moderate, strong, or very strong.

Table 6. An example of access frequency for *Single-Query to Multiple-Files* at time interval $t = 1$

a_f^t	Requestor NodeId	FileId	Number of access frequency
4	97	15	4
10	97	17	10
2	97	21	2

Table 7. An example of access frequency for single query-many files at time interval t = 2

a_f^t	Requestor NodeId	FileId	Number of access frequency
10	97	15	5
5	97	17	2.5
3	97	21	1.5

3.5   Affinity Degree as Dominant Factors

The second dominant factor will be calculated based on the affinity degree between the source node and the destination node. Table 8 shows the *nodeId* and the *fileId* whilst Table 9 shows the discovery layer where the file requested by the source node is found in the destination node. This also refers to the success hit whenever a query file is found in the destination node.

Table 8. An example of *NodeId* and *FileId*

NodeId	FileId
40	23, 6, 34, 36, 17, 30, 15, 29, 19, 22
26	29, 39, 42, 27, 23, 21, 6, 5
39	10, 44, 43, 40, 21, 48
25	10, 44, 43, 40, 18, 3, 6
46	42, 1, 41, 14, 3, 31, 13
27	31, 26, 25, 4, 28, 37
11	6, 43, 38, 24, 19, 23, 7, 32
24	19, 12, 15, 28, 2, 25, 37, 27
97	30, 48, 25, 7, 22, 19
14	23, 17, 36, 34, 40, 29
32	40, 10, 44, 48, 43, 31, 13

Table 9. An example of success hit

Source node	FileId	Destination node
14	15	24, 40
40	1	46
18	1	46
32	21	39
16	23	11, 26
10	3	25
97	17	40
25	21	26, 39
46	21	26, 39
97	15	24, 40
18	21	26, 39

3.5.1   Case-1: Single Query - Single File

In a case of a single query - single file request, only one client node is requesting one file in the system during a period of time. There is no replication and thus affinity degree is not calculated in this case.

3.5.2   Case 2: Single Query - Multi Files

In Sect. 3.2, the definition of affinity and how to calculate the affinity degree has been discussed in detailed. In this section, the affinity degree is calculated based on the formula from 3.1 and 3.2. The affinity degree as the second denominator will be calculated using similar two cases as in Sect. 3.5.

In this case, the same Node is requesting two or more files in a fixed time interval. Prior to this, an average access frequency has been calculated in Sect. 3.5 and the popular files were found. As calculated in Sect. 3.5, only *fileId* 15 and *fileId* 17 are popular whereas *fileId* 21 is below average frequency threshold and therefore is considered as less popular. Next, the affinity degree is calculated between the source node, *NodeId* 97 and the destination node, *NodeId* 40, as shown in Table 9. The affinity degree is calculated as below:

Example 1: Let source/Query node be S_{97} and the destination node be D_{40}. The query file is *fileId* 17.

$$S_{97} = \{30, 48, 25, 7, 22, 19\} \text{ and } D_{40} = \{23, 6, 34, 36, 17, 30, 15, 29, 19, 22\}$$

The affinity is

$$aff_{S_{97}D_{40}}^{S_{97}} = S_{97} \cap D_{40} + \text{Requested File in } D_{40} = \{22, 30, 19, 17\} = 4$$

From equation in 3.2, the affinity degree over S_{97},

$$= \frac{|aff_{S_{97}D_{40}}|}{|S_{97} + f_{qid}(D_{40})|}$$
$$= 4/7$$
$$= 0.57 \text{ (Moderate affinity)}$$

Example 2: Let source node be S_{97} and the destination nodes be D_{24} and D_{40}. The query file is *fileId* 15.

$$S_{97} = \{30, 48, 25, 27, 22, 19\} \text{ and } D_{24} = \{19, 12, 15, 28, 2, 25, 37, 27\}$$

$$aff_{S_{97}D_{24}}^{S_{97}} = S_{97} \cap D_{24} + \text{Requested File in } D_{24} = \{15, 19, 27, 25\} = 4$$

From equation in 3.2, the affinity degree is

$$= 4/7 = 0.57 \text{ (Moderate Affinity)}$$

By calculating the affinity degree of the files between the source nodes and the destination nodes using the proposed affinity formula, the affinity degree in example 1 indicates that the relation is strong. Therefore we can conclude that, *fileId*17 is a popular file and the nodes (the source node and the destination node) has strong relation. Not only *fileId*17 will be replicated but also all the intersection files that represent the affinity data, will be replicated as well to the source node. However, in example 2, the affinity degree calculated indicates "weak affinity". The *fileId*15 will not be replicated since the affinity degree is "low" regardless how popular the File is.

The rationale is that, when a user generates a request for a file, large amount of bandwidth could be consumed to transfer the file from the server to the client. Furthermore, the popular files tend to be accessed more frequently than less popular files in the near future. Therefore to select a popular file in the replica placement strategy is very important. In real world most of the files have affinity with one another. A user searching for one song from "The Beatles", may search for another song from the same music group. A researcher from a university may need more than one related journals or research files from other university. These two examples of searching and accessing files need to be done repeatedly. As a consequence, not only the total access cost is increased but also the total communication cost in accessing the files. However, the increase of both costs can be reduced if related files are copied instead of just one file per request from the client.

Therefore, the idea behind ARSM is to create a set of replicas where affinity and popularity are equally important and very essential criteria in replica selection strategy. Besides, ARSM place the new replicas as close as possible to those clients that frequently request the corresponding files, subject to storage availability. The effectiveness of this ARSM algorithms also depend on the number of accesses threshold value and the proximity threshold value that were used herein to determine the selection of replicas in the distributed systems.

4   Results and Discussion

Figure 1 illustrates the popular files from time interval 1 to time interval 6 whilst the data from time interval 7 to time interval 10. The graph is decreasing towards the end of the intervals. The result indicates that the access frequency that pass the average access frequency threshold were between interval T1 to T6, where from interval T5 onwards, the files queried were less popular. This result illustrates that the files over the time intervals were decreased and the files became less popular. In real scenarios, this reflects that the popularity of files increased in the first dissemination and became less popular after a period of time.

Figure 2 Illustrates the affinity degree calculated based on the files that exceed or equal to the access frequency threshold and the affinity degree that have strong files relatedness. In time interval 4 (T4), there was a slight increase in the number of the replicated files. The replicated files over a period of time in T3 were decreased but gained back in T4 before the pattern is repeated. The graph in Fig. 2 verified that there is a certain access pattern and relatedness of the requested files by the clients in the distributed systems.

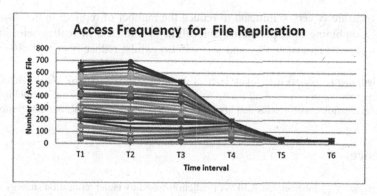

Fig. 1. The relationship between Access Frequency (AF) and Time Interval (T)

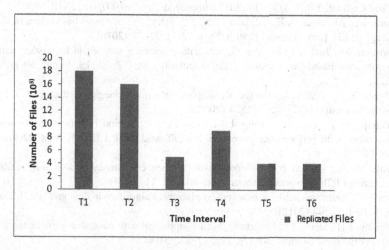

Fig. 2. The relationship between Access Frequency (AF) and Time Interval (T)

5 Conclusion

The demand for the popular and correlated files are high during the first dissemination and then decreased after certain period. Consequently it will gain popularity and correlativity before it decreases hence this pattern will be repeated. In real scenario, in research collaboration for example, a new found technology or research will initially expected to be highly demanded and therefore the number of replicas is increased and copied to the trusted or affine clients. However, this data will decrease over a certain period of time and whenever newer technology is found, the pattern will be repeated.

Generally, other replica selection strategies deal with the quantity of data dissemination. However, ARSM in this thesis deals with the quality over quantity data replication strategy. If we just take popularity as a measure, a system may over replicate. Moreover, in many cases, popularity does not continue. There will be lots of replicas which may not be needed. Therefore, taking affinity into consideration as another measure is very significant to reduce the number of replicas in the distributed systems. Combining both popularity and affinity parameters in replica selection will finally improve data availability and accessibility whilst reduce over replication.

Acknowledgment. We wish to thank internal grant of UNISZA (UniSZA/2017/DPU/72) for financial supporting our work. Also thanks to all team members in reviewing for spelling errors and synchronization consistencies and also for the constructive comments and suggestions.

References

1. Nagarajan, V., Mohamed, M.A.M.: A prediction-based dynamic replication strategy for data intensive applications. J. Comput. Electr. Eng. **57**, 281–293 (2017)
2. Rahmani, A., Azari, L., Daniel, A.H.: A file group data replication algorithm for data grids. J. Grid Comput. **15**(3), 379–393 (2017). https://doi.org/10.1007/s10723-017-9407-1
3. Bsoul, M., Abdallah, A.E., Almakadmeh, K., Tahat, N.: A round-based data replication strategy. IEEE Trans. Parallel Distrib. Syst. **27**(1), 31–39 (2016)
4. Mansouri, M., Javidi, M.M.: An efficient data replication strategy in large-scale data grid environments based on availability and popularity. AUT J. Model. Simul. **50**(1), 39–50 (2018)
5. Rahman, R.M., Alhajj, R., Barker, K.: Replica selection strategies in data grid. J. Parallel Distrib. Comput. **68**(12), 1561–1574 (2008)
6. Nukarapu, D.T., Tang, B., Wang, L., Lu, S.: Data replication in data intensive scientific applications with performance guarantee. IEEE Trans. Parallel Distrib. Syst. **22**(8), 1299–1306 (2011)
7. Meng, C.Z.X.: An ant colony model based replica consistency maintenance strategy in unstructured P2P networks. Comput. Networks **62**, 11 (2014)
8. Fadaie, Z., Rahmani, A.M.: A new replica placement algorithm in data grid. Int. J. Comput. Sci. **9**(2), 491–507 (2012)
9. Abawajy, J.H., Deris, M.M.: Data replication approach with consistency guarantee for data grid. IEEE Trans. Comput. **63**(12), 2975–2987 (2014)
10. Skakowski, K., Sota, R., Król, D., Kitowski, J.: QoS-based storage resources provisioning for grid applications. Future Gener. Comput. Syst. **29**(3), 713–727 (2013)
11. Shorfuzzaman, M., Graham, P., Eskicioglu, R.: QoS aware distributed replica placement in hierarchical data grids. In: Proceedings of the International Conference on Advanced Information Networking and Applications, AINA, 2011, pp. 291–299
12. Jaradat, A., Patel, A., Zakaria, M.N., Amina, M.A.H.: Accessibility algorithm based on site availability to enhance replica selection in a data grid environment. Comput. Sci. Inform. Syst. **10**(1), 105–132 (2013)
13. Hamrouni, T., Slimani, S., Ben Charrada, F.: A data mining correlated patterns-based periodic decentralized replication strategy for data grids. J. Syst. Softw. **110**, 10–27 (2015)
14. Rahmani, A.M., Fadaie, Z., Chronopoulos, A.T.: Data placement using Dewey Encoding in a hierarchical data grid. J. Netw. Comput. Appl. **49**, 88–98 (2015)
15. Mostafa, N., Al Ridhawi, I., Hamza, A.: An intelligent dynamic replica selection model within grid systems. In: 2015 IEEE 8th GCC Conference and Exhibition, GCCCE 2015 (2015)
16. Ranganathan, K., Foster, I.: Identifying dynamic replication strategies for a high-performance data grid. In: Lee, C.A. (ed.) GRID 2001. LNCS, vol. 2242, pp. 75–86. Springer, Heidelberg (2001). https://doi.org/10.1007/3-540-45644-9_8

17. Chang, R.Sh., Chang, H.P., Wang, Y.T.: A dynamic weighted data replication strategy in data grids. In: IEEE/ACS International Conference on Computer Systems and Applications, pp. 414–421 (2008)
18. Abawajy, J.H.: Placement of file replicas in data grid environments. In: Bubak, M., van Albada, G.D., Sloot, P.M.A., Dongarra, J. (eds.) ICCS 2004. LNCS, vol. 3038, pp. 66–73. Springer, Heidelberg (2004). https://doi.org/10.1007/978-3-540-24688-6_11
19. Larbani, M., Chen, Y.W.: A fuzzy set based framework for concept of affinity. Appl. Math. Sci. 3(7), 317–332 (2009)
20. Dancey, C.P., Reidy, J.: Statistics Without Maths for Psychology, 4th edn. Pearson Education Ltd., Harlow (2007)

Does the Operational Model Capture Partition Tolerance in Distributed Systems?

Grégoire Bonin[✉], Achour Mostéfaoui, and Matthieu Perrin

LS2N, Université de Nantes, Nantes, France
{gregoire.bonin,achour.mostefaoui,matthieu.perrin}@univ-nantes.fr

Abstract. In large scale distributed systems, replication is essential in order to provide availability and partition tolerance. Such systems are abstracted by the *wait-free model*, composed of asynchronous processes that communicate by sending and receiving messages, and in which any process may crash. Complexity in local memory has already been studied for several objects, including sets, databases and collaborative editors. However, the literature has focused on a subclass of algorithms, operating in the so-called *operational model*, in which processes can only broadcast one message per update operation and the read operation incurs no communication.

This paper tackles the following question: are the operational model and the wait-free model equivalent from the complexity point of view? We show that, under a weak consistency criterion, implementations in the wait-free model require strictly less local memory than their counterparts in the operational model.

Keywords: Operational model · Eventual consistency ·
Space complexity · Update consistency · Wait-free model

1 Introduction

Eventual Consistency. In large scale distributed systems, replication is essential in order to provide availability and partition tolerance. Problems arise with replication as consistency has to be maintained between the different replicas.

The most natural and intuitive abstraction for the user would be to view a distributed/replicated object as if it is a single physical object shared by all the processes. This means that all the operations on the object, possibly concurrent or interleaving, appear as if they have been executed atomically and sequentially. Such an abstraction has to respect a correctness condition called strong consistency. Unfortunately, the CAP Theorem [6] states that this property is unrealizable in most systems, as it is impossible to combine strong consistency, availability and partition tolerance in asynchronous systems. Eventual consistency was introduced to overcome this issue. It states that, after update operations stop taking place, the different replicas will eventual converge towards an identical state.

© Springer Nature Switzerland AG 2019
V. Malyshkin (Ed.): PaCT 2019, LNCS 11657, pp. 400–407, 2019.
https://doi.org/10.1007/978-3-030-25636-4_31

The Operational Model. In this context, Conflict-Free Replicated Data Types (CRDTs) [11] constitute a family of objects designed to achieve eventual consistency. Those are based on a theorem stating the equivalence between two kinds of objects: the Commutative Replicated Data Types (CmRDTs), in which all update operations commute, and Convergent Replicated Data Types (CvRDTs), the states of which form a lattice. For example, and implementation of the set structure called G-set (grow-only set) provides two different operations: an update operation that inserts an element in the set and a query operation that says whether a specific element belongs to the set. From the CmRDT point of view, the operations "insert x" and "insert y" commute. From the CvRDT point of view, the set inclusion is a lattice order on the states of the set.

The *operational model* has been proposed to abstract the implementation of CRDTs. In the operational model, each replica maintains a local state on which the operations are done. An update operation is divided into two parts. First, the update operation is prepared locally by the replica where the update operation is issued and then a message is broadcast to inform all the other replicas. Then, the local state of each replica is updated at the reception of the update message. Given that the different operations are commutative, all replicas converge to the same state when no update operation is in progress.

As only one message is broadcast per update operation, algorithms in the operational model are, by design, optimal in terms of the number of used messages. The amount of metadata that must be stored on each replica is more problematic and has been widely studied for several objects including sets, counters and registers [5], data stores [2] and collaborative editors [1].

The Wait-Free Model. Despite the fact that algorithms from the operational model are naturally partition tolerant and minimize communication in their implementation, the operational model imposes limitations on the form of its admissible algorithms. It is for example impossible to acknowledge or forward messages, to execute local steps without the reception of a message, or to propagate information during read operations. This prevents algorithms from using more advanced techniques like the message patterns used in checkpointing [3,9].

Such algorithms are usually studied in the *wait-free asynchronous message-passing distributed model*, or simply the *wait-free model*, in which asynchronous processes communicate by sending and receiving messages. Any number of processes may crash: a *faulty* process executes correctly until it *crashes*, it then stops operating. A process that does not crash during an execution is called *correct*. Failure tolerance also captures partition tolerance as it is impossible for a process to wait for an acknowledgement from any other process since all other processes may have crashed.

Processes communicate and synchronize by sending and receiving messages, using the *causal broadcast* abstraction[1] that provides them with a **broadcast**(m) operation and a **receive**(m) event, where m is a message. respecting the following Communication channels are uniformly reliable meaning that all correct processes eventually receive the same set of messages, including their own messages.

However, channels are asynchronous, in the sense that there is no bound on the time it takes for one message to be delivered.

A history in the wait-free model is an abstraction of an execution that contains the information accessible for an outside observer, i.e. the operations that were performed, their invoking process and time, as well as their returned value.

Complexity. We consider deterministic algorithms. This allows us to define a state using an execution or a history. In order to compare the local complexity of algorithms in the different models, we define the *H-complexity* that allows us to compare the efficiency of two algorithms when executing the same history. As the algorithms are deterministic, we can compare equivalent state in the two algorithms (if the states are defined by the same sub-history, then they are equivalent).

More precisely, given a history H that contains a finite number of updates, and an algorithm Λ, we define the H-complexity of Λ as follows. Let S be the set of all local states reachable by any process executing Λ during an execution that can be abstracted by H. We define the H-complexity of Λ as follows:

- if $S = \emptyset$ (i.e. if H is not admitted by Λ), the H-complexity is 0;
- if $|S| = \infty$ (i.e. if S has states of unbounded size), the H-complexity is ∞;
- otherwise, the H-complexity is the maximal size of a state in S.

Problem Statement. The wait-free model is strictly more general than the operational model, as any algorithm from the operational model is also an algorithm in the wait-free model, but the converse does not hold. In particular, this means that the complexity results proven in the operational model may not hold in the wait-free model.

Therefore arises the following question: are the wait-free model and the operational model equivalent in terms of complexity?

Approach. In this paper, we propose a new object, called *update consistent l-countdown-append object*, and compare its wait-free implementations in both models. As its name suggests, the update consistent l-countdown-append object is specified by a sequential specification, that describes the behavior of the object when processes access it sequentially, and a weak consistency criterion, called

---

[1] Note that causal broadcast can be easily implemented in the wait-free model [10]. However, this implementation has a cost in local memory. We choose to include the primitive in the model to isolate the complexity needed to maintain consistency of the shared objects from the complexity needed to ensure causality, and therefore reducing the noise of the complexity results we obtain in the next sections.

update consistency [8], that describe how concurrency affects the sequential behaviour of the object.

The l-countdown-append object, where $l \in \mathbb{N}$, accepts the four update operations given by the set $U = \{a, b, c, d\}$, and one query operation, q. The behavior of the object is divided into two phases: during the first phase, the object counts the number of update operations, starting from l, down to 1, then ε (the empty word). In the second phase, the operation is concatenated at the end of the state. Finally, the query operation returns the local state of the object each time it is executed.

Update consistency is a consistency criterion that strengthens eventual consistency by stating that the convergence state must be obtainable in a sequentially consistent execution. In other words, it can be obtained by a sequential ordering of the update operations.

More formally, a history H is update consistent for an object O if it is in one of the two following cases:

- The processes never stop updating, i.e. H contains an infinite number of update operations.
- It is possible to omit a finite number of query operations such that resulting history has a linearization admitted by the sequential specification of O.

On a computability point of view, it is possible to implement any object with this criterion in both computing models [8].

Contributions. This paper proves that the two models are not equivalent: we prove that $\mathcal{O}(l)$ bits are necessary in the operational model to implement an update consistent l-countdown-append, whereas we give a logarithmic algorithm for the wait-free model.

Organization. Section 2 proves the part of the result for the operational model, and Sect. 3 explores the wait-free model. Finally, Sect. 4 concludes the paper. We could not include all the proofs in this extended abstract, due to space restriction. A complete version of the paper can be found in [4].

2 Lower Bound in the Operational Model

In order to compare these two models, we consider a set of possible histories (executions): the H_v histories. Let $l \in \mathbb{N}$, and let $v = u_1...u_l$ be a word consisting of l update operations of the l-countdown-append object. We denote by H_v any history in which one process performs all updates of v in their order of appearance, and the other processes keep performing the query operation.

We now prove that any algorithm in the operational model has a H_v-complexity of at least $\frac{l}{2} - 1$ bits for some v. Our proof follows the scheme introduced in [5]: we build a family of executions that do not belong to H_v, in such a way that, at some point in the execution, a process p_i performing the operations of v is unable to distinguish between these executions and an execution in H_v. Then, in a later stage of the execution, process p_i must be able to distinguish between enough of them in order to keep convergence possible.

Theorem 1. *For any deterministic algorithm Λ that implements an update consistent l-countdown-append object in the operational model, there exists v such that the H_v-complexity of Λ is at least $\frac{l}{2} - 1$ bits.*

Proof. Let Λ be an algorithm in the operational model implementing an update consistent l-countdown-append object. For each pair of words of update operations (v_1, v_2), where $v_1 \in \{a, b\}^l$ and $v_2 \in \{c, d\}^l$, we define the execution $X_{(v_1,v_2)}$ as follows. Only two processes p_1 and p_2 take steps in $X_{(v_1,v_2)}$. All other processes crash before the beginning of the execution. Initially, process p_1 (resp. p_2) executes sequentially the successive operations of v_1 (resp. v_2). In accordance to the operational model, they broadcast a single message during each operation. In a later stage, processes p_1 and p_2 receive the messages of each other, according to the FIFO order. Finally, both processes perform a query operation. We denote by $\mathcal{X} = \{X_{(v_1,v_2)} | v_1 \in \{a, b\}^l \wedge v_2 \in \{c, d\}^l\}$ the set of all $X_{(v_1,v_2)}$ executions.

Let us first remark that update consistency imposes that both query operations return the same value v_c, that is a suffix of size l, of an interleaving of v_1 and v_2. Let $f(v_1, v_2)$ be the number of c and d operations in v_c. Note that f is well defined because Λ is deterministic.

We now distinguish the executions depending on which process has a majority of operations in the convergence state. We define $\mathcal{X}_1 = \{X_{(v_1,v_2)} \in \mathcal{X} : f(v_1, v_2) \geq \frac{l}{2}\}$ and $\mathcal{X}_2 = \mathcal{X} \setminus \mathcal{X}_1$. As $\mathcal{X}_1$ and $\mathcal{X}_2$ form a partition of $\mathcal{X}$ which has a size 2^{2l}, we have $|\mathcal{X}_1| \geq 2^{2l-1}$ or $|\mathcal{X}_2| \geq 2^{2l-1}$. Without loss of generality, we suppose that $|\mathcal{X}_1| \geq 2^{2l-1}$.

We now partition $\mathcal{X}_1$ based on the value of v_1. For each word $v_1 \in \{a, b\}^l$, let $\mathcal{X}_1(v_1) = \{X_{(v,v_2)} \in \mathcal{X}_1 : v = v_1\}$. There exists a word v_1 such that $|\mathcal{X}_1(v_1)| \geq \frac{|\mathcal{X}_1|}{|\{a,b\}^l|} = \frac{2^{2l-1}}{2^l} = 2^{l-1}$. Let us fix such a v_1.

Let v_2 and v_2' be two words such that $X_{(v_1,v_2)}$ and $X_{(v_1,v_2')}$ belong to $\mathcal{X}_1(v_1)$. By definition of f, if $X_{(v_1,v_2)}$ and $X_{(v_1,v_2')}$ converge to the same state, then v_2 and v_2' differ at most by their $l - f(v_1, v_2) \leq \frac{l}{2}$ first operations. Consequently, there are at least $\frac{2^{l-1}}{2^{\frac{l}{2}}} = 2^{\frac{l}{2}-1}$ different values for v_2 for which $X_{(v_1,v_2)}$ lead to different convergence states. Let $\mathcal{X}'$ be a subset of $\mathcal{X}_1(v_1)$ of size $2^{\frac{l}{2}-1}$, in which all convergence states are different.

In the operational model, the local state of process p_2 at the end of the execution only depends on its local state after executing its own l update operations, and the messages received from p_1 afterwards. In all the executions of $\mathcal{X}'$, the messages received by p_2 are the same in all executions because v_1 is fixed. Moreover, the local state of p_2 at the end of all executions is different. This means that the local state of p_2 after doing its updates is also different in all executions. Consequently, there is a word v_2 such that, after executing all update operations in v_2 (execution X), the local state of p_2 requires at least $\frac{l}{2} - 1$ bits.

Finally, let us consider the execution X' in which only p_2 takes steps, executing the sequence of update operations of v_2. Just after executing its updates, p_2 cannot distinguish between the executions X and X'. Consequently, its local state in X' also requires $\frac{l}{2} - 1$ bits. Moreover, X' is modeled by H_{v_2}. Therefore, the H_{v_2}-complexity of Λ is at least $\frac{l}{2} - 1$ bits.

```
1   var clock_i ∈ Array(ℕ, ℕ) ← [i ↦ 0];
2   var leader_i ∈ ℕ ← i;
3   var countdown_i ∈ {0, ..., l} ← l;
4   var append_i ∈ U* ← ε;
5   operation q()
6   ⌊  if countdown_i = 0 then  return append_i  else  return countdown_i ;
7   operation u() // u ∈ U
8   ⌊  broadcast mUpdate (clock_i[i] + 1, i, u);
9   receive mUpdate (t_j ∈ ℕ, j ∈ ℕ, u ∈ U)
10  │  if clock_i[j] < t_j then
11  │  │  clock_i[j] ← t_j; leader_i ← i;
12  │  │  if countdown_i = 0 then
13  │  │  │  append_i ← append_i · u;
14  │  │  │  broadcast mCorrect (clock_i, i, append_i);
15  │  ⌊  else  countdown_i ← countdown_i − 1 ;
16  receive mCorrect (cl_j ∈ Array(ℕ, ℕ), j ∈ ℕ, a_j ∈ U*)
17  │  if (∀k, clock_i[k] ≤ cl_j[k]) ∧ (j ≤ leader_i ∨ ∃k, clock_i[k] < cl_j[k]) then
18  ⌊  ⌊  append_i ← a_j; clock_i ← cl_j; leader_i ← j;
```

Algorithm 1. The countdown-append object in the wait-free model

3   Upper Bound in the Wait-Free Model

This section exhibits an algorithm (Algorithm 1) that implements an update consistent l-Countdown-append in the wait-free model with a lower H_v-complexity, for any v. This algorithm is a variant of the algorithm UQ_0 proposed in [7].

Each process p_i maintains four variables. Variables $countdown_i$ and $append_i$ represent the current local state at p_i. If $countdown_i > 0$, the l-countdown-append object is in the countdown phase. Otherwise it is in the append phase and its value is $append_i$. Variable $clock_i$ is the equivalent of a version vector, such that $clock_i(j)$ represents the number of operations issued by p_j that are taken into account into the current state of p_i. As p_i does not know the number of participants, it is encoded as an associative array, rather than a vector. Finally, variable $leader_i$ is the identifier of a process such that, $clock_i < clock_{leader_i}$ or p_i and p_{leader_i} are in the same local state.

When a process invokes the query operation q, it computes locally the state of the object based on $countdown_i$ and $append_i$.

When process p_i invokes an update operation a, b, c or d, it increments its local clock $clock_i[i]$ and broadcasts a message mUpdate (Line 8). Upon the reception of such a message, p_i executes the operation (decrements countdown$_i$ if the countdown is still possible, or appends the operation to append$_i$), and answers with a mUpdate message containing its state and its current vector clock.

When a correction message is received, the process checks whether it is more recent according to the vector clock, and if that is the case, it replaces its own data with the received one.

Algorithm 1 is clearly wait-free as its operations contain no loop. It is also update consistent because, (1) all processes constantly maintain a state obtained by a linearization of the operations of their causal past, and (2) after all updates have been performed, all replicas converge towards a common state, that is the state of the correct process with the smallest identifier.

Let $l \in \mathbb{N}$ and $v \in U^l$. In any execution abstracted by H_v, there is a process p_i that performs all l update operations. For all processes p_j, clock$_j$ only contains one entry for p_i, smaller than l. Therefore, clock$_j$ can be encoded in less than $\log(n) + \log(l) = \log(nl)$ bits; The process identifier leader$_i$ can be encoded in $\log(n)$ bits; countdown$_i$ can take at most l different values, so it can be encoded in $\log(l)$ bits and append$_i = \varepsilon$ is a constant value, so it has an encoding of constant size c. Finally, the H_v complexity of Algorithm 1 is $\mathcal{O}(\log(nl))$ bits, which proves the following theorem.

Theorem 2. *There exists an algorithm Λ implementing an update consistent l-countdown-append object in the wait-free model such that, for all $v \in U^l$, Λ has an H_v-complexity of $\mathcal{O}(\log(nl))$ bits.*

We can finally conclude on the non-equivalence between the two computing model in the implementation of update consistency.

Corollary 1. *There exists an object O and an algorithm Λ_{wf} implementing an update consistent O in the wait-free model, such that, for any algorithm Λ_{om} implementing an update consistent O object in the operational model, there is a history H such that Λ_{wf} has a strictly lower H-complexity than Λ_{om}.*

4   Conclusion

In this paper we answered the following question: are the wait-free model and the operational model equivalent in terms of local complexity? We proved that the response to this question is no in the case of update consistency: we proved that there exists an object that has a different complexity in the two models: the l-countdown-append object. In the wait-free model, there is an algorithm for which the complexity required to encode a special state of the object is upper bounded by $\mathcal{O}(\log(nl))$ bits, whereas in the operational model, any algorithm requires at least $\frac{l}{2} - 1$ bits to encode the same state. This means that the operational model does not allow the optimal implementation for update consistency.

The result proposed in this papers shows that the question of whether the operational model is well suited to represent partition tolerance is not simple, especially in the context of determining the complexity in local memory required to implement shared objects. An interesting open question is whether the lower bounds proved for several objects in the operational model can be extended to the wait-free model.

References

1. Attiya, H., Burckhardt, S., Gotsman, A., Morrison, A., Yang, H., Zawirski, M.: Specification and complexity of collaborative text editing. In: Symposium on Principles of Distributed Computing, pp. 259–268. ACM (2016)
2. Attiya, H., Ellen, F., Morrison, A.: Limitations of highly-available eventually-consistent data stores. IEEE Trans. Parallel Distrib. Syst. **28**(1), 141–155 (2017)
3. Baldoni, R., Brzezinski, J., Hélary, J.M., Mostefaoui, A., Raynal, M.: Characterization of consistent global checkpoints in large-scale distributed systems. In: Workshop on Future Trends of Distributed Computing Systems, pp. 314–323. IEEE (1995)
4. Bonin, G., Achour, M., Perrin, M.: Does the operational model capture partition tolerance in distributed systems? extended version (2019)
5. Burckhardt, S., Gotsman, A., Yang, H., Zawirski, M.: Replicated data types: specification, verification, optimality. In: ACM Sigplan Notices, vol. 49, pp. 271–284. ACM (2014)
6. Gilbert, S., Lynch, N.: Brewer's conjecture and the feasibility of consistent, available, partition-tolerant web services. ACM Sigact News **33**, 51–59 (2002)
7. Perrin, M.: Distributed Systems: Concurrency and Consistency. Elsevier, Amsterdam (2017)
8. Perrin, M., Mostefaoui, A., Jard, C.: Update consistency for wait-free concurrent objects. In: International Parallel and Distributed Processing Symposium, pp. 219–228. IEEE (2015)
9. Randell, B., Lee, P., Treleaven, P.C.: Reliability issues in computing system design. ACM Comput. Surv. (CSUR) **10**(2), 123–165 (1978)
10. Raynal, M., Schiper, A., Toueg, S.: The causal ordering abstraction and a simple way to implement it. Inf. Process. Lett. **39**(6), 343–350 (1991)
11. Shapiro, M., Preguiça, N., Baquero, C., Zawirski, M.: Conflict-free replicated data types. In: Défago, X., Petit, F., Villain, V. (eds.) SSS 2011. LNCS, vol. 6976, pp. 386–400. Springer, Heidelberg (2011). https://doi.org/10.1007/978-3-642-24550-3_29

Blockchain-Based Delegation of Rights in Distributed Computing Environment

Andrey Demichev[1]([✉]), Alexander Kryukov[1], and Nikolai Prikhod'ko[2]

[1] Skobeltsyn Institute of Nuclear Physics,
Lomonosov Moscow State University, Moscow, Russia
{demichev,kryukov}@theory.sinp.msu.ru
[2] Yaroslav-the-Wise Novgorod State University, Velikiy Novgorod, Russia
niko2004x@mail.ru

Abstract. The paper suggests a new approach based on blockchain technology and smart contracts to delegation of rights within distributed computing systems, which is fault-tolerant, safe and secure. The implementation of the proposed approach is based on the permissioned blockchains and on the Hyperledger Fabric blockchain platform in conjunction with Hyperledger Composer.

Keywords: Distributed computing · Blockchain · Access rights · Delegation · Hyperledger

1 Introduction

Nowadays, distributed computing systems (DCS) are widely used for solving various problems in scientific, engineering and business areas. The advantage of DCS is the unification and simplification of an access to computing resources, e.g., clouds, supercomputers, databases, and, as consequence, to growth of efficiency of scientific, engineering and business activities. However, using heterogeneous and geographically widely dispersed DCS requires sophisticated and robust solutions for various aspects of the distributed computation in comparison with the case of local resources or more localized DCS. In particular, a reliable but still user-friendly security model for such DCS is of great importance. In this paper we discuss some aspects of the security infrastructure for DCS and suggest possible improvements. Providing the security of DCS implies solving the following basic problems: (1) security of communications: this problem is solved by encrypting the communication channels; (2) authentication: this means confirmation of the truth of the attribute of the data fragment declared by a certain entity as a true one; (3) authorization: this means the granting of access rights according to a policy; (4) delegation: this means delegation of rights from a user or a Web service to another Web service.

This work was funded by the Russian Science Foundation (grant No. 18-11-00075).

V. Malyshkin (Ed.): PaCT 2019, LNCS 11657, pp. 408–418, 2019.
https://doi.org/10.1007/978-3-030-25636-4_32

In this paper we consider the last aspect of the DCS security. We will use grid infrastructures and distributed storages as a reference DCS models for implementation of the security infrastructure. However the same problems are relevant and the suggested solutions are applicable for any DCS which comprises of a set of communicating Web services. The most striking example of grid infrastructure and globally distributed storage is the Worldwide LHC Computing Grid (WLCG) [1,2] which is used for processing and simulation of experimental data from the Large Hadron Collider (LHC) [3]. Other important examples of DCSs are the data storages and processing infrastructures in the area of astroparticle physics [4,5].

The security of most of DCSs, including WLCG, is based on the PKI [6] and X.509 certificates [7] together with proxy certificates [8]. The proxy certificate is a special short-time living certificate used for the purpose of providing restricted rights delegation within a PKI based authentication system. The short lifetime of the proxy certificate is due to security reasons. In DCSs the proxies are used for both user access to computing resources and for processing workflows. A workflow is a composite computational job that must be run sequentially by multiple services, with each service in the sequence receiving requests directly from the previous service. In this case, the delegation of rights from service to service occurs with the help of the proxy certificates. However, the proxies have short lifetimes, while one cannot predict how much time would take request processing especially in the case of the composite jobs. There are special services to support prolongation of proxy lifetime [9], and all this make the security infrastructure overcomplicated and difficult to interact with.

Recently, we proposed an approach [10,11] which allows us to avoid using the proxy certificates in security infrastructures entirely. Roughly speaking, in our scheme each issued request is a pair of a message and individual hash related to it. This single-shot hash has unlimited lifetime so that in our scheme the prolongation service is not needed. At the same time, the security level is not reduced because every hash can be used only once and only for a specific request. Thus hash compromise can only result in the fact that the request has to be processed again. However this approach also requires a central dedicated service, namely validation service, to process requests in DCSs. The point is that upon getting computational request each service checks request's hash against the validation service and continues only if the hash is correct and was not used before. Both the proxy prolongation service and the validation service in the approach suggested in [10,11] being centralized ones are potential points of failure and bottlenecks for the entire distributed systems.

In this work, we suggest a DCS design which allows abandoning the special dedicated centralized services in the DCS security infrastructure and the use instead of them a blockchain-based distributed registry and smart contracts. The very idea of using the blockchain technology for DCS security was expressed in our work [12]. However that paper does not contain any details of the design and is oriented to the Ethereum blockchain platform [13] which is not well suited for DCSs. In the present paper, we propose an approach to solving the problem of delegation on the basis of blockchain technology and smart contracts within the Hyperledger platform [14,15] which is proved to be very suitable for DCS management, in particular for distributed storages [16]. While in the paper [16] we proposed a mechanism for managing provenance metadata and data access rights based on the blockchain technology, in the present work we solved another problem, namely, developing on the same basis a mechanism for delegation of rights in distributed systems. To our best knowledge, the blockchain-based mechanism for delegation of rights in distributed system suggested in this work are completely novel. Other existing blockchain-based suggestions and developments in the field of DCS management are far from the system proposed in this paper, both in their goals and objectives, and in the ways of their implementation. The reader may find discussion of them in the survey [17].

In the next section we shortly consider security infrastructure with the use of proxy certificates and the solution without proxy certificates but with a special central service. In Sect. 3 the blockchain-based delegation of rights in DCS is presented. The Sect. 4 is devoted to conclusions.

2   DCS Security Infrastructure

2.1   Security Infrastructure with the Use of Proxy Certificates

In distributed grid-like systems the security infrastructure is build around Public Key Infrastructure (PKI) that uses asymmetric cryptography. One of the main problem of the security infrastructure is the problem of delegation of rights [18,19]. Let us consider the delegation procedure in DCS for the following workflow (see Fig. 1): a Client asks the Service1 to perform a request; the Service1 sends a subrequest to Service2. It is expected that the Client somehow delegates its rights to Service1 to authenticate it to the Service2 since subrequest is performed on his behalf. Therefore there is a question how this delegation is carried out.

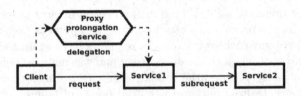

Fig. 1. Delegation of credentials.

The common solution used in grid is to use the proxy certificate with non-critical extension to store information about user rights. The proxy certificate is an extended X.509 public key certificate and has the following properties: it is signed with standard X.509 or another proxy certificate of a user who needs delegation of rights; contains both public and private keys; these are not the original users keys but generated from them; does not require any password (unlike usual PKI certificates); cannot be revoked; is used by grid services, to act on behalf of the proxy issuer. Thus the proxy certificates are essentially less secure objects than standard certificates. To reduce the chance for proxy certificate to be stolen, the proxy must have very short lifetime. This leads to the problem of the renovation of the proxy. The possible solution of the problem is to use certain service that have to manage proxy certificates and renew them if necessary. One of such services is the MyProxy service [20].

The delegation scheme in this case looks as follows: (1) the user creates a proxy certificate; (2) it sends it to the service with a request to perform some action on behalf of the user; (3) from the point of view of any service, having a proxy certificate means that its bearer is authorized to do whatever it likes on behalf of the issuing the proxy. The last item leads to a vulnerability of the proxy certificate approach, namely, the service that received the proxy is given too much leeway on behalf of the entity issuing the proxy certificate. This is in addition to the above mentioned necessity to have the proxy prolongation service which is a potential point of failure, intrusion and bottle neck.

An example of a delegation is copying of a file from Service1 to Service2. For this aim a user transmits to Service1 his proxy certificate and requests it to copy a file to Service2 on his behalf so that the rights to the file will belong not to Service1 or Service2, but to the user. In Sect. 3.2 we will consider this use case for the delegation in the framework of the blockchain-based approach.

2.2   Intermediate Solution: Security Infrastructure Without Proxy Certificates and with Special Central Service

In the papers [10,11] a new security infrastructure model for distributed computing systems was suggested which does not require the proxy certificates. The proposed architecture of the DCS security infrastructure is shown in Fig. 2 on the left hand side.

Each request processed in DCS is accompanied by an accounting information. Accounting information is a triple of the following objects: $\{h, Entity_s, Entity_d\}$, where $h, Entity_s, Entity_d$ are the hash, source and destination entity of the request. This triple means that the entity $Entity_s$ sends a request with the hash h to the entity $Entity_d$ for execution. Complete format of accounting information include some additional objects such as affiliation to a virtual organization and user's roles in it.

Let us consider the processing of a request from the point of view of the credential delegation.

1. The Client generate a request r_1 and the hash $h_1 = H(r_1)$.
2. The Client registers the triple $\{h1, Client, Service1\}$ in the validation service (VS).
3. The Client sends the request r_1 to the Service1 for processing.
4. The Service1 generates the hash from the obtained request r_1 and asks the VS to approve it. If VS approves then Service1 continues.

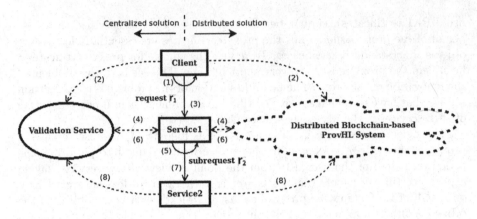

Fig. 2. The architectures of the security infrastructure with the central validation service and with the distributed registry (blockchain).

5. The Service1 generates the new subrequest r_2 that is generated from r_1 and the hash $h_2 = H(r_2)$.
6. The Service1 registers the triple $\{h2, Service1, Service2\}$ in the VS.
7. The Service1 sends the request to the Service2 for further processing.
8. The Service2 generates the hash from the obtained request r_2 and asks the VS to approve it. If VS approves then Service2 continues.

When Service1 registers $\{h2, Service1, Service2\}$, VS, knowing that this is a secondary request generated from the user's one, registers it as a user request. Thus, when accessing it by Service2, it will confirm that the action should be performed on behalf of the user, although received from Service1.

The hash of secondary requests should be calculated not only on the basis of the body of the new request, but also the hash of the primary request (a weak variant of the Merkle tree) from which it is generated. In processing the request, the validation service accumulates chains of accounting information for each request in the DCS. This information can be used for different purposes. In particular, it may be used for revocation of the request at any stage of processing.

One of the possible weak points of the proposed approach is the requirement to have on-line access to the validation service for all other services of the DCS. The simulation using our prototype shows that such an infrastructure is quite stable and works fine at least for the systems with 20 user requests per second. For the critical high-availability systems it is possible to deploy two parallel validation services with on-line database replication. At this case one of the services acts as a master service that processes requests and another is a slave (an inactive full copy of the master). If the master service crashes it would be easy to switch to the slave service immediately with almost no loss of information. An important benefit of the proposed security infrastructure is that during request processing the validation service collects all the information concerning each request in the DCS. This information can be used for monitoring purposes as well as for request revocation at any stage of processing.

3   Use of the Blockchain Technology for Providing Delegation of Rights in DCS

The approach shortly presented in Sect. 2.2 results in essential simplification both registration of new users in the system, and their operations in DCS, in comparison with the most popular infrastructure of public keys (PKI) together with use of the proxy certificates (Sect. 2.1). However the vulnerable point of both the solutions is need of a special fault-tolerant and resistant to malicious operations centralized service in the security infrastructure. In this section, we investigate the possibility to refrain from the special server in the security infrastructure of DCS and to use for this purpose a distributed registry based on the blockchain technology and smart contracts. Since in this case the security infrastructure registry is distributed across a number of nodes in the system, such an approach will lead to increased fault tolerance and level of security of DCS. The basic example of DCS which we use in present work is a distributed storage.

3.1   Distributed Storage with Provenance Metadata Driven Data Management

In the work [16] we proposed a new approach to the construction of data management systems in a distributed environment, based on the integration of the following basic principles and technologies:

- smart contracts [21];
- permissioned blockchains technology [22];
- Hyperledger blockchain platform [14,15] together with Hyperledger Composer [23]; hereafter we shall refer to these two components as HLF&C-platform;
- management of data access rights with the help of special HLF&C-platform tools;
- provenance metadata driven data management: the metadata is written to the blockchain beforehand, and data management systems (DMS) refer to the blockchain and performs the transactions recorded there;
- distributed consensus protocols [22].

Provenance metadata (PMD) contain key information that is necessary to determine the origin, authorship and quality of relevant data, their storage and usage consistency, and for interpretation and confirmation of relevant results of data processing. The need for PMD is especially important when data is jointly processed by several research groups that have their own, although interrelated interests, which is a very common practice in many scientific, engineering, and industrial fields lately. For the details we refer to the work [16] where principles, architecture and operation algorithms have been developed for the PMD management system, entitled ProvHL (Provenance HyperLedger), which is fault-tolerant, safe, reliable in terms of the safety and security of provenance metadata records from accidental or intentional distortion. Moreover, it allows users to perform operations with files and directories in the DCS. The distribution of the main HLF&C modules by administrative domains of the modeled distributed storage environment within the current testbed for the ProvHL system is shown in Fig. 3. Here we shall concentrate on a new blockchain-based method for delegation of rights within distributed computing systems which is free from shortcomings inherent in other solutions.

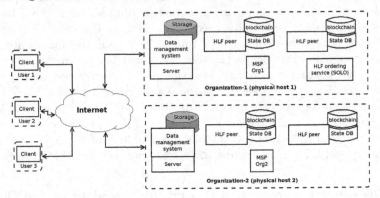

Fig. 3. A simplified scheme of the ProvHL testbed environment.

3.2   Blockchain-Based Delegation in Distributed Storages

The algorithm which we propose for recording transactions with provenance metadata and data management in the framework of ProvHL in a very simplified form reads as follows:

- the owner accesses the chaincode function, which, according to the acl-file ("acl" stands for access control language), allows the owner of the data to grant access rights to these data to another user or group of users;
- a user who is granted access rights by the owner accesses the chaincode with a request to make an operation (Client Request transaction) with data (for example, file download, upload, copy, etc.);
- the chaincode verifies that such a transaction complies with the rules defined in the acl-file and, if it does, sends a request to the HLF&C environment to complete the transaction;
- HLF&C performs transaction processing (transaction workflow: simulation and endorsements → ordering → validation → state updating);
- HLF&C sends a message (event) to the user about the successful transaction and its recording in the blockchain; the message also contains the transaction identification number;
- the user accesses the data management system (DMS) with a request to perform a data operation that contains the number of the corresponding transaction;
- the DMS checks for a record of this transaction in the blockchain;
- if there is a record of the valid transaction, the DMS performs the required operation and, in turn, initiates a transaction record confirming that a data operation was performed (Server Response transaction).

As it can be seen, for each data operation, at least two transaction records are made in the blockchain: one corresponds to the client request, and the second corresponds to the server response. Actually, an operation comprises of even more transactions.

Below we present more details on delegation of rights between services on the example of operation of coping data from one local storage (Storage1) to another (Storage2). Now the Service1 on Fig. 2 stands for the data management system of the Storage1 (DMS_Storage1) and Service2 stands for the data management system of the Storage2 (DMS_Storage2) and we use the right hand side of the figure (distributed solution). Now the content of the request r_1 is: "copy file F from Soragel to Storage2" and that for the request r_2 is: "upload file F to Storage2". In the framework of the ProvHL system, operations with files are defined as assets (alongside with other business network entities) [16] by using the object-oriented modeling language [23] in the so called cto-file. For the delegation mechanism it is important that it contains the obligatory attributes "requester" and "executor". Also it inherits "file owner" attributes from the file asset definition. Upon receiving a request from a User for a file copying the DMS_Storage1 (Storage1 contains the file to be copied) detects the type of the copy operation, namely decides if this is local copying (within the Storage1) or copying to another storage. In the latter case it initiates, on behalf of the User, the operation of uploading the required file to destination Storage2. For this aim it interacts with the chaincode which, among other actions, defines that while for the initial copy operation the value of the requester attribute is equal to the User and the executor is DMS_Storage1, for the induced upload operation the requester is DMS_Storage1 and the executor is DMS_Storage2. In addition, the owner of the file copy on the Storage2 is the same as the owner of source file on the Storage1.

Note that in this case it is not necessary to rely on request hashes, as described in Sect. 2.2. Instead, one can use an arbitrary UUID for the request naming, since an immutability of record for a request sequence is guaranteed by the blockchain structure. The analog of the steps outlined in the Sect. 2.2 reads as follows.

1. The Client (User) generate the request r_1 and UUID for it.
2. The client initiates a transaction to create a copy operation, after which the entire transaction workflow is executed.
3. The Client sends the request r_1 to the DMS_Storage1 for processing. At this stage the 'requester' field of the operation attributes is equal to the Client, and the 'executor' is the DMS_Storage1.
4. The DMS_Storage1 checks that the related transaction is recorded in the blockchain and valid; in the case of positive result it continues carrying out the operation.
5. The DMS_Storage1 generates the new subrequest r_2 to DMS_Storage2 for uploading the file F to Storage2.
6. The DMS_Storage1 initiates recording the corresponding transaction into blockchain. At this stage the 'requester' field of the operation attributes is equal to the DMS_Storage1, and the 'executor' is the DMS_Storage2. It is

worth stressing that the right to initiate this request for the transaction is provided by the appropriate content of the smart contract (chaincode).

7. The DMS_Storage1 sends the request r_2 to the DMS_Storage2 for the file F uploading.

8. The DMS_Storage2 checks that the related transaction is recorded in the blockchain and valid. In the case of positive result it carries out the request.

Thus, the second request r_2 is executed at the initial request of the User, though it is issued by the DMS_Storage1 (source storage), and the file ownership does not change. This means that all goals of a delegation are completed. It is worth mentioning that in contrast to the scheme based on proxy certificates (Sect. 2.1), in the blockchain-based approach, as well as in the mechanism presented in the Sect. 2.2, the delegation is restricted solely to the specified operation. The chain of hashes used in Sect. 2.2 is replaced with a chain of transactions and blocks that make up the history of the copy operation from one storage to another. It is important to note that during the execution of the entire operation, the file F preserves the attribute "file owner" unchanged, that is, the rights to it in the process of the operations carried out by the chain of services (in this case, DMSs) do not change.

The approach proposed in this section allows us to avoid central services that can be bottlenecks, points of failure, and which are controlled by one of the sides of the business process. Instead, a distributed registry (blockchain) is used, which is controlled by all parties of the business process based on a consensus. The flexibility of the proposed mechanism is achieved due to the fact that in smart contracts one can fix any conditions for the delegation of rights. In this paper, we have considered a relatively simple, but in practice, most popular version of such conditions. The proposed mechanism directly extends to the case of arbitrary data processing services. Some technical complications are related to the fact that the result of such services can be an arbitrary number of output files. However, the general approach works in this case too.

The metric values of the developed system are under study and will be presented elsewhere. The preliminary measurements on the testbed depicted on Fig. 3 show that the overheads related to the operation processing by the ProvHL system is of the order of $4 \div 7$ s depending on setup variables such as maximal time of block forming, etc. This is fully consistent with the extensive results of the recent work [24] on the performance of the Hyperledger platform itself, with the measurements in this work were carried out on a testbed similar to ours. In particular, it was shown that for the input transaction rate up to 800 tx/s, the transaction latency is $\lesssim 1$ s, and the transaction throughput is ~ 800 tx/s. If we take into account that each file operation consists of $3 \div 7$ transactions (depending on the type of the operation), we get matching results for the latency, while for the throughput we may expect ~ 100 ops/s. These values, obtained on the testbed with very modest computer facilities, are quite acceptable for operations with files of sufficiently large volumes, the handing time of which (copying, downloading, uploading, etc.) is tens or more seconds. Such volumes of data files are typical for distributed storages intended for large scientific experiments.

4 Conclusion

In this work we have proposed a solution for a security infrastructure and delegation of rights for distributed computing systems based on the blockchain technology and smart contracts in the framework of the Hyperledger Fabric platform. This infrastructure is free from the significant drawbacks inherent to other existing approaches, namely, from the vulnerabilities (bottlenecks, points of failure) associated with the presence of a central services managing the security infrastructure. Due to its distributed nature, the blockchain-based delegation proves to be fully adequate to distributed computing systems. The use of smart contracts, in turn, provides flexibility because they allow one to define various conditions for the delegation of rights in DCSs.

At present, a testbed has been created on the basis of SINP MSU, where a preliminary version of the ProvHL system implementing the developed solution is deployed. Testing of the system has confirmed the correctness of the chosen approach, basic principles and algorithms of work and the preliminary performance measurements showed the suitability of the developed system for large distributed data storages.

The implementation of the suggested solution for delegation of rights in the framework of the ProvHL system of production level will significantly improve the security as well as quality and reliability of the results obtained on the basis of processing and analysis of data in a distributed computer environment.

References

1. Sciaba, A., et al.: Computing at the petabyte scale with the WLCG. Worldwide LHC computing grid. Technical report CERN-IT-Note-2010-006 (2010)
2. WLCG. http://wlcg.web.cern.ch. Accessed 21 May 2019
3. CERN. http://www.cern.ch. Accessed 21 May 2019
4. Berghöfer, T., et al.: Towards a model for computing in European astroparticle physics. arXiv preprint, arXiv:1512.00988 (2015)
5. Kryukov, A., Demichev, A.: Architecture of distributed data storage for astroparticle physics. Lobachevskii J. Math. **39**(9), 1199–1206 (2018)
6. Buchmann, J.A., Karatsiolis, E., Wiesmaier, A.: Introduction to Public Key Infrastructures. Springer, Heidelberg (2013). https://doi.org/10.1007/978-3-642-40657-7
7. Cooper, D., et al.: RFC 5280 - internet X.509 public key infrastructure certificate and certificate revocation list (CRL) profile. Technical report (2008)
8. Tuecke, S., et al.: Internet X.509 public key infrastructure proxy certificate profile. Technical report RFC 3820 (2004)
9. Kouril, D., Basney, J.: A credential renewal service for long-running jobs. In: IEEE/ACM Proceedings of the International Workshop on Grid Computing 13–14, pp. 2–13 (2005)
10. Dubenskaya, J., Demichev, A., Kryukov, A., Prikhod'ko, N.: Special aspects of the development of the security infrastructure for distributed computing systems. Procedia Comput. Sci. **66**, 525–532 (2015)

11. Dubenskaya, J., Demichev, A., Kryukov, A., Prikhod'ko, N.: New security infrastructure model for distributed computing systems. J. Phys. Conf. Ser. **681**, 012051 (2016)
12. Kryukov, A., Demichev, A.: Security infrastructure for distributed computing systems on the basis of blockchain technology. CEUR Workshop Proc. **1787**, 338–342 (2016)
13. A next-generation smart contract and decentralized application platform. White Paper. https://github.com/ethereum/wiki/wiki/White-Paper. Accessed 21 May 2019
14. Hyperledger. https://www.hyperledger.org. Accessed 21 May 2019
15. Androulaki E., et al.: Hyperledger fabric: a distributed operating system for permissioned blockchains. In: Proceedings of the Thirteenth EuroSys Conference, Article No. 30, Porto, Portugal. ACM (2018)
16. Demichev A., Kryukov A., Prikhod'ko N.: The approach to managing provenance metadata and data access rights in distributed storage using the hyperledger blockchain platform. In: Proceedings of Ivannikov ISPRAS Open Conference, Moscow, Russia, pp. 131–136, 22–23 November 2018. IEEE Xplore Digital Library (2019)
17. Salman, T., et al.: Security services using blockchains: a state of the art survey. IEEE Commun. Surv. Tutor. **21**(1), 858–880 (2018)
18. Welch V, et al.: 509 proxy certificates for dynamic delegation. In: 3rd Annual PKI R&D Workshop, vol. 14 (2004)
19. Smirnova O.: Grid computing: delegation and authorisation. http://www.hep.lu.se/courses/grid/2014/Grid-COMPUTE-13.pdf. Accessed 21 May 2019
20. Novotny, J., Tuecke, S., Welch, V.: An online credential repository for the grid: myproxy. In: Proceedings of 10th IEEE International Symposium on High Performance Distributed Computing, pp. 104–111. IEEE (2001)
21. Szabo, N.: The idea of smart contracts (1997). http://szabo.best.vwh.net/smart_contracts_idea.html. Accessed 21 May 2019
22. Baliga, A.: Understanding blockchain consensus models. Technical report, Persistent Systems Ltd. (2017)
23. Hyperledger Composer. https://hyperledger.github.io/composer. Accessed 21 May 2019
24. Baliga, A., et al.: Performance characterization of hyperledger fabric. In: 2018 Crypto Valley Conference on Blockchain Technology (CVCBT), pp. 65–74 (2018)

Participant-Restricted Consensus in Asynchronous Crash-Prone Read/Write Systems and Its Weakest Failure Detector

Carole Delporte-Gallet[1], Hugues Fauconnier[1], and Michel Raynal[2,3(✉)]

[1] IRIF, Université Paris 7 Diderot, Paris, France
cd@irif.fr, hf@irif.fr
[2] IRISA, Université de Rennes, 35042 Rennes, France
michel.raynal@irisa.fr
[3] Department of Computing, Polytechnic University, Hung Hom, Hong Kong

Abstract. A failure detector is a device (object) that provides the processes with information on failures. Failure detectors were introduced to enrich asynchronous systems so that it becomes possible to solve problems (or implement concurrent objects) that are otherwise impossible to solve in pure asynchronous systems where processes are prone to crash failures. The most famous failure detector (which is called "eventual leader" and denoted Ω) is the weakest failure detector which allows consensus to be solved in n-process asynchronous systems where up to $t = n - 1$ processes may crash in the read/write communication model, and up to $t < n/2$ processes may crash in the message-passing communication model. In these models, all correct processes are supposed to participate in a consensus instance and in particular the eventual leader.

This paper considers the case where some subset of processes that do not crash (not predefined in advance) are allowed not to participate in a consensus instance. In this context Ω cannot be used to solve consensus as it could elect as eventual leader a non-participating process. This paper presents the weakest failure detector that allows correct processes not to participate in a consensus instance. This failure detector, denoted Ω^*, is a variant of Ω. The paper presents also an Ω^*-based consensus algorithm for the asynchronous read/write model, in which any number of processes may crash, and not all the correct processes are required to participate.

Keywords: Agreement · Asynchronous system ·
Atomic read/write register · Concurrency · Consensus ·
Eventual leadership · Failure detector · Participating process ·
Process crash · Read/write shared memory · Snapshot object ·
Weakest information on failures

1 Introduction

Concurrent objects. When considering multiprocess programming, concurrent objects are the objects that can be simultaneously accessed by several processes.

© Springer Nature Switzerland AG 2019
V. Malyshkin (Ed.): PaCT 2019, LNCS 11657, pp. 419–430, 2019.
https://doi.org/10.1007/978-3-030-25636-4_33

Examples of such objects are the classical objects encountered in sequential computing (such as stacks, queues, graphs, sets, trees, etc.) now shared by several processes to communicate and cooperate on a common goal, and objects targeting new concurrency-related issues (such as rendezvous and non-blocking atomic commitment objects). When there are no failures the implementation of such objects are usually based on locks, which can be built from base read/write or read/modify/write registers (see concurrency-related e.g., [16,19]).

In a failure-prone system (where a failure is a process crash), the situation is different, and many concurrent objects (as simple as stacks and queues) can no longer be implemented. This impossibility follows from the famous impossibilities to build a consensus object in the presence of asynchrony and process crashes [10, 13] (pedagogical presentations of these results can be found in textbooks such as [3,14,16,19]).

Impossibility Results and Failure Detectors. Several approaches have been proposed and investigated to circumvent the previous impossibilities. One of them, which is system-oriented, consists in enriching the system with failure-related objects providing each process individually with information on failures. This is the *failure detector*-based approach introduced in [6]. More precisely, a failure detector provides each process with one or several read-only local variables, containing information on failures. According to the type and the quality of this information, different failure detector classes can be defined. As a simple example, the class of perfect failure detectors (denoted P) provides each process p_i with a read-only set SUSPECTED$_i$ that (i) never contains a process identity before it crashes, and (ii) eventually contains the identities of all the processes that crashed. It is easy to see that a perfect failure detector allows a process that does not crash not to remain blocked forever because another process is crashed. (The power of perfect failure detectors was investigated in [11].)

A fundamental notion associated with failure detectors is the notion of *weakest failure detector* for a given concurrent object. Intuitively, "weakest failure detector" means that, a failure detector D is the weakest failure detector to implement an object O, if (1) D allows O to be implemented, and (2) any other failure detector, that allows O to be implemented, provides each process with enough information on failures that allow to build D.

This notion was introduced in [5], where it is shown that the "eventual leader" failure detector (denoted Ω) is the weakest failure detector which allows consensus to be implemented in an n-process asynchronous message-passing system where up to $t < n/2$ processes may crash. Ω provides each process p_i with a read-only variable LEADER$_i$ such that, after an arbitrarily long but finite time, the variables LEADER$_i$ of all the non-crashed processes contain the same process identity, which is the identity of one of them. Before this time occurs, the variables LEADER$_i$ can contain different, and varying with time, process identities. (This result was extended in [7] in asynchronous message passing system prone to any number $t < n$ of process crashes and in [12] where it is shown that Ω is the weakest failure detector that allows consensus to be implemented in asynchronous read/write systems prone to any number $t < n$ of process crashes.

Implementations of failure detectors such as Ω in asynchronous crash-prone read/write systems can be found in [8,16]. These implementations rely on underlying behavioral assumptions, which means that the corresponding underlying systems are not fully asynchronous).

Content of the Paper. When one want to solve consensus with the help of Ω, it is implicitly assumed that all processes that do not crash participate in the algorithm. This is due to the fact that Ω may elect any process that does not crash as the eventual leader (a process that does not crash in a run is said to be *correct*). So, if it elects a correct process but this process does not participate in the considered consensus instance, an Ω-based consensus algorithm may never terminate. It follows that, in a model in which correct processes do not participate Ω is too weak to solve consensus.

The system model considered consists of n asynchronous processes, which communicate by reading and writing atomic read/write registers, and where any number of processes may crash. In this setting, the paper considers consensus instances where an a priori unknown subset of processes do not participate. Hence the notion of *participant-restricted consensus*: possibly some processes may crash, but it is possible that, while being correct, some others never participate. In such a context the paper has the following contributions.

- It presents a failure detector (a variant of Ω denoted Ω^*) suited to participant-restricted consensus.
- It then presents an Ω^*-based consensus algorithm, and shows that Ω^* is the weakest failure detector for participant-restricted consensus.

Roadmap. The paper is made up of 5 sections. Section 2 introduces the underlying computing model. Section 3 presents the failure detector Ω^* and shows it is the weakest to solve consensus in the presence of correct processes that do not participate. Then Sect. 4 presents and proves correct an Ω^*-based consensus algorithm. Finally, Sect. 5 concludes the paper. The presentation style used in the paper is voluntarily informal.

2   Basic Computing Model and Consensus

2.1   Process, Communication, and Failure Model

The system is made up of a finite set Π of n sequential asynchronous processes denoted $p_1, ..., p_n$. "Asynchronous" means that each process proceeds to its own speed, which can vary with time and remains always unknown to processes.

The processes communicate by accessing a shared read/write memory made up of atomic read/write registers. From 0 to $(n-1)$ processes can commit crash failures. A process commits a crash when it halts prematurely. Before halting (if it ever halts), a process executes correctly its algorithm. After it crashed, it executes no more steps. Given an execution, a process that crashes in this execution is said to be *faulty*. Otherwise, it is said to be *correct* in this execution.

2.2   High Level Communication Abstraction

The algorithm described in Sect. 4 use high level communication objects, namely snapshot object. Snapshot object can be implemented on top of asynchronous read/write systems in which any number of processes may crash (e.g., [1,2,16,19]). Hence, while they provide processes with a higher abstraction level than atomic read/write registers, snapshot objects do not provide a stronger computational power than registers.

Snapshot Object. A *snapshot* object provides the processes with two operations denoted write() and snapshot() [1,2]. Such an object can be seen as an array of single-writer multi-reader atomic register $SN[1..n]$ such that:

– When p_i invokes the operation write(v), it writes v into $SN[i]$; and
– When p_i invokes the operation snapshot(), it obtains the value of the array $SN[1..n]$ as if it read simultaneously and instantaneously all its entries.

Said another way, the operations write() and snapshot() are atomic (linearizable).

One-Shot Snapshot and Containment Property. A *one-shot* snapshot object SN is such that each process can invoke SN.write() only once.

Let assume an one-shot snapshot object SN initialized to $[\bot, \cdots, \bot]$, where $\bot$ is a default value that cannot be written by a process. The arrays $snap1$ and $snap2$ being the values returned by any two invocations of SN.snapshot(), let us define $snap1 \leq snap2$ as

$$\forall\, x \in \{1, \cdots, n\} : (snap1[x] \neq \bot) \Rightarrow (snap2[x] = snap1[x]).$$

Any one-shot snapshot object SN has the following *containment* property, is an immediate consequence of the fact that each process issues at most one write operation, and the operations can be totally ordered (linearization):

$$(snap1 \leq snap2) \vee (snap2 \leq snap1).$$

2.3   Consensus and Participant-Restricted Consensus

Consensus: Definition. Consensus is one of the most fundamental problems of fault-tolerant distributed computing (see textbooks such as [3,14,16,17,19] for more developments). More precisely, a consensus object is an one-shot object which provides the processes with a single operation denoted propose(). This operation takes a value as input parameter (called input or proposed value) and returns a result (called decided value). A consensus object is defined by the following properties.

– Validity: If a process decides a value v, this value was proposed by some process.
– Agreement: No two processes decide different values.
– Termination: If a process invokes propose() and does not crash, it decides.

When a process p_i invoke propose(v) we say "p_i proposes v". When this invocation terminates and returns value w, we say "p_i decides w".

It is well-known that consensus cannot be implemented in asynchronous crash-prone systems in which the processes can communicate only through atomic read/write registers [10,13]. Hence, it cannot be implemented by using only objects (such as snapshot objects) which can be implemented with read/write registers only.

Participant-Restricted Consensus. A process p_i participates in a consensus instance if it invokes the propose() (from an operational point of view, this corresponds to the first shared memory access invoked by propose()).

The *participant-restricted* consensus is a consensus instance in which not all the correct processes are required to participate. Hence, a non-participating process can be correct or faulty. Moreover the subset of processes that participate is not know in advance. But if a correct process takes one step in the execution then it takes an infinity number of steps.

3   The Failure-Detectors Ω and Ω^*

3.1   The Eventual Leader Failure Detector Ω

The *eventual leader* failure detector, denoted Ω, was introduced in [5], where it is shown to be the weakest failure detector to solve consensus in asynchronous message-passing systems in which a majority of processes do not crash. This failure detector provides each process p_i with a read-only local variable LEADER$_i$, which always contains a process identity, and is such that, after an unknown but finite period, the variables LEADER$_i$ of all the correct processes contain the same identity and this identity is the identity of a correct process (this property is called *eventual leadership*).

An Ω-based consensus algorithm for asynchronous read/write systems in which any number of processes may crash is presented in [12], where it is shown that Ω is the weakest failure detector to solve consensus in asynchronous read/write systems in which any number of processes may crash.

Be the communication medium read/write registers or message-passing, the Ω-based consensus algorithms implicitly assume that all the processes participate in the consensus. This is because the process that is eventually elected as common leader by Ω can be *any* correct process. If this process does not participate, consensus cannot be solved. It follows that Ω is not the weakest failure detector to solve consensus if some correct processes do not participate.

3.2   The Eventual Leader Failure Detector Ω^*

The failure detector Ω^* was introduced in [9,18]. It is used in [9] to boost liveness properties of concurrent objects, and in [18] to solve k-set agreement (a generalization of consensus, which corresponds to the case $k = 1$).

The Failure Detector $\Omega^*(X)$. Given any set X of processes, $\Omega^*(X)$ provides each process p_i with a read-only local variable $\text{LEADER}_i(X)$ such that the following properties are satisfied.

- Validity: At any time, any local variable $\text{LEADER}_i(X)$ contains the identity of a process of X.
- Restricted eventual leadership: There is an unknown but finite time after which the local variables $\text{LEADER}_i(X)$ of the correct processes of X contain the same process identity, which is the identity of a correct process of X.

Hence, given any non-empty set of processes X, there is an arbitrary period during which the processes of X have arbitrary leaders, but this anarchic period is finite. When this period terminates the correct processes of X agree on the same leader, which is one of them. Let us remark that when X is the set of all the processes, $\Omega^*(X)$ boils down to Ω.

The Failure Detector Ω^*. Considering all the non-empty subsets $X \subseteq \Pi$, Ω^* is the failure detector made up of all the corresponding $\Omega^*(X)$.

Failure Detector Reductions. A failure detector D is *weaker than* a failure detector D' (denoted $D \preceq D'$) if there is a reduction algorithm from D' to D, i.e, an algorithm based on D' whose outputs satisfy the properties of D. If D is weaker than D', any problem that can be solved with D can be solved with D'. If $D \preceq D'$ but $D' \npreceq D$, we said that D is *strictly weaker than* D' ($D \prec D'$).

The following theorem follows directly from the definition of $\Diamond P$ (while Ω, Ω^*, and $\Diamond P$ belongs to the family of eventual failure detectors, $\Diamond P$ is the only of them that, after some finite time, behaves as the perfect failure detector P –which was defined in the Introduction–).

Theorem 1. $\Omega \prec \Omega^* \prec \Diamond P \prec P$.

3.3   The Weakest Failure Detector for Participant-Restricted Consensus

Theorem 2. Ω^* *is the weakest failure detector to implement participant-restricted consensus in an asynchronous read/write system in which any number of processes may crash.*

Proof. The fact that Ω^* allows participant-restricted consensus to be solved follows from the existence of the algorithm described in Fig. 1.

The fact it is the weakest results from the following observation. Given an execution, let $part \subseteq \Pi$ be the set of the processes that participate in the consensus (i.e., the set of processes that invoke the operation propose()). In such an execution, it follows from its definition that $\Omega^*(part)$ behaves exactly as Ω in a system of $|part|$ processes. As Ω is the weakest failure detector to solve consensus in a model in which all processes are assumed to participate, it follows that $\Omega^*(part)$ is the weakest when only processes in $part$ participate. $\square_{Theorem\ 2}$

4   An Ω^*-Based Participant-Restricted Consensus Algorithm

This section presents an Ω^*-based consensus algorithm suited to the participating processes model. This algorithm is a round-based algorithm inspired from message-passing algorithms such as the ones described in [4,15,17].

From a notational point of view, shared (snapshot) objects are denoted with uppercase letters. Differently, local variables of each process are denoted with lowercase letters sub-scripted with the index i of the corresponding process p_i.

4.1   Shared Objects and Local Variables

The processes cooperate through a sequence of one-shot snapshot objects, each associated with a specific round. Let $SNAP[r]$ denote the snapshot object associated with round r. The containment property of each of these objects is essential for the correctness of the algorithm. More precisely, the total order on the operations on each one-shot snapshot object, can be seen as replacing both

- the majority of correct process requirement used in message-passing, and
- the requirement that all correct processes must participate.

Local variables at every process p_i. Each process manages the following local variables.

- r_i: local round number.
- $prop_i$: current estimate of p_i's decision value.
- myl_i: current leader of p_i.
- $report_i$: auxiliary variable containing a proposed value or the default value "?" (as $\bot$, "?" cannot be a proposed value).
- $leaderpair_i$: pair made up of a proposed value and a participating process identity.
- $snap_i[1..]$: sequence of one-snapshot objects; $snap_i[r]$ is used at round r.
- $set_i[r]$: set of non-$\bot$ values contained in $snap_i[r]$ (used only at even rounds).

In addition to these local variables, Ω^* provides each process p_i with the read-only variables LEADER$_i(X)$ where X is any non-empty subset of Π.

4.2   Description of the Algorithm

The algorithm is given in Fig. 1. It uses an internal operation myleader() which returns the current Ω^*-based leader of the invoking processes. Operationally, the invocation $SNAP[1]$.snapshot() allows the invoking process p_i to compute the current set of participating processes, denoted $part_i$. Then, myleader() returns the output of the read only local variable LEADER$_i(part_i)$.

The algorithm consists of a sequence of phases, each composed of two consecutive rounds, an odd round followed by an even round.

```
init: ∀ r ≥ 1: SNAP[r] initialized to [⊥, ⋯ , ⊥].

internal operation myleader() is
    snap_i[1] ← SNAP[1].snapshot();
    part_i ← {k such that snap_i[1][k] ≠ ⊥};
    return(LEADER_i(part_i)).     % Current local output of Ω*

operation propose(v_i) is
(1)    prop_i ← v_i; r_i ← 0;
(2)    repeat
       % phase 1: filtering   % odd rounds
(3)        r_i ← r_i + 1;
(4)        myl_i ← myleader();
(5)        SNAP[r_i].write(i, ⟨prop_i, myl_i⟩);
(6)        repeat snap_i[r_i] ← SNAP[r_i].snapshot();
                   leaderpair_i ← snap_i[r_i][myl_i]
(7)        until (leaderpair_i ≠ ⊥) ∨ (myl_i ≠ myleader()) end repeat;
(8)        if (∃ v, ℓ:   % v is the current value of prop_ℓ
              ∀j : (snap[r_i][j] ≠ ⊥) ⇒ (snap[r_i][j] = ⟨−, ℓ⟩) ∧ (snap[r_i][ℓ] = ⟨v, −⟩))
(9)            then report_i ← v else report_i ← ? end if;
           % ∀ i, j : ((report_i ≠?) ∧ (report_j ≠?)) ⇒ (report_i = report_j = v ≠?)
       % phase 2: try to decide   % even rounds
(10)       r_i ← r_i + 1;
(11)       SNAP[r_i].write(i, report_i);
(12)       snap[r_i] ← SNAP[r_i].snapshot();
(13)       let set_i[r_i] be the set of non-⊥ values in snap[r_i];
           % ∀ i, j : set_i[r_i] = {v} where v ≠? and set_j[r_i] = {?} are mutually exclusive
(14)       case (set_i[r_i] = {v} where v ≠?)     then return(v)
(15)            (set_i[r_i] = {v, ?} where v ≠?) then prop_i ← v
(16)            (set_i[r_i] = {?})                then skip %  prop_i keeps its previous value
(17)       end case
(18) end repeat.
```

Fig. 1. Ω^*-based consensus (code for process p_i)

First Round of a Phase. This (odd) round r can be seen as a filtering mechanism, whose aim is to reduce the set of proposed values, to a single value of the default value "?".

To this end, a process p_i first computes its current leader myl_i, and writes the pair $\langle prop_i, myl_i \rangle$ in $SNAP[r]$ (line 3), and enters then in an internal loop (lines 6-7). In this loop, p_i reads $SNAP[r]$, from which it extracts $leaderpair_i$ (which is the pair $\langle prop, leader \rangle$ deposited by p_i'current leader or $\perp$ if this pair has not yet been deposited, line 6). This is repeated until $SNAP[r][myl_i]$ has been written or p_i's current leader changed (predicates of line 7).

When p_i exits the internal repeat loop, it checks if there is a process p_ℓ that is the current leader of all the processes that (up to now) have written in $SNAP[r]$

(line 8). If this is the case, p_i reports the proposal v of p_ℓ in $report_i$. The idea is here to decide the value v. Otherwise, it writes "?" in $report_i$, whose meaning is "during this phase, p_i cannot help deciding".

When the first (odd) round of a phase terminates (i.e., after line 9), the following predicate is satisfied, where v is a proposed value:

$$\mathcal{PR}1 \equiv \forall\, i,j : \big((report_i \neq ?) \wedge (report_j \neq ?)\big) \Rightarrow (report_i = report_j = v \neq ?).$$

Second Round of a Phase. When a process p_i enters this (even) round, it writes it report in $SNAP[r+1]$, reads its content, locally saves it in $snap_i[r+1]$, and computes $set_i[r+1]$ (lines 10-13). When this is done, the following predicate is satisfied:

$$\mathcal{PR}2 \equiv \forall\, i,j : set_i[r+1] = \{v\} \neq \{?\} \text{ and } set_j[r+1] = \{?\} \text{ are mutually exclusive.}$$

Then, there are three cases according to the value of $set_i[r+1]$. If $set_i[r+1] = \{v\} \neq \{?\}$, p_i decides v. If $set_i[r+1] = \{v,?\}$, p_i adopts v as new proposed value ($prop_i$). Otherwise, p_i keeps its previous proposal. In the last two cases, p_i starts a new phase.

4.3   Proof of the Algorithm

Lemma 1. *If, during an odd round r, p_i and p_j execute line 9, the predicate* $\mathcal{PR}1 \equiv \forall\, i,j : \big((report_i \neq ?) \wedge (report_j \neq ?)\big) \Rightarrow (report_i = report_j = v \neq ?)$ *is satisfied, where v is the value defined at line 8.*

Proof. Let r be an odd round executed by p_i, at the end of which p_i writes $v \neq ?$ in $report_i$ (line 9). As process p_i obtained v from the predicate of line 8, it follows from the second part of this predicate that v is the value written in $SNAP[r]$ by p_ℓ at line 3. Moreover, as p_i read atomically $SNAP[r]$ for the last time at line 6, $SNAP[r]$ contained no pair with a leader different from p_ℓ (first part of the predicate of line 8).

Let us assume, by contradiction, that a process p_j writes $v' \neq v, ?$ in $report_j$ at line 9. For the same reason as before, there is a process $p_{\ell'}$ that wrote at line 3 the pair $\langle v', -\rangle$ in the snapshot object $SNAP[r]$, i.e., $SNAP[r][\ell'] = \langle v', -\rangle$.

Let τ_i (resp., τ_j) be the time at which p_i invoked $SNAP[r]$.snapshot() for the last (line 6). As the snapshot object $SNAP[r]$ is atomic, and $SNAP[r]$ did not contain $\langle v', -\rangle$ at time τ_i (otherwise, p_i would not have written v in $report_i$), it follows that $\tau_i < \tau_j$. It then follows, from the containment property of $SNAP[r]$ that, at time τ_j, $SNAP[r]$ contains both the pair $\langle v, -\rangle$ and the pair the pair $\langle v', -\rangle$. The predicate of line 8 is consequently not satisfied by the last value of $SNAP[r]$ read by p_i. It follows that p_j assigns the default value $\perp$ to $report_j$ at line 9. A contradiction. $\square_{Lemma\ 1}$

Lemma 2. *If, during an even round r, p_i and p_j execute line 13-17, it is not possible to have $set_i[r] = \{v\}$, where $v \neq ?$ and $set_j[r] = \{?\}$.*

Proof. Let us assume, by contradiction, that $set_i[r] = \{v\}$ (where $v \neq ?$) and $set_j[r] = \{?\}$. It follows from the atomicity of the read of $SNAP[r]$ by p_i at line 10, and the definition of $set_i[r]$ at line 13, that, when it read it, $SNAP[r]$ contained at least one v, possibly $\bot$, and no other values. This atomic read of $SNAP[r]$ occurred at time τ_i.

Similarly, it follows from the atomic read of $SNAP[r]$ by p_j at line 10 that, when read by p_j, $SNAP[r]$ contained at least one ?, possibly $\bot$, and no other values. This atomic read of $SNAP[r]$ occurred at time τ_i.

As the operation $SNAP[r]$.snapshot() is atomic, we have either $\tau_i < \tau_j$ or $\tau_j < \tau_i$. Without loss of generality, assume $\tau_i < \tau_j$. Due to its containment property, at time τ_j, $SNAP[r]$ contains v, and we have $set_j[r] \neq \{?\}$. A contradiction.
$$\square_{Lemma\ 2}$$

Lemma 3. *A decided value is a proposed value.*

Proof. A process p_i decides a value v when it executes line 14. This occurs during an even round r during which $set_i[r] = \{v\}$, where $v \neq \bot, ?$ (lines 13-14). The proof consists in showing that $SNAP[r]$ contains only $\bot$, "?", or a proposed value. This is an immediate consequence of Lemma 1; and the fact that, initially, and then by induction on the updates of $prop_i$ executed at line 15, any update of $prop_i$ assigns it a previous $prop_j$ value.
$$\square_{Lemma\ 3}$$

Lemma 4. *No two processes decide different values.*

Proof. Let r be the first (even) round at which a process p_i decides, and v the value it decides. Hence, we have $set_i[r] = \{v\}$ (line 14). Let p_j be another process that executes round r. There are two cases.

- p_j decides during round r. Let us assume it decides v'. Hence, $set_j[r] = \{v'\}$. It then follows from the containment property of $SNAP[r]$ that $v = v'$.
- p_j does not decides during round r. If follows from Lemma 2, that $set_j[r] = \{v, ?\}$. Hence, p_j adopts v as new proposed value. It follows that all the processes p_j that progress to the next round are such that $prop_j[r] = v$. Consequently, v is the only value that remains in the execution, and consequently no other value can be decided.
$$\square_{Lemma\ 4}$$

Lemma 5. *If a process that invokes* propose() *and does crash decides.*

Proof. Let us assume by contradiction that no correct participating process terminates. Let $PART$ be the set of participating processes. Let τ denote a time after which no more participating process crashes and Ω^* returns forever the same correct participating process identity (say ℓ) to the participating processes.

Claim C. Assuming no correct participating process terminates, none of them blocks forever in the internal repeat loop (lines 6-7).

Proof of the claim. Let r be the first round at which a correct participating process loops forever in the internal loop. This means that the predicate $(leaderpair_i \neq \bot) \vee (myl_i \neq$ myleader()) (line 7) is never satisfied. From time τ, the invocation of myleader() by p_i (line 7) always returns the value of $leader_i(PART)$, that is ℓ. There are two cases.

- If $myl_i \neq \ell$, p_i exits the internal loop (second predicate at line 7).
- If $myl_i = \ell$, the predicate $leaderpair_i \neq \bot$ (first predicate at line 7) must be satisfied for p_i to exit the loop. But in this case, $leaderpair_i$ was previously assigned the pair $\langle prop_i, myl_i \rangle = \langle prop_i, \ell \rangle$ (line 6), which allows p_i to exit the loop.

It follows that r is not the smallest round during which a correct participating process loops forever, contradicting the Claim assumption. End of the proof of the claim.

Assuming no process decides, it follows from Claim C that the correct processes of $PART$ execute rounds forever. Moreover, after time τ they all have the same correct participating leader p_ℓ. Let r be an odd round executed after time τ by the correct processes of $PART$. During r, they all assign ℓ to their local variables myl_i, and each process p_i writes the pair $\langle prop_i, \ell \rangle$ in $SNAP[r]$. Moreover, the predicate $myl_i \neq$ myleader() is never satisfied when they evaluate the predicates of line 7.

Due the claim C, all processes progress, which means that, for p_i, the predicate $leaderpair_i \neq \bot$ is eventually satisfied, which means that p_ℓ wrote a pair $\langle v, - \rangle$ in $SNAP[r]$ when it executed line 3 of round r. It follows that, all p_i are then such that (i) $snap_i[r][\ell] = \langle v, - \rangle$, and (ii) $(snap_i[r][j] \neq \bot) \Rightarrow (snap_i[r][j] = \langle -, \ell \rangle)$. Consequently, they all report v (the current proposal of the leader) in their local variables $report_i$. It then follows that they all decide during the next even round $r + 1$. A contradiction. $\qquad \Box_{Lemma\ 5}$

Theorem 3. *The algorithm described in Fig. 1 implements a participant-restricted consensus object in an asynchronous read/write model, in which any number of processes may crash.*

Proof. The proof follows from Lemma 3 (Validity), Lemma 4 (Agreement), and Lemma 5 (Termination). $\qquad \Box_{Theorem\ 3}$

5   Conclusion

This paper was on the implementation of consensus when a subset of correct processes only participate in the consensus instance, hence the name *participant-restricted consensus*.

After having introduced the failure detector Ω^* (a straightforward generalization of Ω), the paper has presented an Ω^*-based algorithm that solves participant-restricted consensus, and shown that Ω^* is the weakest failure detector to solve this problem.

Acknowledgments. This work was partially supported by the French ANR project DESCARTES (16-CE40-0023-03) devoted to layered and modular structures in distributed computing. We want to thank the referees for their constructive comments.

References

1. Afek, Y., Attiya, H., Dolev, D., Gafni, E., Merritt, M., Shavit, N.: Atomic snapshots of shared memory. JACM **40**(4), 873–890 (1993)
2. Anderson, J.: Multi-writer composite registers. Distrib. Comput. **7**(4), 175–195 (1994)
3. Attiya, H., Welch, J.L.: Distributed Computing: Fundamentals, Simulations and Advanced Topics, 2nd edn. Wiley-Interscience, p. 414 (2004). ISBN 0-471-45324-2
4. Ben-Or, M.: Another advantage of free choice: completely asynchronous agreement protocols. In: Proceedings of 2nd ACM Symposium on Principles of Distributed Computing (PODC 1983), pp. 27–30. ACM Press (1983)
5. Chandra, T., Hadzilacos, V., Toueg, S.: The weakest failure detector for solving consensus. J. ACM **43**(4), 685–722 (1996)
6. Chandra, T., Toueg, S.: Unreliable failure detectors for reliable distributed systems. J. ACM **43**(2), 225–267 (1996)
7. Delporte-Gallet, C., Fauconnier, H., Guerraoui, R.: Tight failure detection bounds on atomic object implementations. J. ACM **57**(4), 32 (2010). Article 22
8. Fernández, A., Jiménez, E., Raynal, M., Trédan, G.: A timing assumption and two t-resilient protocols for implementing an eventual leader service in asynchronous shared-memory systems. Algorithmica **56**(4), 550–576 (2010)
9. Guerraoui, R., Kapalka, M., Kuznetsov, P.: The weakest failure detectors to boost obstruction-freedom. Distrib. Comput. **20**(6), 415–433 (2008)
10. Fischer, M.J., Lynch, N.A., Paterson, M.S.: Impossibility of distributed consensus with one faulty process. J. ACM **32**(2), 374–382 (1985)
11. Hélary, J.-M., Hurfin, M., Mostéfaoui, A., Raynal, M., Tronel, F.: Computing global functions in asynchronous distributed systems with perfect failure detectors. IEEE Trans. Parallel Distrib. Syst. **11**(9), 897–909 (2000)
12. Lo, W.-K., Hadzilacos, V.: Using failure detectors to solve consensus in asynchronous shared-memory systems. In: Tel, G., Vitányi, P. (eds.) WDAG 1994. LNCS, vol. 857, pp. 280–295. Springer, Heidelberg (1994). https://doi.org/10.1007/BFb0020440
13. Loui, M., Abu-Amara, H.: Memory requirements for agreement among unreliable asynchronous processes. In: Preparata, F.P. (ed.) Advances in Computing Research, vol. 4, pp. 163–183. JAI Press, Greenwich (1987)
14. Lynch, N.A.: Distributed Algorithms, p. 872. Morgan Kaufmann Pub., San Francisco (1996)
15. Mostéfaoui, A., Raynal, M.: Solving consensus using Chandra-Toueg's unreliable failure detectors: a general quorum-based approach. In: Jayanti, P. (ed.) DISC 1999. LNCS, vol. 1693, pp. 49–63. Springer, Heidelberg (1999). https://doi.org/10.1007/3-540-48169-9_4
16. Raynal, M.: Concurrent Programming: Algorithms, Principles, and Foundations, p. 515. Springer, Heidelberg (2013). https://doi.org/10.1007/978-3-642-32027-9
17. Raynal, M.: Fault-Tolerant Message-Passing Distributed Systems: An Algorithmic Approach, p. 492. Springer, Switzerland (2018). https://doi.org/10.1007/978-3-319-94141-7
18. Raynal, M., Travers, C.: In search of the holy grail: looking for the weakest failure detector for wait-free set agreement. In: Shvartsman, M.M.A.A. (ed.) OPODIS 2006. LNCS, vol. 4305, pp. 3–19. Springer, Heidelberg (2006). https://doi.org/10.1007/11945529_2
19. Taubenfeld, G.: Synchronization Algorithms and Concurrent Programming, p. 423. Upper Saddle River, Pearson Education/Prentice Hall (2006)

Capture on Grids and Tori with Different Numbers of Cops

Fabrizio Luccio[✉] and Linda Pagli

Department of Informatics, University of Pisa, Pisa, Italy
luccio@di.unipi.it

Abstract. This paper is a contribution to the classical cops and robber problem on a graph, directed to two-dimensional grids and tori. We apply some new concepts for solving the problem on grids and apply these concepts to give a new algorithm for the capture on tori. Then we consider using any number k of cops, give efficient algorithms for this case yielding a capture time t_k, and compute the minimum value of k needed for any given capture time. We introduce the concept of *work* $w_k = k \cdot t_k$ of an algorithm and study a possible *speed-up* using larger teams of cops.

Keywords: Cops · Robber · Capture time · Grid · Tori · Work · Speed-up

1 Introduction

The problem of cops and robber on a graph has received considerable attention. Started as a pure pursuit-evasion game it has shown interesting theoretical implications and importance in graph searching, network decontamination, motion planning, security and environment control. As a consequence many versions of the problem have been studied, typically depending on the type of graph, the knowledge of the actors on the positions of the others, the type and speed of movements allowed.

In the basic version of the problem cops and robber stay on the vertices of a graph and can move to adjacent vertices or stay still, starting from initial positions chosen first by the cops, then by the robber. The chase proceeds in rounds, each of which is composed of a parallel move of the cops followed by a move of the robber which is captured when a cop reaches its vertex and the game terminates. We study the problem in this basic version if the graph is a toroidal grid, also revisiting some known results on grids.

1.1 A Brief Analysis of the Literature

The cops and robber problem was defined by Quillot [19] and Nowakowski and Winkler [18] as a pursuit-evasion game with one cop, to generate a complex

© Springer Nature Switzerland AG 2019
V. Malyshkin (Ed.): PaCT 2019, LNCS 11657, pp. 431–444, 2019.
https://doi.org/10.1007/978-3-030-25636-4_34

theory in the following years. After studying graphs where the game can be won by a single cop, the attention was directed to solve the problem on different classes of graphs with a minimum number of cops, called the *cop number*. A general survey in this direction can be found in the comprehensive book by Bonato and Nowakowski [5] which brings together the main structural and algorithmic results on the field known when the book appeared. In particular they thoroughly discuss the still open Meyniel conjecture on the sufficiency of $\sqrt{n}$ cops for capturing the robber in an arbitrary graph of n vertices.

Many variants of the basic problem exist for general graphs or for particular classes of graphs, such as considering more than one robber; or cops and robber moving at different speeds; or a robber being invisible for some rounds; or, more recently, the robber escaping surveillance if it maintains a given distance from the cops [6].

With specific reference to two dimensional grids and toroidal grids studied in this paper, the proof that the cop number is 2 in a two-dimensional grid was originally given in [14], and the capture time was determined in [15]. The cop number 3 for tori can be derived from the results of [17] where the robber capture is studied for products of graphs. Several variations were proposed, in particular if the visibility of each cop is limited to edges and vertices of its row or column. In [7, 16] the cops win if they can see the robber, and in [20] it is shown that the problem with limited visibility has application in motion planning of multiple robots. The study of [20] has been revisited in [8] and algorithms for the capture using one, two or three cops having constant maximal speed are given. In [3] the cop number is determined if the robber can move at arbitrary speed. A more recent work [9] assumes that the initial positions of cops and robber are chosen randomly. In [2] the study is extended to n-dimensional grids.

Other important problems with a relation with ours were born in the field of distributed computing with moving agents, see the survey in [1]. Reachability issues are also connected to thr cops and robber problem.

Finally we recall some studies on complexity issues for the cops and robber problem for arbitrary graphs. In [12] general results on the EXPTIME-completeness of determining the cop number are given. In [10] it is proved that computing such number is NP-hard. Changing the perspective, in [4] it was shown that determining the number of cops needed for the capture in no more than a given capture time is NP-hard.

1.2 Our Contribution

Among a wealth of possibilities, we limit our treatment to the standard game on 2-dimensional grids starting from the results of [15], and then extend it to toroidal grids. In Sect. 2 we introduce some new concepts on the capture valid for general graphs, to be used in Sect. 3 for showing how the results of [15] can be found with a new different approach. In Sect. 4 this approach is applied to toroidal grids, for which we give efficient algorithms for the capture using three cops, together with a new proof that these numbers are the minimal possible. In Sect. 5 we treat the capture problem as a function of any (hence not necessarily minimum) number of cops.

For any given capture time t^* we also determine the minimum number of cops needed for the capture in at most t^* rounds using our algorithms. In this context we introduce the concept of *work* $w_k = k \cdot t_k$ of an algorithm run by k cops in total time t_k inherited from parallel processing, discussing the *speed-up* that emerges using a larger number of cops.

2   Basic Model and Properties

In the basic model of the problem one or more cops and one robber are placed on the vertices of an undirected and connected graph $G = (V, E)$. The game develops in consecutive rounds, each composed of a cops turn followed by a robber turn. In the cops turn each cop may move to an adjacent vertex or stay still. In the robber turn, the robber may move to an adjacent vertex or stay still. The game is over when a cop reaches the vertex of the robber.

The initial positions of the cops are arbitrarily chosen, then the initial position of the robber is chosen accordingly. The aim of the cops is capturing the robber in a number of rounds as small as possible, called *capture time* t; while the robber tries to escape the capture as long as possible. If needed two or more cops can stay on the same vertex and move along the same edge. All agents are aware all the time of the locations of the other agents. k, the *cop number*, denotes the smallest number of cops needed to capture the robber.

We will direct our study to 2-dimensional grids or tori. However, first we extend some known preliminary properties valid for all undirected and connected graphs. For a vertex $v \in V$, let $N(v)$ denote the set of neighbors of v, and let $N[v] = N(v) \cup \{v\}$ denote the closed set of neighbors. We pose:

Definition 1. *A siege $S(v)$ of a vertex v is a minimum set of vertices containing cops, such that at least one vertex $u \in S(v)$ is in $N(v)$, and $\bigcup_{w \in S(v)} N[w] \supseteq N(v)$. Among all the sieges of v, $\bar{S}(v)$ denotes one of these sets of minimal cardinality.*

Definition 1 depicts the situation shown in Fig. 1, where black and white circles on the graph denote vertices occupied by the cops, or by the robber, respectively. Let the robber be in v, and assume that the cops have just been moved into the vertices of $S(v)$. Now the robber has to complete the current round, but whether it moves or stands still it will be captured in the next round. In fact the condition $\bigcup_{w \in S(v)} N[w] \supseteq N(v)$ indicates that all the escape routes for the robber have been cut. We immediately have:

Lemma 1. *The robber is captured in round i if and only if at round $i - 1$ the robber is in a vertex v and there is a siege $S(v)$.*

Lemma 2. *Let v be a vertex for which $\bar{S}(v)$ has minimal cardinality among all the vertices of the graph. Then $k \geq |\bar{S}(v)|$.*

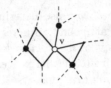

Fig. 1. A minimal siege $\bar{S}(v)$ with the robber (white circle) in v and three cops (black circles) in $\bar{S}(v)$.

Based on the definition of siege we can also establish a lower bound on the capture time t based on the initial positions of the cops and the moving strategy of the robber. First we pose:

Definition 2. *For a graph $G = (V, E)$ and an integer $e \geq 4$, an e-loop L is a chordless cycle of e vertices where each vertex of $V \setminus L$ is adjacent to at most one vertex of L.*

Note that a single cop would chase forever a robber that moves inside an e-loop. We then have (for the proof see [13]):

Lemma 3. *Let the initial positions of the k cops $c_1, c_2, \ldots, c_k$ be established; let v be the initial position of the robber; let $d_1 \leq d_2 \leq \cdots \leq d_k$ be the distances (number of edges in the shortest paths) of $c_1, c_2, \ldots, c_k$ from v; and let h be the cardinality of a minimal siege for G, $2 \leq h \leq k$. We have:*

(i) $t \geq d_1$;
(ii) if v belongs to an e-loop, $t \geq d_h - \lfloor \frac{e}{2} \rfloor$.

3 Capture on Grids

An elegant approach to studying the capture on an $m \times n$ grid has been presented in [15], where it is proved that two cops are needed, the capture time is $t = \lfloor \frac{m+n}{2} \rfloor - 1$, and this result is optimum. We examine this problem under a different viewpoint, as a basis for studying robber capture on tori.

Formally an $m \times n$ *grid* $G_{m,n}$ is a graph whose vertices are arranged in m rows and n columns, where each vertex $v_{i,j}$, $0 \leq i \leq m - 1$ and $0 \leq j \leq n - 1$ is connected to the four vertices $v_{i-1,j}, v_{i+1,j}, v_{i,j-1}, v_{i,j+1}$, whenever these indices stay inside the closed intervals $[0, m - 1]$ and $[0, n - 1]$ respectively. The vertices are obviously divided into *corner vertices*, *border vertices*, and *internal vertices*, having two, three, and four neighbors respectively.

If two vertices u, w of a grid are adjacent, the set $N(u) \cap N(w)$ is empty. If w is at a distance two from u, the set $N(u) \cap N(w)$ contains one or two vertices. This implies that the siege $S(u)$ has cardinality three if u is an internal vertex, or cardinality two if u is a border or corner vertex, see Fig. 2. Since the minimal siege for a grid has cardinality two, the number of cops needed to capture the robber is $k \geq 2$ by Lemma 2, and in fact two cops suffice as already proved

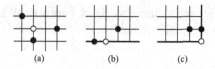

Fig. 2. Examples of a siege $S(u)$ in a grid, if u is an internal vertex, a border vertex, or a corner vertex.

in [15]. Lemma 1 shows that any algorithm using two cops must push the robber to a border or to a corner vertex to establish a siege around it, as three cops would be needed for a siege around an internal vertex. Furthermore, wherever the cops are initially placed, there is a vertex v where the robber can be placed that is at a distance $d_1 \geq \lfloor \frac{m+n}{2} \rfloor - 1$ from the closest cop, or at a distance $d_2 \geq \lfloor \frac{m+n}{2} \rfloor + 1$ from the other cop. Since all the vertices of a grid belong to an e-loop consisting of square cycles of $e = 4$ vertices we have $\lfloor \frac{e}{2} \rfloor = 2$, hence $t \geq \lfloor \frac{m+n}{2} \rfloor - 1$ by Lemma 3 case (i) or (ii), that confirms the lower bound of [15].

Consider now the *shadow cone* of a cop c, namely a zone of the grid from where the robber is impeded by c to escape. A similar concept was proposed in [20] for different instances of the problem. Let c be in vertex $u = v_{i,j}$ and consider two straight lines at $\pm 45°$ through u that divide the grid into four zones whose borders contain vertices placed on the two lines (or *edges*), and vertices placed on the border of the grid, see Fig. 3. The shadow cone of c is the zone containing the robber which is said to stay *within* the cone if it stays *in* the cone but not on one of its edges.

Fig. 3. The shadow cone of cop c with the the robber *within* it (x), or *in* it (y).

W.l.o.g. let the shadow cone of c lay "below" the cop. Two cases may occur to which the following Cone Rule applies, whose role is to keep the robber in the cone, possibly moving the cone to compensate the robber's movement (point 2.*iv* of the rule). We have (for the proof see [13]):

Lemma 4. *By applying the Cone Rule, if the robber reaches an edge of the cone it will never be able to reach the opposite edge.*

Note that if at each round the robber stands on an edge of the cone, the cop would not be able to reach the robber. The following algorithm GRID runs in $t = \lfloor \frac{m+n}{2} \rfloor - 1$ rounds as for the algorithm of [15], but is useful for the discussion that follows. We report only the cops' moves, under the standard assumption that the robber will properly move to delay capture. W.l.o.g. we let $m \leq n$, $m_1 = \lfloor \frac{m-1}{2} \rfloor$, $m_2 = \lceil \frac{m-1}{2} \rceil$, $n_1 = \lfloor \frac{n-1}{2} \rfloor$, $n_2 = \lceil \frac{n-1}{2} \rceil$.

CONE RULE

The cop c is in vertex $v_{i,j}$, and the robber is in the shadow cone of c.

1. Let be the cop's turn to move. (i) If the robber is within the cone (vertex x in Figure 3) the cop moves "down" to vertex $v_{i+1,j}$ thereby reducing the size of the cone while keeping the robber in it. (ii) If the robber is on an edge of the cone (e.g. in vertex y) the cop does not move.
2. Let be the robber's turn to move. (iii) If the robber remains in the cone the subsequent cop's move takes place as specified in points 1.i or 1.ii whichever applies. (iv) If the robber moves out of the cone from one of its edges (e.g. from vertex y, moving "up" or "to the right"), the cop moves across to vertex $v_{i,j+1}$ thereby shifting its shadow cone by one positions to keep the robber in the cone.

---

algorithm GRID(m,n)

1. initial positions of the cops c_1, c_2:
   for m even and n odd ($e \,|\, o$), or for $e \,|\, e$, **place** c_1 in v_{m_1,n_1} and c_2 in v_{m_2,n_1}; for $o \,|\, o$, **place** c_1 is in v_{m_1-1,n_1} and c_2 is in v_{m_1,n_1}, see figure 4.a; for $o \,|\, e$, **place** c_1 is in v_{m_1,n_1} and c_2 is in v_{m_1,n_2};
   // assume to work on an $o \,|\, o$ grid (the others are treated similarly)
   // the shadow cones are chosen so that at least one of them will contain
      the robber; assume that they lay below the cops as in Figure 4

   initial position of the robber: any vertex not adjacent to a cop;

2. **repeat**
2.1   **if** (the robber is in the two cones)
            move both cops one step down
2.2   **else** (the robber is on the edge of a cone but outside the other
                cone) **or** (the robber is outside the two cones)
            move both cops horizontally in the direction of the robber;
3. **until** the robber makes its last move inside a siege;
// the siege is established with the robber in a grid corner (figure 4.c)
4. **capture** the robber;

The cops are initially adjacent and their cones have a large portion in common, see Fig. 4. Then, up to the round in which the siege is established, they move in parallel so their mutual positions do not change. At the beginning the robber is in at least one of the shadow cones and is kept in this condition after each cops' move. Note that to delay the capture as much as possible the robber must eventually escape from one side of a cone. By Lemma 4, however, it is forced to escape always from the same side until it ends up in a siege, in a grid corner. We state the following Theorem 1 (for the proof, based on the analysis of algorithm GRID, see [13]).

Theorem 1. *In a grid $G_{m,n}$ two cops can capture the robber in $t = \lfloor \frac{m+n}{2} \rfloor - 1$ rounds.*

4   Capture on Tori

We now extend our study to 2-dimensional grids in the form of *tori* $T_{m,n}$ with closure in both dimensions. That is each vertex $v_{i,0}$ is connected with $v_{i,n-1}$, $0 \leq i \leq m-1$, and each vertex $v_{0,j}$ is connected with $v_{m-1,j}$, $0 \leq j \leq n-1$.

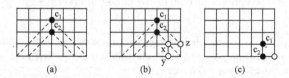

(a)   (b)   (c)

Fig. 4. (a) Initial placement of the two cops for grids with m odd and n odd, denoted as $o\,|\,o$. A similar procedure applies to grids $e\,|\,o$, $o\,|\,e$, $e\,|\,e$. (b) The cops push the robber towards the border. Vertices x, y, z indicate particular positions of the robber. Vertices x, y are in the two cones as in step 2.1 of GRID. Vertex z is in one of the two possible conditions indicated in step 2.2. (c) The final siege.

From the known results on the cop number for the capture on products of graphs proved in [17] we have that 3 cops are needed for tori. Based on Lemma 2 we confirm these numbers as lower bounds, give an algorithms for the capture that uses 3 cops, and compare the time required with the lower bound given in Lemma 3. Note that no explicit algorithm was given in [17] and the number of moves was not computed.

The capture on tori $T_{m,n}$ is more difficult as there are no borders where to push the robber. All the vertices now admit a siege of cardinality 3, then at least three cops are needed, see Fig. 2. The following capture algorithm TGRID calls the procedures GUARD and CHASE and uses three cops c_1, c_2, c_3 with shadow cones $\gamma_1, \gamma_2, \gamma_3$. Without loss of generality we define the algorithm for $n \geq m$ (simply exchange rows with columns if $m > n$), and let $m \geq 6$ and $n \geq 6$ to avoid trivial cases. Place all the cops c_1, c_2, c_3 in row 0, and in columns 0, $\lceil \frac{2n}{3} \rceil$, and $\lceil \frac{n}{3} \rceil$, respectively (see Fig. 5). Note that initially there is a cop-free gap of $\lceil \frac{n-3}{3} \rceil$ columns between c_1 and c_3, and a cop-free gap of $\lceil \frac{n-3}{3} \rceil$ or $\lfloor \frac{n-3}{3} \rfloor$ columns between c_3 and c_2 and between c_2 and c_1 around the torus. Starting with the cops in any row will be the same because we work on a torus.

The strategy is to bring a cop to *guard* the robber r (procedure GUARD), that is the cop will reach the column of r and then follow r if it moves horizontally, so to build a *virtual border* if it moves horizontally, so to build a *virtual border* along the row of the guard that prevents r from traversing it. When the guard is established, the other cops start chasing r (procedure CHASE). Without loss of generality we assume that the initial position of the robber is such that c_2 or c_3 becomes the guard.

To understand how algorithm TGRID works observe the following:

– After step 1, the algorithm is divided in a phase GUARD to establish the guard c_g, with $g=2$ or $g=3$, followed by a phase CHASE of chasing.

---

algorithm TGRID(m,n)

1. initial positions of the cops c_1, c_2, c_3:
   place c_1 in $v_{0,0}$; **place** c_2 in $v_{0,\lceil \frac{2n}{3} \rceil}$; **place** c_3 in $v_{0,\lceil \frac{n}{3} \rceil}$;
   let $\gamma_1, \gamma_2, \gamma_3$ be the shadow cones of c_1, c_2, c_3;
   initial position of the robber r: any vertex not adjacent to a cop;
   // w.l.o.g let the column of r lie in the closed interval $[\lceil \frac{n}{3} \rceil : \lceil \frac{2n}{3} \rceil$-1$]$

2. GUARD;
   // c_2 and c_3 move to establish the guard; upon exit c_g is the guard
   // and c_h is in column $\lceil \frac{n}{2} \rceil$ to start chasing r together with c_1,
   // with $g=2$, $h=3$, or $g=3$, $h=2$

3. CHASE;
   // r is captured by c_1, c_h with an extension of algorithm SGRID
   // c_g is the guard

---

phase GUARD

1. **let** y_0, y_1, y_2, y_3 be the columns of r, c_1, c_2, c_3 respectively;
   $g = 0$;
   // $g=0$, $g=2$, $g=3$ respectively denote that: the guard has not
   // yet been established, or c_2 is the guard, or c_3 is the guard;

2. **repeat**   // establishing the guard
2.1   **if** ($y_2 == y_0$) $\{g = 2$; **move** c_3 to the right $(y_3=y_3 + 1)$;$\}$
2.2   **else if** ($y_3 == y_0$) $\{g = 3$; **move** c_2 to the left $(y_2=y_2 - 1)$;$\}$
2.3   **else** $\{$**move** c_3 to the right; **move** c_2 to the left;$\}$

3. **until** $g \neq 0$;

4. **if** ($g == 2$) $h = 3$ **else** $h = 2$;   // now c_g is the guard

5. **repeat**   // cop c_h reaches the initial chasing position
5.1.   **move** c_g horizontally to follow r;
5.2.   **move** c_h horizontally towards column $\lceil \frac{n}{2} \rceil$;

6. **until** c_h reaches column $\lceil \frac{n}{2} \rceil$;

– GUARD is repeated until c_h reaches the column $\lceil \frac{n}{2} \rceil$ to start the chase together with c_1.
– In the CHASE phase, the shadow cones $\gamma_1, \gamma_2, \gamma_3$ lie below the cops c_1, c_2, c_3. As before the robber r must start on the edge of a cone to delay the capture as much as possible, but now the best position for it is not below row $\lfloor \frac{m}{2} \rfloor$ (Fig. 5), otherwise c_1 and c_h would chase it "from the bottom".

– When c_1 and c_h have established a pre-siege, r must move down. The novelty here is that c_g moves towards r in step 1.6, reducing its distance from r hence the number of rounds for the capture.

---

phase CHASE

1. **repeat**   // chasing r with cops c_1 and c_h, while c_g is the guard
1.1   **if** (r is outside γ_1 and γ_h) **move** c_1 and c_h horizontally towards r
1.2   **else if** (r is within γ_1 and/or within γ_h) **move** c_1 and c_h down
1.3   **else if** (r is on an edge of γ_1 (resp. γ_h)
                and outside γ_h (rep. γ_1))
                move c_h (resp. c_1) horizontally towards that edge
1.4   **else if** (r is on an edge of γ_1 and on an edge of γ_h)
                {**if** (the cops are in different rows)
                        move down the cop in the highest row
                    else move down one of the cops};
1.5   **if** (c_g and r are in different columns) **move** c_g to the column of r
1.6   **else if** (c_1, c_h build a pre-siege)
                move c_g from its row z to row $(z - 1) \bmod m$;
1.7   **move** the robber in any way to try to escape from the cones;

2. **until** the robber makes its last move inside a siege;
// the siege is established with the robber adjacent to c_g

3. **capture** the robber;

We state the following Theorem 2 (for the proof, based on the analysis of algorithm TGRID, see [13]).

Theorem 2. *In a torus $T_{m,n}$ three cops can capture the robber in time t such that:*

(i) $\frac{2n}{3} + \frac{5m}{4} - \frac{9}{2} \leq t \leq \frac{2n}{3} + \frac{5m}{4} - \frac{25}{12}$, *for $m \leq \lceil \frac{n}{2} \rceil$;*

(ii) $\frac{25n}{24} + \frac{m}{2} - \frac{9}{2} \leq t \leq \frac{25n}{24} + \frac{m}{2} - \frac{17}{8}$, *for $\lceil \frac{n}{2} \rceil < m \leq n$.*

For $T_{7,15}$ of Fig. 5, case (i) of Theorem 2 applies and we have $11.75 < t < 16.67$, that is $12 < t < 16$ since t must be an integer. Computing t without approximation, using the exact values shown in the proof of the theorem, we have $t_1 = 2, t_2 = 11, t_3 = 1$ hence $t = 15$. In the following Lemma 5 we establish a lower bound on the capture time (for the proof see [13]).

Lemma 5. *The capture time in a torus $T_{m,n}$ is such that $t \geq \lfloor \frac{n}{2} \rfloor + \lfloor \frac{m}{2} \rfloor - 2$.*

Corollary 1. *In $T_{m,n}$ the ratio ρ between the upper and lower lower bound on t is such that $\rho \to 4/3$ for $n/m \to \infty$, and $\rho \to \sim 37/24$ for $n/m \to 1$.*

Letting $n < m$, the new upper bounds for t are the ones of Theorem 2 exchanging n with m, while the lower bound of Lemma 5 holds unchanged. So the first statement of Corollary 1 is rephrased as: $\rho \to 4/3$ for $m/n \to \infty$.

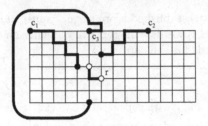

Fig. 5. Chase with three cops in $T_{7,15}$ up to a pre-siege, assuming that c_3 becomes the guard. The first two moves of c_2, c_3, and r take place in the GUARD phase, that ends when c_2 reaches column $\lceil \frac{n}{2} \rceil = 8$.

5   Using Larger Teams of Cops

We now take a new approach to the problem, discussing how the capture time decreases using an increasing number of cops, and conversely which is the minimum number of cops needed to attain the capture within a given time. This has a twofold purpose. On one hand, the possibility of employing the cops immediately in a new chase when they have completed their previous job. The second purpose is completing a job within a required time when a smaller team of cops cannot meet that deadline.

For this new approach we inherit the concept of *speed-up* introduced in parallel processing. The the *work* w_k of a process carried out by k agents in time t_k is defined as $w_k = k \cdot t_k$, and the speed-up between the actions of j over $i < j$ agents to catch the robber is defined as w_i/w_j. If the algorithms run by the two teams of i and j agents are provably optimal, the speed-up is an important measure of the efficiency of parallelism. Referring to the cops and robber problem, the speed-up is a measure of the gain obtained using an increasing number of cops with the best available algorithms.

In algorithm GRID-K the robber may be captured on a left or on a right corner of the grid by the leftmost or by the rightmost pair of cops; or it may be captured on the top or on the bottom border by two cops, one from each pair, in a vertex between the two pairs. We have (for the proof, based on the analysis of algorithm GRID-K, see [13]):

5.1 k Cops on a Grid

Let us consider the case of $k > 2$ cops on a grid $G_{m,n}$, with $m \geq 4, n \geq 4$ to avoid trivial cases. W.l.o.g let $m \leq n$. The following algorithm GRID-K is designed as an extension of algorithm GRID, taking k even. GRID-K is limited to its main lines, however sufficient for computing the capture time.

The cops $c_1, \ldots, c_k$ start in $h > 1$ pairs of adjacent vertices, $k = 2h$, with the cops of each pair placed in rows $\lfloor \frac{m}{2} \rfloor - 1$ and $\lfloor \frac{m}{2} \rfloor$. The pairs are almost equally spaced, with $\lceil \frac{n-h}{h} \rceil$ and $\lfloor \frac{n-h}{h} \rfloor$ cop-free columns between them except for the